T0093885

Framing Global Mathematics

Norbert Schappacher

Framing Global Mathematics

The International Mathematical Union between Theorems and Politics

Norbert Schappacher
IRMA
Strasbourg, France

ISBN 978-3-030-95685-1 ISBN 978-3-030-95683-7 (eBook)
https://doi.org/10.1007/978-3-030-95683-7

Mathematics Subject Classification: 01A55, 01A60, 01A61, 01A65, 01A72, 01A74

This Springer imprint is published by the registered company Springer Nature Switzerland AG
The registered company address is: Gewerbestrasse 11, 6330 Cham, Switzerland

Elif için

Preface

я бы, впрочем, не пускался в эти весьма нелюбопытные и смутные
объяснения и начал бы просто-запросто без предисловия:
понравится - так и так прочтут; но беда в том, что
жизнеописание-то у меня одно, а романов два.
Главный роман второй.
....
Обойтись мне без этого первого романа невозможно,
потому что многое во втором романе стало бы непонятным.

I would not, in fact, venture into these rather vague and uninteresting
explanations but would simply begin without any introduction
—if they like it, they'll read it as it is—
but the trouble is that while I have just one biography,
I have two novels. The main novel is the second one.
......
It is impossible for me to do without that first novel,
or much in the second novel will be incomprehensible.

Fyodor Dostoevsky, from the preface to *The Brothers Karamazov*[1]

This book is about international relations in mathematics over the last two hundred years, since the 1820s. It focusses on institutions and organizations that were created with a view to framing the international dimension of mathematical research.

[1] Translation by Richard Pevear & Larissa Volokhonsky.

Nowadays, the organization that first comes to mind in this context is the *International Mathematical Union* (IMU), a non-governmental organization (NGO) with more than eighty member countries, whose most visible recurring activity is the orchestration of an *International Congress of Mathematicians* (ICM) once every four years. Indeed, the idea of the present book arose in 2019 from the observation that the initial variant of the IMU was founded in 1920 in Strasbourg on the occasion of the first postwar ICM. At the time it was mostly called *Union mathématique internationale* (UMI), or one of several other French names.

However, the Strasbourg Congress was not a particularly successful event in the series of ICMs, nor was it possible to celebrate in 2020 a full century of the foundation of the UMI in Strasbourg, because that *Union* had disbanded in 1932. A new, different IMU was created after World War II, in 1952. It is this second IMU which today manages the quadrennial ICMs, the Fields Medals and several other awards.

The series of ICMs had started as early as 1897, before the creation of the first IMU, and were also organized in the 1930s, when there was again no IMU. This long continuity of the congresses, which was interrupted only by the two World Wars, shows that today's IMU is the eventual outcome, but not the origin of the story of international relations in mathematics. Indeed, this story has its roots in two developments that emerged during the nineteenth century: the broad professionalization of the sciences, in particular mathematics, at universities on the one hand, and the competitive behavior of Nation states on the other.

The nineteenth century also paved the way for other crucial ingredients of today's global mathematics: the creation of research facilities and institutes outside of the universities. Their effect on international mathematical contacts and careers first became apparent between the two World Wars, in particular through Rockefeller travel grants and the founding of centers such as the *Institut Henri Poincaré* in Paris and the *Institute for Advanced Study* in Princeton.

These are the elements which place this preface—with all due respect, and in spite of its utter literary negligibility—in a position vaguely analogous to what Dostoevsky expresses in the above quote with respect to his double novel on the Karamazov Brothers. While my book converges on a portrait of today's *International Mathematical Union*, it was impossible for me to do without an exploration of its prehistory, or much in the last part would be incomprehensible.

This is why the book is organized according to three historical periods:

Part I	Chap. 1–3	The long nineteenth century until the end of World War I
Part II	Chap. 4–7	The interwar period 1919–1949
Part III	Chap. 8–10	The past seventy years.

Readers who, for instance, do not want to go through the long nineteenth century before reading about the (first) IMU, may start reading with Chapter 4; cross-references will suggest sections from Part I for additional background.

Part I traces the historical roots of scientific, and especially mathematical internationalism. The notion of internationality which is at work here originated in nineteenth century Europe. Its introduction into the world of science was conditioned and shaped by the broad professionalization of scientific research and teaching in the nineteenth century. This is the reason why Part I focusses largely on Europe, even though Japanese and Indian mathematics are also discussed.

Chapter 1 unfolds the panorama of national vs. international activities in mathematics that took place during the nineteenth century. Side glances to other sciences—chemistry, for instance, but also history—are occasionally offered to get a clearer picture of the specific situation of mathematics. In the late nineteenth century, the search for international organizations involved even the scientific Academies inherited from the age of the enlightenment. Felix Klein's project of an *Encyclopedia of the Mathematical Sciences* assumed certain kinships between nations and mathematical disciplines. Chapter 2 looks at different views of the foundational debate in mathematics before, during and after World War I. The effects of that War on mathematics and their national as well as international dimension is discussed in Chapter 3.

Part II dwells on what may have been both the most productive and the most brutally disruptive period of the twentieth century. As a consequence of World War I, the history of this intermediary period was dominated by the USA and Europe. The life and death of the first IMU, which took place between the two World Wars, is detailed in Chapter 4.

Chapter 5 describes the way in which the Rockefeller Foundation and the Institute for Advanced Study (IAS) in Princeton contributed to opening up the international perspective of many mathematicians, especially young researchers. Chapter 6 chooses Emmy Noether's personal approach to mathematics to explain why and in which way the 1920s and in particular the 1930s gave birth to many of the most characteristic features of twentieth century mathematics. We refer to this as a process of consolidation and unification of mathematics, which strongly manifested itself in the 1930s, and we trace its effects through international conferences, book publications and review journals. Chapter 7 turns to the forced displacement of scholars as well as the profound and lasting effects of World War II on global mathematics.

Part III looks at the past seventy years, the establishment of the second IMU, and the way in which this NGO has framed the worldwide community of mathematicians. It was not our intention to duplicate or to extrapolate the well-documented, partly personal account of the first half century of the second IMU by Olli Lehto.[2]

We first survey the past seventy years in Chapter 8, highlighting the effects of the evolution of world politics on the international mathematical scene. This sketch of contemporary history is then complemented, in the final Chapter 10, by a condensed portrait of the principal commissions of the IMU. The subsequent final part of Chapter 10 is dedicated to an analysis of data relating to the ICMs since 1950, which

[2] See [Lehto 1998], Chapters 4–12.

has been realized by Birgit Petri in Darmstadt. The objective is to explore in some detail the image of mathematical excellence which the ICMs of the past seventy years have projected, for the mathematical community and the interested public.

The brief Chapter 9 in the middle of Part III gives a very condensed account of the history of ICMI, the *International Commission on Mathematical Instruction*. This Commission has a longer and more continuous history than the two International Mathematical Unions, which justifies the separate chapter, if not its brevity.

Framing. We use this term in the title as well as in the text of the book. It alludes on the one hand to common expressions such as "the framers of the constitution"—see for instance the title of Section 4.1—, and on the other to framing models from the field of sociology, psychology, and communication inspired by the works of Erving Goffman, Charles Fillmore, and Marvin Minsky. Without entering into methodological arguments about this concept, suffice it to say that the phenomenon of framing assesses how individuals or institutions may use, modify, or challenge ways to apprehend activities or situations.

By managing the organization of ICMs and the way in which Fields medalists and other prize winners are selected, the International Mathematical Union is instrumental in framing a public image of the mathematical sciences. One of the goals of this book is to describe this process from different vantage points.

"Mathematics International." Throughout this book I shall occasionally use the expression *Science International* to allude to the underlying phenomenon whose history I am writing about.[3] Sometimes this is narrowed down to *Mathematics International*. I like these crisp formulas where the word 'international' cannot quite make up its mind whether it is an adjective or a noun. They allow me to allude to a whole spectrum of possible organizational structures for scientific, respectively mathematical cooperation between different nation states. The book traces the historical evolution of concrete realizations of these vague notions.

Convention. As a rule, when persons are mentioned in the narrative for the first time, they are given with their full names and their years of birth and death (if applicable). If the mention is inside a quote, the additional information is added in square brackets. These data come from publicly available sources, and have been checked as best I could. (In rare individual cases I relied on personal communication.)

[3] This formula has been employed by others in the same context; for instance as the title of Frank Greenaway's book on the history of ICSU—see [Greenaway 1996]. Also ICSU's Newsletter Magazine was called *Science International* after the 1985 meeting at Schloss Ringberg. (This is not to be confused with any of the two online scientific journals that carry the same name today.) – There was even a Canadian Television series in the 1970s—created among others by the Berkeley differential geometer Michael Spivak—which was initially called "Science International", although it would soon be renamed to: "What will they think of next?"

Credits, Permissions, and References. The author has made all reasonable efforts to trace all rights holders to any copyrighted material used in this work. In cases where these efforts have not been successful the author welcomes communications from copyright holders, so that the appropriate acknowledgements can be made in future editions, and to settle other permission matters.

Credits for those illustrations which are reproduced thanks to a special permission requested and granted are briefly indicated in the caption. For *every* illustration reproduced in this book, we spell out the rights, to the best of our knowledge, in the first part of the References section at the end of the book.

This reference section also contains: the list of the *archives* consulted; the *publications* referred to, as well as the list of *websites* referred to in the text.

Acronyms, initialisms and shortcuts used in this book are reviewed right after this preface.

Acknowledgements. I am first of all grateful to the instances of the International Mathematical Union that entrusted this book project to me and assisted its realization step by step. The central person to be mentioned here is Helge Holden, the Secretary General of the IMU. His extremely well-informed, sober advice and helpful suggestions in numerous discussions have found their way into the final version of this book in more ways than could be enumerated here. My thanks also go to the other readers of consecutive versions of the manuscript for their comments: to Nalini Joshi, Paolo Piccione, and especially to Reinhard Siegmund-Schultze.

The staff at the IMU headquarters in Berlin was more than welcoming. Special thanks go to the IMU Archivist Birgit Seeliger, who went out of her way to suggest and facilitate my access, both to archival and published material.

I thank the former IMU secretary Martin Grötschel for sharing his reminiscences with me in a long conversation that we had in Berlin in October 2019.

Hearty thanks to Fulvia Furinghetti, Livia Giacardi, and Bernard Hodgson for opening my eyes to the world of ICMI, the International Commission on Mathematical Instruction. I am equally grateful to June Barrow-Green and Craig Fraser for sharing valuable insights with me about the International Commission on the History of Mathematics (ICHM).

I am deeply indebted to Birgit Petri for her relentless and painstaking work on the data-analysis which is presented in the final chapter, Sections 10.3–10.5. Part of this study exploits the generous access to zbMATH Open data that was arranged by Olaf Teschke via a contract signed with FIZ Karlsruhe/Leibniz Institut für Informationsinfrastruktur.

Thanks to Springer Verlag for the friendly cooperation with the production crew, which was managed by Leonie Kunz and Martin Peters. It was also Martin who suggested Barnaby Sheppard as copy-editor. Working with Barnaby was a delight.

Many other persons contributed to the final outcome of this book project in ways that are too diverse to be recalled here in detail. I have thus received valuable information or help from John Ball, Jean-Pierre Bourguignon, William McCallum, Hans-Joachim Dahms, Thomas Delzant, Vladimir Dragović, Christophe Eckes, Danielle Fauque, Catherine Goldstein, Johannes Huebschmann, Tinne Hoff Kjeldsen, Harald Kümmerle, Markus Messling, Jan Nekovář, Michael Rapoport, Volker Remmert, David Rowe, Marie-Françoise Roy, Peter Schöttler, Peter Schneider, Brigitte Stenhouse, Rossana Tazzioli, Renate Tobies, Dirk Werner, and Yuri Zarhin.

This book was generously funded by the Klaus Tschira Stiftung. The IMU and the author are particularly grateful to Frau Beate Spiegel for her enthusiastic support of the project.

Strasbourg and Berlin, January 2022 Norbert Schappacher

Acronyms, Initialisms and Shortcuts

ACM	Association for Computing Machinery – Sec. 10.1.2.3.
APS	American Physical Society – Sec. 8.2.4.1.
CC	Consultative Committee (of an ICM) – Sec. 10.2.
CCC	Copyright Clearance Center – References.
CDC	Commission for Developing Countries – Sec. 10.1.3.
CDE	Commission on Development and Exchange – Sec. 10.1.3.
CFRS	Committee on Freedom and Responsibility in the Conduct of Science – Sec. 8.2.2.3.
CI	Institution of the Cupola – Sec. 10.4.2.
CIEC	Committee for Electronic Information and Communication – Sec. 10.1.1.
CIMPA	Centre international de mathématiques pures et appliquées, Nice, France – Sec. 10.1.3.
CMI	Clay Mathematics Institute – Sec. 8.2.4.2, 8.2.5.1.
CoD	Committee on Diversity – Sec. 10.1.
COSTED	Committee on Science and Technology in Developing Countries – Sec. 8.1.2.
CTI	Confederation of Intellectual Workers – Sec. 6.2.2.
CWM	Committee for Women in Mathematics – Sec. 10.1.2.3.
DCSG	Developing Countries Strategy Group – Sec. 10.1.3.
DHST	Division of History of Science and Technology – Sec. 10.1.4.
EIM	Einstein Institute of Mathematics, Jerusalem – Sec. 7.1.
ETH	Swiss Federal Institute of Technology, Zürich – *passim*.
FNP	Fields Medal, Nevanlinna Prize, or Plenary Speaker – Sec. 10.4.2.
GAMM	Gesellschaft für angewandte Mathematik und Mechanik – Sec. 10.1.

HCM	Hausdorff Center of Mathematics, Bonn, Germany – Sec. 10.4.2.
IAA	International Association of Academies – Sec. 1.3.2.
IAU	International Astronomical Union – Sec. 1.4, 4.1.1.
ICC	International Criminal Court – Sec. 8.2.4.1.
ICIAM	International Council for Industrial and Applied Mathematics – Chap. 9, Sec. 10.1.
ICIC	International Committee on Intellectual Cooperation – Sec. 4.2.
ICHM	International Commission on the History of Mathematics – Sec. 10.1.4.
ICHS	International Committee of Historical Sciences – Sec. 8.2.3.1.
ICM	International Congress of Mathematicians – *passim*.
ICMI	International Commission on Mathematical Instruction – Chap. 9.
ICS	International Commission on Stratigraphy – Sec. 10.1.
ICSU	International Council of Scientific Unions – Sec. 8.1.2.
ICTP	International Centre for Theoretical Physics, Trieste, Italy – Sec. 10.1.3.
IEB	International Education Board – Sec. 5.1.
IEUS	International Encyclopedia of Unified Science – Chap. 6, opening.
IHES	Institut des Hautes Études Scientifiques, Bures-sur-Yvette, France – Sec. 8.3.
IHP	Institut Henri Poincaré, Paris – Sec. 5.1, 8.3.
IMA	Institute for Mathematics and its Applications – Sec. 10.1.
IMKT	International Mathematical Knowledge Trust – Sec. 10.1.1.
IMPA	Instituto Nacional de Matematica Pura e Aplicada, Rio de Janeiro, Brazil – Sec. 8.3.
IMU	International Mathematical Union – *passim*.
IMUK	Internationale Mathematische Unterrichtskommission (the German name of ICMI) – Chap. 9.
IRC	International Research Council – Sec. 4.1.
ISA	International Seismological Association – Sec. 1.4.
ISC	International Science Council – Sec. 8.2.2.3.
ISSC	International Social Science Council – Sec. 8.2.2.3.
IUBS	International Union of Biological Sciences – Sec. 1.4, 4.1.1, 10.1.2.3.
IUGG	International Union of Geodesy and Geophysics – Sec. 4.1.1, 10.1.
IUGS	International Union of Geological Sciences – Sec. 10.1.
IUHPST	International Union of History and Philosophy of Science and Technology – Sec. 10.1.2.3.
IUPAC	International Union of Pure and Applied Chemistry – Sec. 1.4, 4.1.1, 10.1.2.3.
IUPAP	International Union of Pure and Applied Physics – Sec. 10.1.2.3.
KIAS	Korea Institute for Advanced Study, Seoul, Korea – Sec. 8.3.
KWU	The German *Kaiser-Wilhelms-Universität* at Strasbourg – Sec. 4.3.

LGTRS	Locally-grounded Transnational Research Site for mathematics – Sec. 5.2; 8.3.
MPC	Minor Planet Center – Sec. 10.1.
MPIM	Max Planck Institut für Mathematik, Bonn, Germany – Sec. 8.3.
MPIMN	Max Planck Institute for Mathematics in the Natural Sciences, Leipzig, Germany – Sec. 8.3.
MSC	Mathematics Subject Classification – Sec. 10.3.
MSRI	Mathematical Sciences Research Institute, Berkeley, California – Sec. 8.3.
NGO	Non-governmental Organization – *passim*.
OC	Organizing Committee (of an ICM) – Sec. 10.2.
ONR	Office of Naval Research – Sec. 7.2.3.
OWSD	Organization for Women in Science for the Developing World – Sec. 10.1.2.3.
PC	Program Committee (of an ICM) – Sec. 10.2, 10.4.1.2.
RIMS	Research Institute for Mathematical Sciences, Kyoto, Japan – Sec. 8.3.
SC	Structure Committee (for the ICMs) – Sec. 10.2.
SCFCS	Standing Committee on the Free Circulation of Scientists – Sec. 8.2.2.3.
SIAM	Society for Industrial and Applied Mathematics – Sec. 10.1.
SMAI	Société de Mathématiques Appliquées et Industrielles – Sec. 10.1.
TIFR	Tata Institute for Fundamental Research (TIFR), Mumbai, India – Sec. 8.3.
URSI	International Union of Radio Science – Sec. 1.4, 4.1.1.
UAI	International Academic Union – Sec. 4.2.
UIA	Union of International Associations – Sec. 1.4.
UIR	*Ufficio Invenzioni e Ricerche*, Italian Office for Inventions and Research – Sec. 3.3.1.
UNESCO	United Nations Educational, Scientific and Cultural Organization – Sec. 4.2, 8.1.2.
URSI	International Union of Radio Science – Sec. 1.4, 4.1.1.
WDM	World Directory of Mathematicians – Sec. 10.1.1.
WIAS	Weierstraß-Institut für Angewandte Analysis und Stochastik, Berlin – Sec. 8.2.5.2.
WTO	World Trade Organization – Sec. 8.2.4.1.

Contents

Part I The Long Nineteenth Century that Made the IMU Possible: 1800–1918

1 Nationalism, Internationalism and the Sciences in the Long Nineteenth Century 3
1.1 Jobs and Journals 8
1.1.1 The Humboldt Brothers 11
1.1.2 Adolphe Quetelet and Mathematical Statistics 13
1.1.3 England 17
1.1.4 Paris and France 19
1.1.5 Italy 21
1.1.6 Gösta Mittag-Leffler 28
1.1.7 A Woman Mathematician with International Connections 29
1.2 Nation Branding through Science 30
1.2.1 Chemical Elements 30
1.2.2 Nation, Culture, Science 34
1.2.3 Nation and Mathematics 36
1.3 Felix Klein, a Sample of Projects he was Involved in 38
1.3.1 Attempts to Federate Pure and Applied Science 40
1.3.2 The Unlikely Resurrection of Scientific Academies as (Inter)National Agents of Science 40
1.3.3 The Encyclopedia of the Mathematical Sciences Including Their Applications 48
1.4 World Mathematics before World War I 56
1.4.1 International Congresses 58
1.4.2 Japan Goes West 63
1.4.3 Mathematical Associations 66
1.4.4 India's Entry onto the World's Mathematical Stage 70

2 Unfirm Foundations ... 73
2.1 Mathematics Meets Literature ... 75
2.1.1 Mathematics and Name Worshipping ... 75
2.1.2 Robert Musil ... 76
2.2 Hermann Weyl's Changing Attitudes to the Foundational Crisis ... 77
2.2.1 The First Phase: Before the War ... 77
2.2.2 The Second Phase: 1917–1918 ... 78
2.2.3 The Third Phase: After the War ... 83

3 World War I ... 85
3.1 To the Civilized World! ... 86
3.2 Intellectual Warfare ... 88
3.3 Mathematic(ian)s during World War I ... 90
3.3.1 Vito Volterra and Mauro Picone ... 91
3.4 International Congresses during the Great War ... 93

Part II Mathematical Consolidation in Times of Tempest: 1919–1949

4 The First IMU: Triumph and Demise ... 99
4.1 The Framers of the Council, the IRC ... 100
4.1.1 The First Scientific Unions within the IRC; Preparing for the IMU ... 105
4.2 The UAI; the League of Nations; the ICIC ... 108
4.3 Strasbourg ... 111
4.3.1 Maurice Fréchet in Strasbourg ... 114
4.3.2 The IMU Founded in Strasbourg ... 115
4.3.3 The 1920 ICM in Strasbourg: "la grande manifestation patriotique et scientifique" ... 117
4.4 The Waning Influence of the IMU ... 121
4.4.1 John Charles Fields ... 123
4.4.2 "The Disagreeable Tempest which Raged at Toronto" ... 126
4.4.3 Bologna and the Marginalization of the IMU ... 130

5 Philanthropic Capital for Mathematics ... 139
5.1 The Rockefeller Philanthropies ... 140
5.2 The Institute for Advanced Study, Princeton ... 148

6 Mathematical Consolidation and Unification in the 1930s ... 153
6.1 Emmy Noether's Legacy ... 157
6.1.1 What is 'Modern' about 'Modern Algebra'? ... 159
6.1.2 Emmy Noether's *Auffassung* and its Influence ... 165
6.1.3 Emmy Noether's International Network ... 167
6.2 Encounters, Workshops and Congresses in the 1930s ... 170
6.2.1 Specialized Conferences ... 171
6.2.2 ICMs of the Thirties ... 174

6.3 Books, Journals; *Zentralblatt* and *Mathematical Reviews* 180
6.3.1 Books 180
6.3.2 Journals and Politics 182
6.3.3 Review Journals and Politics 185
6.4 Three Journeys to the West 187

7 Forced Migration and World War II 195
7.1 Global Redistribution of Scientists in the 1930s and 1940s 196
7.2 What World War II Meant for Mathematics 202
7.2.1 Searching for the Hiding Place of the IMU 203
7.2.2 Mathematics for the War 205
7.2.3 How World War II Reshaped the World – the Case of Mathematics 208

Part III Seventy Years of Globalization: 1950–2020

8 Seventy Years, Eighteen ICMs, and One IMU 217
8.1 A New IMU and an ICM in Another World 217
8.1.1 United Nations, International Tribunals 218
8.1.2 UNESCO and ICSU 220
8.1.3 The New IMU 224
8.1.4 Gathering "a Very Large Part of the Mathematical World" .. 227
8.2 IMU Time Intervals 230
8.2.1 Gearing up to Run Mathematics International: The New IMU 1950–1962 230
8.2.2 From Moscow to Helsinki: 1966–1978 241
8.2.3 New Horizons: 1982–1990 251
8.2.4 Mathematics Without Borders? 1994–2002 259
8.2.5 Global Reach from a New Homebase: 2006–2018 264
8.3 A World Wide Web of Institutes 271

9 ICMI, The Resilient Nucleus of the IMU 275

10 Framing Mathematical Excellence 289
10.1 The Infrastructure of the IMU 289
10.1.1 The Committee for Electronic Information and Communication (CEIC) 291
10.1.2 Women in Mathematics 294
10.1.3 The Commission for Developing Countries (CDC) 304
10.1.4 The International Commission on the History of Mathematics (ICHM) 308
10.2 Framing ICMs 310
10.3 The Database 314
10.4 The Cupola of the ICMs 317
10.4.1 Parts of the mathematical world 317
10.4.2 Institutions of the Cupola 326

10.5 Framing Domains of Mathematics 339
10.5.1 Mathematical Subdomains 339
10.5.2 Fields Medalists 341
10.5.3 Plenary Speakers 342
10.5.4 Filtering the Mathematical Production 344

References 347

Index 377

Part I
The Long Nineteenth Century that Made the IMU Possible: 1800–1918

Der Freihandel der Begriffe und Gefühle steigere
ebenso wie der Verkehr in Produkten und Boden-
erzeugnissen den Reichtum und das allgemeine
Wohlsein der Menschheit. Dass das bisher nicht
geschehen sei, liege an nichts anderm als daran,
dass die internationale Gemeinsamkeit keine
festen moralischen Gesetze und Grundlagen habe.

The free trade of concepts and sentiments,
just like the commerce of products and crops,
increases the wealth and general well-being
of humanity. The fact that this has not yet
happened is due solely to the fact that international
commonality has no solid moral laws and foundations.

Johann Wolfgang von Goethe[4]

Recalling the century which preceded the founding of all the big international scientific unions which are still going strong today is more than a chore that one has to get out of the way before moving on to the real story; it is the only way to understand what these international unions could be in the first place, at the time of their creation, and how they could develop thereafter. The new international scientific unions created after World War I, and the IMU in particular, were shaped by closely intertwined, diverse and contradictory tendencies. These influences—political, economic, technical, cultural, scientific; romantic or rational—are reflected on in the following pages.

[4] Conversation with the Polish poet Antoni Edward Odyniec on 25 August 1829, three days before Goethe's eightieth birthday. Quoted from [Brunner et al. 2004], vol. 3, p. 375; my translation.

Chapter 1
Nationalism, Internationalism and the Sciences in the Long Nineteenth Century

This book is about *Science International*, and what it meant for mathematics over the past two centuries. The emergence of the particular kind of *Science International* at work in the twentieth century hinges on the concept of Nation States, i.e., states which claim to be political and cultural units at the same time. *Science International* thus originated from European, especially continental European developments in the nineteenth century.

Archimedes. Years ago, on the way to dinner during a small workshop in Strasbourg in the 1970s, I asked René Thom (1923–2002) what the Fields Medal looked like. He told me with a smile that he had unfortunately mislaid it a while ago and did not remember too well. It was only in 2019, at the Berlin offices of the IMU, that I saw a replica of the medal for the first time. One side of it shows a fantasy portrait of Archimedes (d. 212 BCE), identified as such by his name, which is written in Greek (in the genitive form, as was often done on ancient coins).[1] Let us pick up the cue, if only in a few lines, to orient ourselves.

Fig. 1.1 The two sides of the Fields Medal (Courtesy IMU).

[1] For more information about the graphic conception of the Fields Medal, see [Riehm & Hoffman 2011], pp. 184–186.

N. Schappacher, *Framing Global Mathematics*,
https://doi.org/10.1007/978-3-030-95683-7_1

In the third century BCE, when Archimedes wrote from Syracuse (on the East coast of Sicily, nowadays Italy) to Dositheus in Alexandria (nowadays Egypt) about his latest findings, he was evidently hoping to connect to a network of competent colleagues, such as Conon of Samos and Eratosthenes, beyond the borders of his home city state. Moreover, the fact that Archimedes wrote his letters and treatises in the Sicilian Doric dialect, instead of the *Koine* Greek used by others for global exchange across the hellenistic world, may conceivably[2] reflect King Hieron's ardent desire to elevate Syracuse to a dominating metropolis on the Mediterranean, sandwiched as it was between Rome and Carthage. Alexandria on the other hand, the capital of the multi-ethnic Ptolemaic Kingdom, was home to the unique central research hub of the hellenistic world, the *Museion*, which was run by Greeks who had settled there. Archimedes himself may or may not have profited from a research training sojourn at the *Museion* early in his career; we do not know. At any rate, publications by Archimedes in the form of letters are evidence of scientific networking across the Mediterranean at the time. Yet, even if his Doric idiom actually did have local political connotations, calling this exchange *international* in today's sense would be a serious anachronism.

To see why, we have to pin down what made the situation at the turn from the nineteenth to the twentieth century so profoundly different from Archimedes's time and world. Doing this will actually show us how the late nineteenth century differed also from other historical moments, in particular from the eighteenth century.

National scientific communities around 1900. First of all, the numbers of individuals involved were of course not the same. For fear of being mauled by expert historians of antiquity, I will not venture to gauge how many learned men around the Mediterranean may have engaged in scientific networking at the time of Archimedes. But at any rate, this whole group was surely not of the same order of magnitude as, for instance, the roughly 2000 "mathematicians and mathematical physicists" who were sent invitations to participate in the first International Congress of Mathematicians in 1897 in Zürich. Those invitation letters, by the way, were centrally printed in Switzerland, but then dispatched by national correspondents to invitees of their own or neighboring countries. The dispatchers were located in Woolwich (UK), Palermo (Sicily, Italy), Halle, Göttingen (Germany), West-Nyack (New York, USA), Paris (France), Gent (Belgium), St. Petersburg (Russia), Vienna (Austria-Hungary), Stockholm (Sweden), Groningen (The Netherlands), Athens (Greece), and Porto (Portugal).[3] The international mathematical world of 1897 was thus managed from Europe and the USA; and targeted countries had to have recognizable communities of "mathematicians and mathematical physicists" in order to be on the map of the organizers.

The Republic of Letters. Not only did this scientific specialization allude to a ramification and richness of scientific theories unheard of in antiquity, but the scientific communities of the late nineteenth century had a sort of national and professional

[2] This possibility is suggested in [Schneider 1979], p. 7, once one excludes mere personal linguistic limitations as the reason for Archimedes's choice of his native dialect.

[3] See Section I.A: *Vorgeschichte des Kongresses*, in Proceedings ICM 1897, pp. 3–21.

grounding which was unlike anything that had existed even one hundred years earlier. Indeed, characteristic of the eighteenth century organization of science across Europe (and to a certain extent, the USA) was what contemporaries then called the *Republic of Letters*. How this proud self-organization of *savants* differed from what one encounters later, in the nineteenth century, transpires for instance from a paper which Lorraine Daston once wrote for a special volume devoted to the concept of *style*, in particular national style, in the history of science. The author then found herself in the peculiar situation where she had to explain why this notion is essentially out of place as far as the sciences of the age of enlightenment are concerned:

> The Republic of Letters of the late seventeenth and eighteenth centuries teaches us two lessons about style in science. First, the bearer of style—individual, nation, institution, religious group, region, class—depends crucially on historical context. When the organization and values of intellectual life are self-consciously cosmopolitan, and when allegiances to other entities (e.g., Protestant versus Catholic, or urban versus rural) are culturally more compelling than those to the nation-state, distinctively national styles are far to seek. This was largely the case for the Republic of Letters, that immaterial (it lacked location, formal administration, and brick and mortar) but nonetheless real (it exercised dominion over thoughts and deeds) realm among the sovereign states of the Enlightenment. Second, that form of objectivity which made science seem so curiously detached from scientists, and therefore so apparently unmarked by style at any level, also has a history. The unremitting emphasis on impartial criticism and evaluation within the Republic of Letters encouraged its citizens to distance themselves first from friends and family, then from compatriots and contemporaries. . . [4]

The *Republic of Letters* was held together by an impressive web of correspondences, and it was institutionally represented by scientific Academies which, among other things, published papers (*memoirs*). Writing and receiving scholarly letters—that "peculiar hybrid of the personal and the public, composed with both a particular reader and a general readership in mind"[5]—had grown since Archimedes into a daily routine, which continued its relentless acceleration, clocked by the timetable of postal collections and deliveries, as history moved into the nineteenth century. For the *Republic of Letters*, letters were still almost as important as publications, and publishing (freely copy-edited) letters in print would remain common practice throughout the nineteenth century even as scientific journals independent of Academies were establishing themselves as the principal medium of scientific communication.

The scientific Academies of the seventeenth and eighteenth centuries "faithfully reflected" the non-national character of the *Republic of Letters*:

> Almost all included provisions for foreign or 'corresponding' members. [Jean-Baptiste] Colbert [1619–1683] invited the Dutch physicist Christiaan Huygens [1629–1695] to head the newly established Paris *Académie des Sciences* in the 1660s, and he installed the Italian astronomer [Giovanni Domenico] Cassini [1625–1712] as head of the observatory. Friedrich II of Prussia invited the French physicist [Pierre Louis de] Maupertuis [1698–1759] to lead the Berlin Academy; the French astronomer [Joseph-Nicolas] Delisle [1688–1768] was only one of several illustrious savants of French, Swiss, and German origin to spend years at the Russian Academy of St. Petersburg, the prizes of the Parisian *Académie des Sciences* (and

[4] [Daston 1991], p. 367.

[5] [Daston 1991], p. 371.

> also of less lofty academies such as those of Bordeaux, Amsterdam, Stockholm, Vienna, and about thirty others . . .) were explicitly open to foreigners, and the 1719 rules governing the *Académie des Sciences* competitions offered foreign contestants the option of writing in Latin rather than French, with the costs of translation to be borne by the Perpetual Secretary . . . [6]

Leonhard Euler. In mathematics, we have the outstanding example of the immensely productive Leonhard Euler (1707–1783).[7] He was born in Basel, Switzerland, and grew up nearby. When he could not get a job at Basel University, he left his home country, hardly 20 years old, and traveled to St. Petersburg where he would soon become one of the most prolific and visible members of the tsarist Academy. In 1741, during an unstable period of Russian politics, he shifted to the Berlin Academy under the recently installed King of Prussia Frederic II ('the Great'), while keeping excellent relations with the Academy in St. Petersburg, to which he would finally return for good in 1760, over increasing tensions with the Prussian King. Euler's massive correspondence fluently oscillates between languages, mostly Latin, French and German (with more or less pronounced Swiss-German elements, depending on the correspondent). Euler also mastered Russian—which cannot be said of all foreign members of the St. Petersburg Academy—, and while in Berlin he translated/reworked a British treatise on Artillery for the benefit of the Prussian military. The overwhelming majority of his equally polyglot research papers are not in scientific journals—of which only few existed in the eighteenth century—but published as memoirs of the Academies with which he was in liaison. A proud list of seminal books, in several languages and published in various European countries, added to the scientific authority of Euler well beyond the inner circle of mathematical research.

A number of academic teaching jobs at universities did exist in the eighteenth century, but Euler's career as a European researcher illustrates the *Republic of Letters* which, however dependent Academy positions could be on the local political situation, transcended states and kingdoms by appealing to a would-be cosmopolitan ideal of reason. This is all the more important to note as Euler's lifelong attachment to the belief of the Reformed Christian Church, in which his father had been a pastor, would otherwise make him an unlikely representative of the age of enlightenment.

Jeremy Bentham. Not only did the Academies thus incorporate a kind of *Science International* before the nineteenth century, but also the very word 'international' goes back to the eighteenth century. According to the findings of a voluminous historical research project[8], the English philosopher of law Jeremy Bentham (1748–1832) may well have been the first author to propose this new expression as he

---

[6] [Daston 1991], p. 372. For references given by Daston, see the original article.

[7] For a well-written concise overview of Euler's scientific career, I like to refer to [Weil 1987], Chap. III, §II, pp. 162–169. Yet the reader will not fail to appreciate that my main argument in Part I of the present book goes precisely *against* Weil's tongue-in-cheek comment just before his biographical sketch of Euler, to wit: "In short, scientific life, by the turn of the [eighteenth] century, had acquired a structure not too different from what we witness to-day." [Weil 1987], p. 162.

[8] See [Brunner et al. 2004], Vol. 3, p. 369.

was trying to "mark out" the "leading points ... in respect of which the laws of all civilized nations might, without inconvenience, be the same," and by extension distinguish the

> *political quality* of the persons whose conduct is the object of the law: These may, on any given occasion, be considered either as members of the same state, or as members of different states: in the first case, the law may be referred to the head of *internal*, in the second case to that of *international*[9] jurisprudence.

Dating a newly coined expression precisely is, more often than not, just as hopeless as trying to pin down the very first occurrence of a mathematical notion or theorem in the history of mathematics. The fact that the word *international* is claimed as a newly created term by Bentham , and that the historical project which we quoted also found no earlier occurrence, may therefore be worth mentioning. It happens to be in line with a string of attention that the author, who postulated 'the greatest happiness of the greatest number', has received in recent years.[10] From the philosophy of law, the word gradually spread first to commercial and economic contexts. From there it was, on the one hand, increasingly generalized, like in Goethe's comment which we chose as epigraph to Part I. On the other hand it became, as of 1864, a technical term in the history of the workers' movement.[11]

Having very briefly recalled earlier types of international scientific structures, we now have to unravel what changed in the course of the long nineteenth century, and which new meaning of the word *international* was taken for granted about one hundred years after Bentham. This new meaning would then become the frame of reference for the kind of *Science International* among highly developed nation states which would foster foundations like the IMU.

[9] Here Bentham attaches a footnote which begins: "The word *international*, it must be acknowledged, is a new one; though, it is hoped sufficiently analogous and intelligible. It is calculated to express in a more significant way, the branch of law that goes commonly under the name of *law of nations*: an appellation so uncharacteristic, that, were it not for the force of custom, it would seem rather to refer to internal jurisprudence. ..." These quotes are from page cccxxiv of [Bentham 1780/1789]. Not surprisingly considering Bentham's great admiration for French figures of the Enlightenment, his treatise appeared in French translation in 1802 (*Traité de législations civile et pénale*, E. Dumont Paris), helping the spread of the French word *international*.

[10] See for instance the European (British/French/Continental) approach to Bentham in [de Champs 2015]. Another context in which Bentham is regularly mentioned today, as an exceptional thinker of the eighteenth century, is the footnote in the same treatise in which he reorients the basic questions for the philosophy of animals: "... the question is not, Can they reason? nor, Can they talk? but, Can they *suffer*?" [Bentham 1780/1789], p. cccix.

[11] Cf. [Brunner et al. 2004], Vol. 3, pp. 370–397.

1.1 Jobs and Journals

New academic jobs, especially in universities, and new journals were the key innovations that the nineteenth century brought to the organization of science within society. Europe led the way for both mathematical jobs and journals, although—as usual with the political patchwork of this continent—in a medley of different national histories.

Some jobs go along with buildings. If you are thinking of planting a university campus today, you will need a lot of structures to accommodate the broad variety of disciplines which your university is supposed to make thrive. Not every single speciality may need a separate building; historians (especially of more remote history) may not complain if you make them move in with archaeologists. But mathematicians nowadays tend to have quarters of their own, more or less separated from other scientists. All this seems natural enough to everyone who is familiar with today's universities, whose architecture often reflects the independent disciplines which have established themselves. However, on the historic scale, it is a fairly recent phenomenon. Traditionally, the first specialized institutes that could convincingly argue for a separate building were anatomy (because of the smell) and astronomy (in order to be at a safe distance from distracting lights). This was accepted even before those institutes belonged to something one could call a university.

Once there is an independent building dedicated to a specific discipline, it will come equipped with at least two jobs: a director and a caretaker. This is the structural reason why, at the beginning of the nineteenth century, i.e., before the creation of substantial numbers of teaching and research university positions in mathematics, highly qualified candidates could sometimes more easily find an astronomical observatory to direct. Cases in point are Carl Friedrich Gauss (1777–1855) in Göttingen and Wilhelm Bessel (1784–1846) in Königsberg (nowadays, the Russian exclave of Kaliningrad).

Gauss's chair(s). Gauss had become a celebrity in 1801 for two very different achievements: on the one hand he fundamentally changed the scope of Higher Arithmetic—and in passing could prove the constructibility of the regular 17-gon by ruler and compass—in his seminal *Disquisitiones arithmeticae*, finally published in 1801. On the other hand, he successfully predicted the orbit of the minor-planet Ceres by dint of massive least-squares-approximations, based on rather scarce observational data. It was this astronomical success, together with his related theoretical astronomical publications, which would qualify him to become in 1807 the director of the Göttingen astronomical observatory, and in this way Professor at Göttingen University. Gauss died in 1855. Already by the time of the death of his immediate successor, Peter Gustav Lejeune Dirichlet (1805–1859), who did not work in astronomy at all, the situation of universities had evolved sufficiently to endow the subsequent successor Bernhard Riemann (1826–1866) with a mathematical chair independent of the observatory. And as of 1868—two years after Göttingen along with the Kingdom of Hannover had been swallowed up by Prussia—the astronomical observatory would itself be split into two departments, each with its own director, one for theoretical, the other for practical astronomy. The first director of the theoretical branch was Ernst Schering (1833–1897), who would also be the first managing editor of Gauss's Collected Papers.

Königsberg. The big Prussian city of Königsberg had already had a university for more than two and a half centuries when Wilhelm Bessel was appointed director of the university's observatory there in 1810, during Wilhelm von Humboldt's (1767–1835) reform of Prussian universities. Bessel's achievements as astronomer are tremendous, and his mathematical innovations, like the functions that today go under his name, are all closely related to mathematical physics and astronomy. But considering his methodology, he may well be counted also among the Königsberg mathematicians, the first of an impressive list which continues with Carl Gustav Jacob Jacobi (1804–1851), Franz Ernst Neumann (1798–1895), Heinrich Weber (1842–1913), Carl Louis Ferdinand Lindemann (1852–1939), Adolf Hurwitz (1859–1919), David Hilbert (1862–1943), Hermann Minkowski (1864–1909), Franz Meyer (1865–1934), and Kurt Reidemeister (1893–1971).[12] This list of mathematicians whose careers were marked by the University of Königsberg is an outstanding example of the impact that University mathematics would gather over the nineteenth century. In addition, certain seminal reforms originated there:

> A notable innovation of Jacobi's was the adoption of the research seminar (taken over from linguistics, another subject enjoying a period of growth in Germany). This spread from Königsberg to Berlin and beyond. Here for the first time advanced students were introduced to research, helped to find out what to read, invited to follow the professor closely and to discuss ideas of their own. Whenever possible a room was set aside for the purpose, and provided with a modest list of books, journals and reprints. Topics could even be assigned. This structured education, in mathematics as in other subjects, did much to bring about the German integration of science into the industrial process.[13]

But we are getting ahead of ourselves; local and national differences have to be taken into account more carefully.

Charles Babbage, Ada Lovelace, Mary Somerville. Charles Babbage (1791–1871) is probably best remembered and renowned today for his *Difference* and *Analytical Engines*, i.e., as a father of the programmable computer. Together with John F.W. Herschel (1792–1871) and George Peacock (1791–1858) he also "initiated a renewal of mathematics which forced the adoption of Leibnizian notation in the examinations at Cambridge University and essentially built up a new conception of algebra, as the language of symbolic reasoning."[14]

---

[12] See the list of chapters dedicated to mathematicians in the memorial volume [Rauschnigg & v. Nerée 1995], pp. 459–575. Note that it is not implied here that all these men actually held teaching positions at Königsberg.

[13] From Jeremy Gray's introduction to the last part of [Goldstein, Gray, Ritter 1996], p. 350.

[14] From the abstract of Marie-José Durand-Richard's Chapter 20 in [Goldstein, Gray, Ritter 1996], pp. 447–477; the quote is from p. 446. While focussing on the Cambridge network of Analytics, this chapter aptly highlights the transformation of the British educational system in response to the social and technological transformation of the industrial revolution.

Mentioning Babbage invites us to insert a remark on Ada Lovelace, i.e., Augusta Ada King, Countess of Lovelace, née Byron (1815–1852). Her 1843 English translation, with copious additions, of Luigi Federico Menabrea's (1809–1896) paper *Notions sur la Machine Analytique de M. Charles Babbage*, besides being a document of international scientific transmission, has attracted a great amount of admiration, along with historical interest and debate. "As well as broader speculation about the potential of the machine, for example to do algebra or compose music, the paper contains a large table setting out the calculation of the seventh Bernoulli number, often called the 'first computer program'."[15]

At the beginning of her scientific instruction, Ada Lovelace received some guidance from the Scottish mathematician and scientist Mary Somerville (1780–1872). Somerville "was remembered on her death as 'one of the most distinguished astronomers and philosophers of the day.' In her lifetime she published four books, which cumulatively went through 17 editions (not including the many pirated editions published in the United States of America), as well as appearing in translation in French, German, and Italian. Somerville also had papers published in the *Philosophical Transactions* and the *Quarterly Review*, and extracts from her letters were published in the *Comptes Rendus de l'Académie des Sciences* and the Edinburgh *New Philosophical Journal*."[16]

Mary Somerville, Ada Lovelace—and also Sophie Germain (1776–1831) in Paris[17]—lived in a world that did not offer women professional opportunities as scientists. Not only did they have to rely on family fortunes for their subsistence, but even entering into scientific networking required personal recommendations, or other arrangements. Sophie Germain for instance would first enter into correspondence with Gauss by using the male pseudonym Antoine Auguste Le Blanc. Mary Somerville's case is particularly interesting in this respect because of the role played by her husband.[18]. Somerville also stands out among the three women because of her extensive international sojourns in France and especially in Italy. However, her prolonged stay in Italy also strained her contacts with the British community.[19] In spite of all the difficulties that stood in the way of Germain or Somerville, their biographies demonstrate that women have long had not only interest in mathematics, but also the ability and resilience to contribute to mathematical knowledge at a high level.

Let us turn to Charles Babbage's international connections: Ever ready to go public with grouchy remarks, he published in 1830 a small book, the title of which leaves no doubt as to its message: *Reflections on the Decline of Science in England.*[20] His

---

[15] [Hollings et al. 2017a], p. 203. See also [Hollings at al. 2017b].

[16] From the second page of [Stenhouse 2020].

[17] Both Sophie Germain's historical situation and the way her role was seen at the end of the nineteenth century are analyzed by Boucard in her chapter in [Kaufholz-Soldat, Oswald 2020], pp. 186–230.

[18] This is shown in [Stenhouse 2020].

[19] See [Patterson 1983], p. 193, where a planned meeting with Babbage in Siena in 1840 is discussed. I am grateful to Brigitte Stenhouse for this information and the reference.

[20] See [Babbage 1830].

argument seems to be directed in the first place against the *Royal Society*.[21] Babbage criticized the allegedly cavalier way in which its fellows were recruited, and he also thought that there were simply too many of them. The Royal Society, which dates back all the way to the 1660s, is for Great Britain what scientific Academies were for their respective political entities on the continent. So we see that Babbage's argument naturally started from the emblematic institutions of the *Republic of Letters*. But then, comparing Great Britain with the continent in 1830, he detected a new trend: "Perhaps, at the present moment, Prussia is, of all the countries in Europe, that which bestows the greatest attention, and most unwearied encouragement on science."[22]

1.1.1 The Humboldt Brothers

It is one thing to create or reform universities, but another to make them work in a way that could impress an observer from the British Isles. In 1810, in the middle of a few years of 'peace' with Napoleon, Wilhelm von Humboldt—albeit himself a somewhat unlikely candidate for this job, which he would actually quit after only 16 months—oversaw a profound reorganization of the whole system of higher education in Prussia in a *neohumanist* spirit. Aside from renovating Königsberg University, he also first created a university in Prussia's capital Berlin. Starting from a reading of ancient Greek civilization as a model for humanity and succeeding to inspire ideas of personal perfection and emancipation in the broader bourgeois strata of the society, 'Humboldt's reform' has become to this very day a household name for a "notion, a program, which ... was inconsistent in itself, has never been fully realized, possibly could never have been realized, but has had extraordinary impact and consequences..."[23], at least in Germany.

It was, however, Wilhelm's brother, the world explorer Alexander von Humboldt (1769–1859), who would finally take it upon himself to deliver on the initial promise also for the exact sciences and actually turn Berlin University, and in fact the whole Prussian system of higher education, into an enterprise that scientists abroad would notice.

When the publication of the results of his South American expeditions was nearing completion, Alexander von Humboldt was obliged by the Prussian King to leave his beloved Paris after 19 years of splendid scientific and social exchange and return to Berlin in 1827. Humboldt chose to travel from Paris to Berlin via London. There, on 26 April 1827, he seized the occasion to inspect the construction site of the Thames Tunnel from a diving bell in the company of the chief engineer and two other men; one of them was 79-year-old Jeremy Bentham.[24]

---

[21] Cf. [Levere 2001], p. 122.

[22] [Babbage 1830], p. 31.

[23] My translation from [Vierhaus 1987], p. 63. Cf. [Pyenson 1983].

[24] Cf. the Digital Humboldt Edition, in particular its searchable chronology, at [URL 01].

Once installed in Berlin—with a population of about 230,000, about one third the size of Paris at the time—Alexander von Humboldt set up what Herbert Pieper has appropriately called 'a network of scientific philanthropy' in which mathematicians with particular interest in pure mathematics, especially number theory, such as Gauss, Dirichlet, Jacobi, Ernst Eduard Kummer (1810–1893), Gotthold Eisenstein (1823–1851), and others played an important role. Already in Paris he had socialized with mathematicians of the previous generation such as Joseph-Louis Lagrange (1736–1813), Pierre-Simon Laplace (1749–1827), Adrien-Marie Legendre (1752–1833), Jean-Baptiste Joseph Fourier (1768–1830), as well as Siméon Denis Poisson (1781–1840) and Augustin-Louis Cauchy (1789–1857). According to his own testimony, this had allowed him to develop a "certain flair for the relative quality of various mathematicians"[25] in spite of his freely avowed total ignorance of mathematics.[26] In this way,

> ... a perpetually increasing circle of number theorists developed around Gauss and Humboldt, and remained in close contact. They progressively constructed a network of relations, of friendships, of links, a communication network which was used sometimes to exchange mathematical knowledge, sometimes also to support newly-discovered young mathematicians. At times ... also Leopold Crelle [1780–1855], the editor of the *Journal für die reine und angewandte Mathematik*, Christian Schumacher, the editor of the *Astronomische Nachrichten*, Franz Encke, the Secretary of the mathematical (resp. mathematical-physical) class of the Berlin Academy of Sciences, Wilhelm Bessel, the Director of the Königsberg Observatory, and others, were involved in this network. ... Humboldt and his colleagues operated their philanthropic activities in the interest of mathematics, using each other's judgments and arguments to secure financial means and positions for the newcomers.
>
> As far as positions were concerned, Humboldt had mainly in view the Prussian universities. His declared objective on his return to Berlin was to act in favour of the development of mathematics and the natural sciences in Prussia, and first of all in Berlin. He also succeeded in avoiding several potential departures to non-Prussian universities by improving the financial situation of the scientist in question. He wrote recommendation letters to Gauss, to the King, to the ministers, but also gratifying and encouraging letters to young mathematicians with advice on efficient behaviour to adopt...[27]

Charles Babbage came to Berlin in September 1828 to meet Alexander von Humboldt on the last leg of an extensive tour of the European continent (especially Italy). Upon his arrival in Berlin he learned about, and then participated in, the scientific congress *Naturforscherversammlung* that Humboldt was hosting in Berlin later that same month. Babbage was invited twice to brunch at Alexander von Humboldt's house. Both times, Lejeune Dirichlet was among the guests. On the first occasion, Dirichlet's future brother-in-law Felix Mendelssohn-Bartholdy joined the party. The second time, Babbage was invited because Gauss, whom Humboldt had brought to Berlin for the Congress, had expressed the wish to make Babbage's acquaintance.

---

[25] See [Biermann 1988], p. 37, both for the list of French mathematicians and for the quote (my translation).

[26] Cf. [Pieper 2007], especially the synthesis at the end of the chapter, pp. 227–228.

[27] [Pieper 2007], p. 228.

Crelle. August Leopold Crelle was already mentioned in the quote above as belonging to the Berlin circle around Alexander von Humboldt. As a matter of fact, Crelle had started preparing the ground for his momentous *Journal für die reine und angewandte Mathematik* several years before Humboldt returned to Berlin.[28] He wrote the preface to the first volume in December 1825, almost two years before the two men first met (November 1827). In the following years, both would complement each other in many ways, lobbying with the government while entertaining personal relations with scientists to consolidate Berlin as a scientific, and especially mathematical center.

Crelle originally was a government building officer specializing in roadworks, but he had a keen interest in mathematics, in particular applied mathematics. He also had his own considerable international network of mathematicians, including Frenchmen such as Joseph Diez Gergonne (1771–1859)—who conceived and edited the first French journal exclusively devoted to mathematics, the *Annales de mathématiques pures et appliquées* which appeared from 1810 through 1832[29] —and Legendre, but also younger colleagues like Louis Poinsot (1796–1874), Poisson, Jean-Victor Poncelet(1788–1867), and eventually Joseph Liouville (1809–1882), who would start his influential *Journal de mathématiques pures et appliquées* in 1836. In the core of Crelle's group of authors we find from the very start the geometer Jakob Steiner (1796–1863), Jacobi, and the Norwegian genius Niels Henrik Abel (1802–1829). Crelle would translate Abel's papers from French into German for publication in his *Journal*; the four issues of the journal which appeared in the first year of its output already include five of Abel's mathematical papers and two of his notes on mechanics.

1.1.2 Adolphe Quetelet and Mathematical Statistics

Among Crelle's correspondents we also find the Belgian astronomer and statistician Adolphe Quetelet (1796–1874), who had founded in 1825 the Belgian[30] journal *Correspondance mathématique et physique*, which had a broader scientific and literary

---

[28] Cf. [Eccarius 1976], especially pp. 9–10.

[29] As we go along, we will occasionally mention examples of mathematical journals that were newly founded in the nineteenth century. Even if we do not discuss them in greater detail, we trust that the reader will appreciate—as suggested by the title of Section 1.1—that these new means of publication, which were independent of scientific Academies, were part and parcel of the new kind of professionalization of mathematics. For an in-depth study of the world wide history of the mathematical press we refer to the harvest of the historical research project CIRMATH; see for instance the special issue 19–2 (2017) of *Philosophia scientiae*, especially the introduction [Nabonnand et al. 2015], and the special issue 45–4 (2018) of *Historia Mathematica*. Cf. [URL 02].

[30] To be precise, published in the *Royaume des Pays-Bas*, until the Belgian revolution of 1830.

Fig. 1.2 Alexander von Humboldt (top), August Leopold Crelle (left), Adolphe Quetelet (right).

spectrum than Crelle's. As editor in chief of his journal he had his own international network of correspondents, including for instance Mary Somerville and her husband.[31]

In his notes of a journey to Berlin in the Summer of 1829, Quetelet praised Crelle in particular for the *véritable service* delivered by Crelle's journal. He also acknowledged the subsidy from the Prussian government, "always ready to support scientific endeavors," which had enabled Crelle to launch the journal.[32] This German trip further led Quetelet to Weimar in time for Goethe's eightieth birthday, after which he briefly met Gauss in Göttingen,[33] and thereafter continued to Heidelberg for the 1829 edition of the *Naturforscherversammlung*, i.e., the sequel of the 1828 Congress that Alexander von Humboldt had hosted in Berlin. About the final discussion there preparing the subsequent meeting for 1830, he wrote in particular:

[31] See [Stenhouse 2020].

[32] See [Quetelet 1830], p. 144; my translation.

[33] See the second part of [Quetelet 1830].

> The question also came up whether the meeting ought to be extended to all parts of Europe, or whether it should remain specifically attached to Germany. The question was decided affirmatively in the latter sense, and I think with good reasons. The national *amour-propre* now tends to give it the highest possible solidity and luster. One does not reject foreigners, on the contrary: one admits them with the same respect, the same rights as the other members. But one does not want the place of the meeting to wander anywhere within the limits of Europe. Also, the variety of languages, let alone the indifference which would spring from such an extension, would soon destroy its unity, which is the condition of its existence. After all, the various countries can have their special meetings; as a matter of fact, the German conference was created emulating a Swiss model. However, it would be useful if such meetings, if and when they are initiated, could relate to each other and communicate.[34]

Quetelet's reaction highlights the significance of national identity for the professionalization of the exact sciences at the time. We will discuss this more extensively for the later nineteenth century in Sections 1.2 and 1.3 below. At first glance it could seem to contradict, or at least constrict the idea of *Science international*. However, what actually happened was more subtle. All the international scientific unions for which the long nineteenth century prepared the stage were in fact based neither on a candid internationalist program nor on purely nationalistic ideas; but on a remarkable entanglement of national vs. international options and priorities. Cooperation and competition were two equally important and central modes of interaction when scientists from different nations got to work together. And what counted as success in such a cooperation and competition could depend on what scientists considered to be culturally relevant (see Section 1.3.3.1 below), or in the public interest.[35]

What we quoted above is of course not what Adolphe Quetelet is chiefly remembered for today.[36] His distinguished place in the history of science is due to his project of a new science of 'social physics' based on mathematical statistics, and more precisely on the normal distribution. The latter he had transferred from his work as an astronomer, and more precisely from the Laplace–Gauss theory of errors of multiple measurements. To quote Alain Desrosières's concise characterization:

> On the one hand, the regularity of the annual rates of births, deaths, marriages, crimes, or suicides in a given country, opposed to the contingent and random nature of each of these occurrences, suggested that these additions were endowed with properties of consistency quite different from those of the occurrences themselves. On the other hand, the striking resemblance between the forms of distribution of large numbers of measurements—whether these were repeated measurements of the same object, or one measurement of several different objects (for example, a group of conscripts)—confirmed the idea that these two processes were of the same nature, if one assumed the existence, beyond these individual contingent cases, of an *average man* [*homme moyen*], of which these cases were but imperfect copies. Thus each of these two ways of relying on statistical records and their totals in order to create a new being involved the centralized collection of a large number of cases. The task of reconciling and orchestrating these two different ideas—and also of organizing the censuses

[34] See [Quetelet 1830], p. 231; my translation.

[35] This is analyzed for the example of expeditions to the South pole in [Soutschek & Nickelsen 2019].

[36] Speaking of what he is or is not remembered for, let us recall in passing that Quetelet was apparently the first to have recognized the remarkable stability of the ratio of weight by the square of the height in adult humans, even though he certainly did not envisage the normative way in which the *Body Mass Index* is used nowadays in medical practice. See [Quetelet 1833].

and the national and international statistical systems needed to produce these figures—was all accomplished by one man, Adolphe Quetelet (1796–1874), a Belgian astronomer who was the one-man band of nineteenth-century statistics.[37]

It was this scientific project of Quetelet's that would usher in the first series of truly international scientific congresses with an essentially mathematical substance. The event and its consequences inform us about the sense that the notion of *Science international* had acquired by the middle of the nineteenth century, and mark the first intersection of our story with that of the World Fairs.

The need for the principal States to get together and exchange their knowledge and experience was felt in England where on the initiative of Prince Albert (the consort of Queen Victoria) the Great (or Universal) Exhibition was held in London in 1851.[38]

At this Exhibition there was a visitor, Adolphe Quetelet, the distinguished Belgian mathematician and astronomer, and it was on his initiative that, after consultation with the many foreign delegates present at the Exhibition, it was decided to hold an international statistical congress, and the Central Statistical Commission of Belgium called such a congress for September 1853 in Brussels. In December 1860 Babbage (one of the founders of the Statistical Society of London, later the Royal Statistical Society) in a letter to [William] Farr [1807–1883] the distinguished statistician and President of the Society in 1871–3, relates the part played by Quetelet: "At length, the conviction of the importance of the value of Statistical Science becoming widely extended in other countries, M.[onsieur] Quetelet saw that a fit time had arrived for summoning a European Congress. The results of such meetings are invaluable to all sciences but more particularly to statistics in which names have to be defined, signs to be invented, methods of observation to be compared and rendered uniform; thus enhancing the value of all future observations by making them more comparable as well as more expeditiously collected."[39]

The Congresses initiated by Quetelet would eventually result in the founding of the *International Statistical Institute* ISI in 1885, which is still thriving today. We saw above that Quetelet had a definite mathematical model to treat data. However, the international congresses and the encouragement that they gave to create national statistical commissions in all participating countries, with a view to unifying criteria for the collection and treatment of data, show that the internationality of this domain was chiefly a data-driven collaboration on national and international levels. Indeed, by 1885 the nature of mathematical statistics was already steering away from Quetelet's fixation on mean values, under the influence of Francis Galton (1822–1911) and others who would marry a new kind of variational statistics (linear *regression*, correlation analysis, etc.) with the idea of eugenics.

[37] [Desrosières 1998], p. 73–74.

[38] *Comment by N.Sch.:* Although it was preceded by a series of National Industrial Exhibitions, in particular in Paris, the 1851 event in London's Crystal Palace generally counts as the first World Fair.

[39] Quoted from [Nixon 1960], p. 5–6.

1.1.3 England

Particularly in mathematics, the British educational system remained largely dominated throughout the nineteenth century by the Colleges and the famous examinations which they trained their students for. A few biographies shed light on the (non-)availability of jobs for mathematical talent. The tumultuous life of James Joseph Sylvester (1814–1897) is on this account as instructive as it is distressing.

> During the course of his life, James Joseph Sylvester repeatedly experienced a sense of sliding on 'the world's slippery path' . . . , but his footing seemed perhaps most unsure in the years prior to 1850 as he sought to establish himself professionally while pursuing his mathematical interests. The balance proved difficult to strike for a number of reasons. First and foremost, the twentieth century concept of a 'professional mathematician' as someone who earns a living by proving theorems did not exist in Victorian Britain. The universities had mathematics chairs, fellows, and tutors, but undergraduate teaching exclusively defined these positions. Moreover, as a Jew and thereby a non-Anglican, Sylvester could not subscribe to the Thirty-nine Articles of the Church of England and so was ineligible for any degree from or post associated with an Anglican institution. How, then, could he make his way doing what he did best, namely, mathematics? The answer was unclear, and he took a number of false steps in trying to find it.[40]

This is not the place to recount Sylvester's biography. Let us just note that the first university-like British institution where he did obtain a teaching job was the Royal Military Academy in Woolwich in 1855. In 1876, he became inaugural professor of mathematics at Johns Hopkins University in Baltimore, Maryland. At the time this was a very recent (1873) philanthropic foundation with Humboldtian inspirations. In 1878, Sylvester founded the *American Journal of Mathematics*. And in 1883 he finally had the opportunity to return to the UK to succeed Henry J.S. Smith (1826–1883)—whose work on Higher Arithmetic was atypical for Britain at the time but well perceived on the continent[41]—on the Savilian Chair of Geometry at Oxford University, one of the rare traditional mathematical posts in the UK, instituted by donation in 1619.

Sylvester and Arthur Cayley (1821–1895) met in 1847 when they both studied law to become potentially independent of the meagre academic job market. At the same time they continued their mathematical research, which for both of them would soon turn to algebraic invariant theory. Cayley, or as Sylvester would call him some twenty years later: "Our Cayley, the central luminary, the Darwin of the English school of mathematicians"[42], was one of the most prolific mathematicians of the nineteenth century. His approach would tend to privilege the purely mathematical aspect even of those questions which were motivated by physical problems. In 1863, he would be appointed the first Sadleirian Professor of Pure Mathematics at Cambridge University. He held this post until his death. The transformation of the

[40] This is the first paragraph of the first chapter: *Negotiating "The slippery path"* in [Parshall 1998], p. 1.

[41] Cf. the remarks on Smith in Sloane Despeaux's chapter in [Parshall & Rice 2002], p. 82.

[42] Quoted from Sylvester's Presidential Address to the British Association in 1869; see [Sylvester 1908], p. 655.

Sadleirian Lectureship into a professorship linking teaching and research was an institutional acknowledgement of the new nineteenth century needs in University mathematics.

Stokes. There is another famous mathematical professorship in England which, like the Savilian chair, goes back to the seventeenth century: the Lucasian chair at Cambridge. Charles Babbage held it between 1828 (after his European journey, which was mentioned in Section 1.1.1) and 1838. Yet he neglected his teaching, a practice which was continued by his immediate successor on the chair, the President of Queen's College Joshua King (1798–1857). In fact, King did not really publish either.[43]

And so it was George Gabriel Stokes (1819–1903) who, as of 1849, restored the dignity of this professorship that had once been held by Isaac Barrow (1630–1677) and Isaac Newton (1642–1726). At first, Stokes did extra teaching at the London Government School of Mines to make up for the scanty salary of the Lucasian chair. He continued to hold the chair for the rest of his life. Clearly a mathematical physicist, Stokes tailored "his papers on pure mathematics to his requirements for solving physical problems."[44] In this respect, he was a natural complement at Cambridge to Cayley's outlook on mathematics, and he would have probably been amused if someone could have told him that "his theorem" would become the ultimate destination of a major rewriting of analysis undertaken in the twentieth century by a group of pure mathematicians publishing under the name of Bourbaki.[45]

> In the early years of his career, through the Cambridge Philosophical Society, his teaching, and the examinations he composed, Stokes was a pivotal figure in furthering the dissemination of French mathematical physics at Cambridge. Partly because of this, and because of his own researches, Stokes' was a very important formative influence on subsequent generations of Cambridge men, including [James Clerk] Maxwell [1831–1879]. With [George] Green[1793–1841], who in turn had influenced him, Stokes followed the work of the French, especially Lagrange, Laplace, Fourier, Poisson, and Cauchy. This is seen most clearly in his theoretical studies in optics and hydrodynamics; but it should also be noted that Stokes, even as an undergraduate, experimented incessantly. Yet his interests and investigations extended beyond physics, for his knowledge of chemistry and botany was extensive, and often his work in optics drew him into those fields.[46]

We cited this aspect of Stokes's scientific orientation because a similar tendency would continue to characterize the situation of mathematics at Cambridge on its way into the twentieth century:

> Perhaps paradoxically, it was the natural sciences (applied mathematics and physics) at Cambridge that offered a greater threat to the place of mathematics at Cambridge. Under J.[oseph] J.[ohn] Thomson's [1856–1940] leadership, the School of Natural Sciences

[43] Of all the mathematicians we have named so far, he is the first who is not mentioned once in [Gillispie 1970–1980].

[44] [Gillispie 1970–1980], Vol. XIII, p. 78.

[45] For a first historical orientation about "Stokes's Theorem", a special case of which he set as an examination question in 1854, see [Katz 1979].

[46] Quoted from E.M. Parkinson's article on Stokes in [Gillispie 1970–1980], Vol. XIII, pp. 74–79.

grew in the 1880s and 1890s, and students switched in increasing numbers from mathematics to science. Matters came to a head in 1906 when, after much debate, it was decided to change the mathematics syllabus. Leading the debate was E.[rnest] W.[illiam] Hobson [1856–1933], who noted that the ideas of many Continental mathematicians 'had never permeated the teaching of Cambridge to a sufficient degree to form a real school of mathematics.' Cambridge mathematics thereafter proceeded with a strong tendency towards applied mathematics.[47]

1.1.4 Paris and France

On more than one occasion we have listed the names of French mathematicians with whom our actors were in touch. Paris was obviously the cultural and scientific capital of Europe at the beginning of the nineteenth century. The French academic landscape was marked by the Napoleonic reforms in the wake of the French Revolution. Institutionally the *École Polytechnique* stands out, relegating universities to a subordinate level. But putting it like this conveys only imperfectly the profound difference between French higher education and that of the other European countries. In fact, even speaking of 'universities' is inadequate. When addressing institutions of higher learning outside the elitist *Grandes Écoles*—specifically the *École Polytechnique* and the *École Normale Supérieure* (ENS), the latter emerging towards the end of the nineteenth century as another center of mathematical excellence in Paris—the French considered faculties (*Facultés*), rather than universities, to be the primary objects. Furthermore, the professors teaching at these faculties share the French word *professeur* with the teachers at secondary schools. Those admitted and trained at the *École Polytechnique*, on the other hand, would typically graduate from this institution to join the technocratic[48] elite of military or civil servants running the country. The fact that mathematics was valued so highly in the supreme institutions of the educational system transferred a special kind of splendor to this field of knowledge which seems unmatched in other countries.

The leading role of mathematics in the technocratic culture of the nineteenth century results in fact from a long historical process with both military and academic roots. It was the science of soldiers and engineers on the one hand, and of *savants* on the other. Associating in an organic way the academic world with the technocratic universe, the *École Polytechnique* achieves the definitive coalescence of these two conceptions regarding the role of mathematics.

... But this two-sided orientation, joining theoretical and applied mathematics, is not sufficient to adequately characterize the teaching of mathematics at *École Polytechnique*. One also has to take into account the decisive role that mathematics played in the grading of the students, all the way from admission to the final exam. It is this selective role of the field, even more than the usefulness of its applications, which conveyed to mathematics its predominant position at the *École*.[49]

---

[47] From Jeremy Gray's introduction to the last part of [Goldstein, Gray, Ritter 1996], p. 353.

[48] I follow this deliberate anachronism proposed for the case at hand in [Belhoste 2003], *passim*; see in particular p. 13.

[49] [Belhoste 2003], p. 174; my translation.

In the second half of the nineteenth century, the French system found itself confronted more and more painfully with the Prussian model. After the French defeat in the German-French war of 1870–1871 and the ensuing proclamation of the German Empire at the Palace of Versailles, restoring the glory of French science in the face of the German competitor became part of the French national agenda, and was generally accepted to require a reshaping of universities. But given the structural differences between the German and the French system, this task was also paradoxical from the outset, and copying the German model was downright impossible.[50]

> The university reform was initiated by scientists who suffered more than others from the limits of the Napoleonic system. But their respective ambitions were driven by three distinct projects. Some of them, typically with a position in Paris, were eager to catch up with their foreign competitors and thus regain the predominant position which French science had enjoyed at the beginning of the nineteenth century in Europe. The others, typically teaching in the province, were first of all trying to obtain the resources they were lacking, simply in order to be able to work without having to try and get promoted to a position in the capital. And the third and most ambitious colleagues wanted to propagate their scientistic ideal . . . in all of the (re-established) universities, and thereby transform in a lasting manner the opinions and attitudes of the French elites whose failure in 1870, they were convinced, was due to a lack of method and scientific spirit. These different goals, if they were all inspired by the same idea, popular at the time, to the effect that Science, i.e. essentially, the application of an experimental and rational method could solve any problem did not call for the same collective behavior.[51]

Thus, after 1870, French science found itself trapped in an awkward position. This general, institution-oriented description of the situation is worth highlighting. It is part of the crucial background information one needs to appreciate the politics surrounding French mathematicians and the founding of the IMU in 1919–1920. But we will not go further into the history of French educational reforms around the turn of the century.

Charles Hermite. Rather, we propose to look at one French mathematician of the period who stands out, not only in view of his scientific excellence, but also because of his unusual career, research interests, and his tremendous international networking.

> By general consensus Charles Hermite (1822–1901) was one of the most important mathematicians of the nineteenth century. He was at the center of French mathematics in the second half of the century: elected member of the Academy in 1856, *maître de conférences* at the École Normale Supérieure, and then professor both at the École Polytechnique and at the Sorbonne in the 1870s. He wove an enormous web of international correspondents and visitors including Italy, the US, Germany, Russia, Sweden, Bohemia, etc. He is also the only French mathematician whom André Weil chose to count among 'the great number theorists' of the nineteenth century.[52]

---

[50] Cf. [Charle 1994] for a detailed prosopographic study of the question, across all scientific disciplines.

[51] [Charle 1994], p. 136–137; my translation.

[52] [Goldstein 2011], p. 129; my translation. The reference to Weil alludes to the beginning of the Introduction to [Kummer 1975].

Hermite's international contacts are particularly noteworthy as they began at a time when "the belief that there was mathematical activity abroad with which French mathematicians should be acquainted and involved was very much a minority view [in France] during the 1860s and 1870s. Indeed, the work of men like Gaston Darboux [1842–1917] and Charles Hermite was quite exceptional."[53] This exceptional role is even more remarkable when Hermite's networking with Germany—which included both an early journey to Berlin in the 1850s and his participation in the Göttingen commemoration of Gauss's centenary in 1877, i.e., after the German-French war which had turned Hermite's birthplace into a German town—is seen against the backdrop of his conservative, anti-Republican political convictions.[54]

Young Hermite had dropped out of the *École Polytechnique*, at least partly for reasons of health. Early on, he was interested in (pure) mathematics for its own sake, not with a view to applications, nor as a stepping stone toward a brilliant civil servant or military career in his home country. To him, however, pure mathematics was a natural science, with given objects that have to be classified.[55] He transmitted this fundamental orientation to his disciples, in particular to Henri Poincaré (1854–1912).

1.1.5 Italy

Unlike France, Italy did not exist as a national entity during the first half of the nineteenth century and her academic landscape was anything but centralized, in spite of widely visible academic centers like Torino, or Napoleonic foundations like the *Scuola Normale Superiore* in Pisa. And unlike what happened in the German states that finally coalesced in the German *Reich*, the *Risorgimento*, i.e., the building of the Italian Nation through the struggle for independence and reunification (1849–1870), was in a peculiar way conceptualized through statistics, and a remarkable number of mathematicians fought in the battles for national unity or appeared on the political scene of the new Italian Nation.

1.1.5.1 Nation Building through Statistics

That "Italy, in a sense, was a creature of statistics"[56] has been established by Silvana Patriarca.[57] The word 'statistics' here seems to refer first and foremost to what was also called 'political arithmetic' at the time, i.e., the sort of descriptive stately, administrative exercise of collecting and presenting data for politics and economics. Some parts of Italy experienced this practice when they had belonged to Napoleonic

[53] See [Gispert 2002], p. 105. On Darboux, cf. [Rowe 2018].

[54] See [Goldstein 2007], p. 379–380; cf. [Archibald 2002].

[55] See [Goldstein 2007], [Goldstein 2018].

[56] [Osterhammel 2014], p. 29.

[57] See [Patriarca 1996].

France. However, the author shows that the sort of statistics that helped to format a unified concept of the Italian Nation, across regional variations, were tightly interwoven with nineteenth century *mathematical* statistics, specifically with Adolphe Quetelet's project of 'social physics' and the related time-honored notion of 'moral statistics', in particular criminal statistics.

To cite just one example, "the first rigorous application of the language and methodological procedures of Queteletian social statistics by an Italian scholar"[58] was an elaborate essay[59] by Angelo Messedaglia (1820–1901), a professor of political economy and statistics at Padova University, in which he meticulously deconstructed, "with the authority of an already well-known scholar", the criminal statistics of the region of Lombardy-Venetia that had been published by the Austro-Hungarian authorities. His analysis was presented "to the academicians gathered in the *Imperial Regio Istituto Veneto di scienze, lettere e arti*, the most prestigious scientific institute of Venetia which had just elected him member." When he published the complete report, Messedaglia could proudly recall that Venice too had in the meantime been able "to integrate the common fatherland, liberated as she was from foreign rule."[60]

Pride about the new Italian nation apart, the scientific essence of Messedaglia's political argument is the quintessential question of any work in applied mathematical statistics: does the number crunching establish a certain disposition as the relevant cause of observed events.

Indeed, the Austro-Hungarian statistics of the region of Lombardy-Venetia had interpreted the potentially violent southern passions (*passioni meridionali*) of the inhabitants—what is North of Rome may still be far South from Vienna—as the essential reason for their criminal record. This is what Messedaglia managed to invalidate through painstaking arguments. In 1866 Messedaglia "was elected representative to the Lower House, where he sat in the center-right, and was appointed senator in 1884."

In a completely different context, we will revisit this same problem, of establishing a causal relation via statistics, once more when we look at Bartel L. Van der Waerden's (1903–1996) would-be "proof" of the fact that women are naturally less gifted in mathematics than men—see Section 10.1.2.1.

1.1.5.2 Mathematicians as Politicians

The immediate explicit link between mathematics and politics that we saw in Messedaglia's work was of course not available to predominantly pure mathematicians like Enrico Betti (1823–1892), Francesco Brioschi (1824–1897), Luigi Cremona (1830–1903), Eugenio Beltrami (1835–1900), and others. While all of them actively fought in battles of the *Risorgimento* and upheld their "scientific pa-

[58] Unless stated otherwise, the quotes in this little section are from [Patriarca 1996], p. 156.

[59] See [Messedaglia 1865/66].

[60] [Messedaglia 1865/66], p. VII; my translation.

triotism"[61] all their lives, once national unity was achieved in 1870, they tended to approach politics rather as a "secondary profession"[62] whereas teaching and research continued to be their primary job.

Thus, while there can be no doubt that for many mathematicians of the *Risorgimento*, "science and politics represented two complementary aspects of a larger renovation project of Italian culture"[63], one may wonder how this complementarity worked out in individual cases.[64] Antonin Durand[65] explores this by first investigating how the mathematical jobs in academia and the hiring procedures changed during the *Risorgimento*, and then the national and local conditions of political careers in Parliament. We cite just one example, Betti's student Ulisse Dini (1845–1918), who had been mathematically influenced first by his year in Paris, where he studied with Joseph Bertrand (1822–1900) and Charles Hermite and worked on the differential geometry of surfaces, and later, as of the 1870s, as a seminal contributor to the refoundation of real analysis, inspired by what Karl Weierstrass (1815–1897) taught at Berlin. Dini managed to get elected to the Lower house in 1880 as a deputy from Pisa, where he was teaching, and was reelected three times.[66] In the small town of Pisa, with its important university and the *Scuola Normale*, it was possible to build Dini's election campaign specifically on his mathematical, academic reputation, and to appeal to Dini's authority with respect to his students as a voucher also for his political competence.[67] It is unlikely that this could have been done elsewhere.

In the context of the present book, we are particularly interested in understanding how the patriotic engagement of the Italian mathematicians affected their attitude towards international scientific relations. It turns out that both ideas went quite a long way hand in hand.

> We have seen that even in the pamphlets of his election campaign Ulisse Dini justified his political legitimacy by his journey to France in 1864, his Parisian contacts and the recognition he had earned there. More generally, the scientists, and in particular the mathematicians, are early and essential actors of transnational exchanges. . . . The international experience thus acquired could be converted into political legitimacy. It could both inform Italy about foreign models of universities, administration and politics, and give to the mathematicians an international reputation which would be accepted by their colleagues in the political arena.[68]

---

[61] [Durand 2018], p. 311.

[62] As [Durand 2018], p. 26, aptly points out with a reference to Max Weber's (1864–1920) reflections from the end of World War I, about different types of political attitudes that members of academia can adopt.

[63] See [Bongiorno & Curbera 2018], p. 55

[64] On a broad scale, not restricted to mathematics, but presented from the point of view of Torino, this is investigated in the rich study [Roero 2013]; see for instance the simultaneously scientific and political appreciation, p. 374, of an international scientific gathering in 1840, with the participation of the British mathematicians Babbage and Hamilton, among others.

[65] See [Durand 2018].

[66] See [Gillispie 1970–1980], Vol. 4, p. 102.

[67] Cf. [Durand 2018], pp. 205–209.

[68] [Durand 2018], p. 216–217; my translation.

1.1.5.3 Betti, Brioschi, Casorati

In the Fall of 1858, Betti from Pisa, Brioschi from Milano, and Felice Casorati (1835–1890) from Pavia set off together on a private (non-commissioned) mathematical reconnaissance mission which took them to Zürich, Munich, Leipzig, Dresden, Berlin, Göttingen, Heidelberg, Karlsruhe, Strasbourg, and Paris. This journey would have a lasting influence on Italian mathematics, and it is a memorable example of Science International in the making.[69]

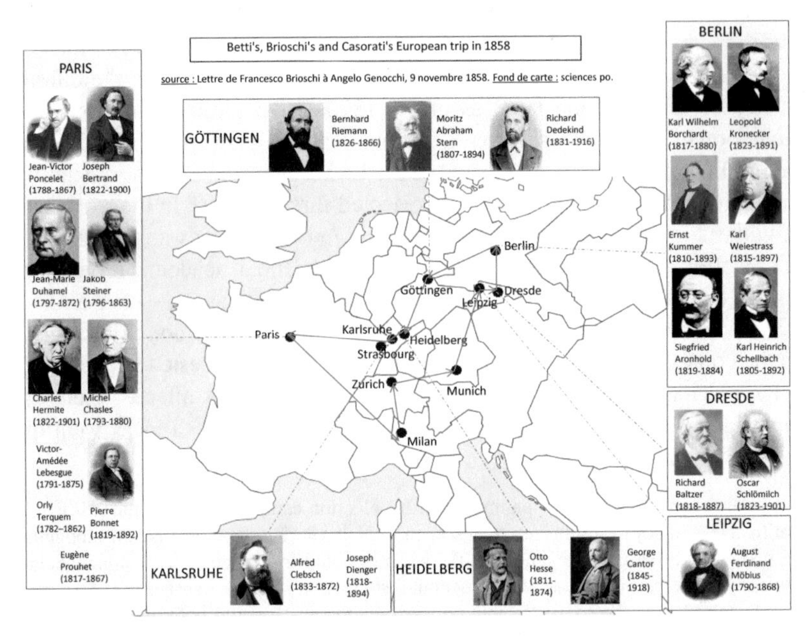

Fig. 1.3 The trip of Betti, Brioschi, and Casorati (Courtesy A. Durand, cf. [Durand 2018], p. 220).

On their trip they met—among others—Ernst Eduard Kummer, Leopold Kronecker (1823–1891), Karl Weierstrass , Bernhard Riemann, Richard Dedekind (1831–1916), Georg Cantor (1845–1918), Alfred Clebsch (1833–1872), Joseph Bertrand, Michel Chasles (1793–1880), and Charles Hermite

Earlier the same year, the mathematical journal *Annali di Scienze matematiche e fisiche*—which had been created in 1850 by Barnaba Tortolini (1808–1874) in Rome, inspired by Liouville's *Journal des mathématiques pures et appliquées*, the English *Quarterly Journal*, and *Crelle's Journal*—was taken over by Betti, Brioschi and Angelo Genocchi (1817–1889), and its name was changed to *Annali di matematica pura ed applicata* on this occasion. One goal of the journey was to acquire new authors and works for this journal, thus meeting the challenge that Betti had posed: to inform its readers also about new mathematical developments abroad. After their

[69] Cf. [Durand 2018], pp. 217–221.

return, Betti would translate Riemann's doctoral thesis, i.e., Riemann's account of the theory of complex analytic functions of one variable, and publish it in the December 1859 issue of the *Annali*.[70]

These research-oriented endeavors for up-to date mathematical exchange were soon complemented by political activities for the renovation of the Italian educational system, to which mathematicians contributed substantially, and in which the discussion of various international models, in particular the German vs. the French one, played an important part. Once, in 1890, a mathematician, Luigi Cremona, was even Minister of Education (*Ministro della Pubblica Istruzione*), and the important post of under-secretary charged with communication between this ministry and the Parliament was held, at different time periods, by four other mathematicians.[71]

The journey was a major "turning point in the history of Italian mathematics and their appearance on the international scene."[72] The joint effects of the journey and the educational policy opened the newly unified Italy towards the German model of higher education.

A German bifurcation. Let us note in passing that Clebsch, whom the three travelers met in Göttingen, was a leading figure of a network of mathematicians which would in the long run result in the mathematical center at Göttingen being an antipode to Berlin:

> Throughout the 1850s and 1860s there were few signs of open conflict in the rivalry between the two leading centers for mathematics in Germany, but by 1870 clear signs of division had emerged. Berlin's dominance was challenged by Alfred Clebsch, a product of Königsberg who taught in Göttingen from 1868 to 1872. Together with Carl Neumann [1832–1925], Clebsch founded the *Mathematische Annalen*, which served as a counterforce to the Berlin-dominated journal founded by Crelle, edited after 1855 by Carl Wilhelm Borchardt [1817–1880]. Other leading representatives of this Königsberg tradition during the 1860s and 1870s included Otto Hesse [1811–1874], Heinrich Weber, and Adolf Mayer [1839–1908]. Along with Clebsch and Neumann they operated on the periphery of the Berlin school and its associated Prussian network. These mathematicians had very broad and diverse interests, making it difficult to discern striking intellectual ties. What they shared, in fact, was mainly a sense of being marginalized, and they looked up to Clebsch as their natural leader."[73]

The further development of this network, after Clebsch's sudden death in 1872, was marked by Felix Klein (1849–1925), whose overwhelming influence will be discussed in Section 1.3 below.

---

[70] Cf. [Tazzioli 2018a], p. 28–29, as well as the interesting comparison between French and Italian journals in [Gispert 2001].

[71] Cf. [Durand 2018], pp. 230–260.

[72] See [Bottazzini 2001], p. 37–38.

[73] See [Rowe 2008].

1.1.5.4 Algebraic Geometry, a New International Domain?

At the opening session of the Second International Congress of Mathematicians, in 1900 at Paris—on the grounds of the Universal Exhibition, in front of an audience which exceptionally included as guests many women and young women in bright Summer dresses (*aux claires toilettes*), as the Proceedings chose to record[74]—Vito Volterra (1860–1940) commemorated the famous journey in his lecture on "Betti, Brioschi, Casorati: three Italian analysts and three ways to deal with the questions of analysis." Here too, the triptych of different characters that Volterra sketches for his audience involves both a variety of mathematical approaches and various attitudes towards administration, politics, and teaching. Moreover, the central character Betti is portrayed by Volterra as being split between the algebraic approach (like in his rendering of Galois theory) on the one hand, and Riemann's function theory inspired by electrophysics on the other. But in spite of his focus on analysis, in the title and the major part of his lecture, Volterra did not fail to also mention the importance of (various kinds[75] of) geometry in nineteenth century Italian mathematics.

In the twentieth century it became common—and it still is today—to refer to 'Italian Algebraic Geometry' to characterize a methodologically fairly coherent corpus of mathematical production one of the highlights of which was the classification of algebraic surfaces by Guido Castelnuovo (1865–1952) and Federigo Enriques (1871–1946) at the turn of the century. As of the 1930s, this 'Italian Algebraic Geometry' would be increasingly criticized for an alleged lack of rigor. This criticism reflected the first major rewriting of this branch of mathematics in the twentieth century at the hands of Van der Waerden, Oscar Zariski (1999–1986), André Weil (1906–1998), and others.[76]

Considering this sequence of events it is not superfluous to recall that, certainly at the outset, this "geometric Risorgimento"[77] was an essentially international enterprise. The original international character of 'Italian Algebraic Geometry' is marvellously illustrated by Corrado Segre (1863–1924) in Torino, the first *maestro* of this branch of mathematics, who would for instance begin one of his lecture courses by putting "down, as principal works of reference, books in four different languages, and remarked that those of his hearers who could not read English, French and German must certainly make up the deficiency in the course of the year."[78] Among the influential mathematicians listed in the first obituary of Corrado Segre we find,

[74] See Proceedings ICM 1900, p. 14.

[75] This is not the place to go, for instance, into the history of nineteenth century projective geometry, whose results and scientific values seem so unusual, if viewed from the twentieth century perspective. Cf. [Gray 2008], Sec. 2.1.1 and 3.1.2.

[76] Cf. [Schappacher 2010], Section 5; [Schappacher 2015a].

[77] See the obituary [Coolidge 1927], p. 352.

[78] Coolidge's recollection, quoted from [Casnati et al. 2016], p. 39.

next to the three Italians Giuseppe Veronese (1854–1917), Eugenio Bertini (1846–1933), and Cremona, also Steiner, Clebsch, Cayley, Alexander Brill (1842–1935), Max Noether (1844–1921) , and Felix Klein.[79]

1.1.5.5 Palermo, Guccia, and his Medal

Nations, even if unified, are rarely homogeneous. In Italy the cleavage between North and South (*mezzogiorno*) continues to affect society and culture even today. This makes it all the more remarkable that the first Italian mathematical journal with an international editorial board was created in ... Palermo, Sicily, by the wealthy Giovanni Battista Guccia (1855–1914): the *Rendiconti del Circolo Matematico di Palermo*. The first volume was issued in 1887. As the name suggests, Guccia had first established the *Circolo Matematico di Palermo* at his home, the *Palazzo Guccia*, in 1884. This was the first international mathematical society in history.[80] Guccia had been trained in algebraic geometry by Cremona in Rome. This also brought him into contact with mathematical Europe at large. Coming from a wealthy family of Sicilian nobility, with their business in local water management, Guccia was able to travel extensively in Europe.

Eventually Guccia became full professor at Palermo University, overcoming obstacles created by Corrado Segre.[81] The two mathematicians apparently had a somewhat shaky relationship,[82] but Segre finally accepted the invitation to be the Italian member on the committee for awarding the *Medaglia Guccia* at the ICM in Rome in 1908. The two other members on the committee were Max Noether and Henri Poincaré. The medal was awarded to Francesco Severi (1879–1961) for his (early) work in algebraic geometry. Guccia had endowed this medal with his personal funds at the 1904 ICM in Heidelberg. The medal showed the *Trinacria*-logo of Guccia's *Circolo* on one side, and an idealized portrait of Archimedes—the most famous scientist from Sicily—on the other, identified as such in Italian. It was awarded only this one time in 1908. One naturally wonders if the presence of Archimedes on the Fields Medal consciously echoes this early predecessor of 1908. I do not know.[83]

---

[79] See *Annali di matematica pura ed applicata* Ser. 4, Vol. 1 (1924), pp. 319–320. For more material on the international character of early Italian Algebraic Geometry, see [Casnati et al. 2016], especially the lavishly documented chapter written by Erika Luciano & Clara Silvia Roero, pp. 93–241. Cf. Aldo Brigaglia's chapter: "The creation and persistence of national schools: the case of Italian algebraic geometry" in [Bottazzini & Dahan 2001], pp. 187–206.

[80] See the review [Tazzioli 2020]; cf. the extensive study [Bongiorno & Curbera 2018]; see also Aldo Brigaglia's dense Chapter 10 in [Parshall & Rice 2002], pp. 179–200.

[81] See [Bongiorno & Curbera 2018], Section 5.2.

[82] For instance, in Segre's correspondence with Castelnuovo, his way of referring to Guccia as *amicone*, i.e., his 'special friend', might be read as an ironic exaggeration. Cf. [Casnati et al. 2016], p. 127, note 104.

[83] Cf. [Bongiorno & Curbera 2018], p. 154; [Tazzioli 2020], p. 79. Also [Riehm & Hoffman 2011], pp. 98–99 and 184–186, remain silent about a possible link of inspiration.

1.1.6 Gösta Mittag-Leffler

Let us leave the South to catch up with the North of Europe. In Djursholm (near Stockholm), Sweden, another energetic man about town, businessman, organizer, and enthusiastic mathematician Gösta Mittag-Leffler (1946–1927) founded his mathematical journal, the *Acta Mathematica* in 1882—i.e., five years before the first volume of the *Rendiconti*—and also in other respects played a similar role for the European dimension of mathematics of his time as Guccia did from Sicily. Early in his career, Mittag-Leffler had visited Paris where he exchanged with Hermite, and Berlin where he was a student of Weierstrass's. One of the first authors he recruited for his new journal was Poincaré. The founding of his journal required diplomatic skills:

> Mittag-Leffler tried to keep all of this planning work hidden from his German friends; not even [Carl Johan] Malmsten's [1814–1886] son-in-law Schering[84] knew about the activities. Mittag-Leffler wanted first to ensure that the journal was actually established before surprising his German friends with a publication offer that they wouldn't be able to refuse. Nor did Mittag-Leffler want to affect the situation among international mathematics journals before he had his own journal in place. The two leading European journals—*Borchar[d]t's (= Crelle's) Journal* in Berlin and *Liouville's Journal* in Paris—had ceased being international publications after the war in 1870–71. And both *Mathematische Annalen* and the *American Journal of Mathematics* were in the early phases of becoming established. The American journal, founded in 1878, was primarily centered around Johns Hopkins University. The fact that Weierstrass and Kronecker had just taken over the editorial responsibility (in 1880) for *Crelle's Journal* made the situation even more delicate. Mittag-Leffler was reluctant to appear to be a competitor to his German friends and teachers. On the other hand, he had doubts about the administrative abilities of Weierstrass and Kronecker, and he was worried about what might happen to the well-respected German journal. The blessing of Oscar II and Malmsten's position were both important to Mittag-Leffler's plan for establishing his own publication (the name *Acta Mathematica* had not yet been chosen) without antagonizing the Germans. Malmsten wrote a letter to Weierstrass, Kronecker, Kummer, and Schering in which he, on behalf of King Oscar II, asked them to contribute articles to Stockholm. Hence it was through a royal communiqué that the German mathematicians learned of the new journal. Everyone except Kummer promised at once to send something. Also important to this plan was the fact that all four of these German mathematicians, more or less on Malmsten's initiative, had been awarded a royal Swedish order. When Mittag-Leffler received a positive reply from Weierstrass, he immediately wrote to Malmsten with the words: 'This was truly a master coup.'[85]

We have mentioned above the international board of editors of Guccia's *Rendiconti*. In fact, the non-Italian members whom Guccia invited to his board of editors were Henri Poincaré and Gösta Mittag-Leffler. Meanwhile the editorial board of *Acta Mathematica* remained Scandinavian.[86] And yet it was an international journal, as was the list of authors and contributions that it published:

---

[84] Cf. the paragraph about Gauss's chair(s) at the beginning of Section 1.1.

[85] [Stubhaug 2010], pp. 275–276.

[86] In the sense that its members were either Scandinavian by birth or had jobs in a Scandinavian country, like Kovalevskaya—see below.

1.1.7 A Woman Mathematician with International Connections

Mittag-Leffler not only put Swedish mathematics on the European map as a visible player. He was also the first man to arrange for a woman, Sofya Kovalevskaya (1850–1891), to be promoted to one of those teaching and research positions that the nineteenth century created.

> Mittag-Leffler heard Kovalevskaya present a paper at a conference in St. Petersburg in 1880 and from then on devoted considerable energy to championing her as a mathematician. Eventually overcoming substantial opposition from many of his colleagues, he arranged a teaching position for her in Stockholm. She arrived there at the end of 1883, and shortly afterwards, Mittag-Leffler invited her to join the editorial board of *Acta*. Time spent in both Berlin and Paris meant that she was well-placed to communicate with the mathematicians there, as well as those in her native Russia. Her main responsibility at *Acta* was to act as a sort of international liaison officer, although she also took over some of the fund-raising activities.
>
> One of Kovalevskaya's assignments was to try to persuade the Russian Academy of Sciences to give both institutional and financial support to the journal. Although she made strenuous efforts on this account during her visits to Russia, she never managed to succeed. It seems that the lack of support was essentially political and stemmed from *Acta*'s endorsement of the Finnish mathematical establishment as a separate entity. This endorsement, which derived from the fact that Mittag-Leffler had spent four years as a professor at the University in Helsinki prior to his return to Stockholm in 1881, was construed by the Russians as support for the Finnish movement for independence from Russia. As a result, Russian mathematicians thought it wiser to avoid a formal association with the journal.
>
> Kovalevskaya had rather more success with her task of obtaining Russian manuscripts for *Acta*. [Pafnuty Lvovich] Chebyshev [1821–1894], in particular, was keen to contribute and was grateful to her for translating the first of his five papers to appear in the journal. She was also popular with several other Russian colleagues, who felt that she would treat their work fairly. To a great extent, they considered her as their own representative in the West and relied on her to publicize their results. This led to a large amount of correspondence—mainly in the form of requests of one sort or another (reprints, preprints, technical queries, etc.)—and kept a channel of communication open between the two countries. As a result of Kovalevskaya's efforts on behalf of *Acta*, Russian mathematicians were provided both with a means to make their work known in Europe and a conduit for contact with European mathematical developments. Conversely, just as Mittag-Leffler had hoped when he originally involved Kovalevskaya with the journal, *Acta* itself benefited from having such a direct Russian connection. Kovalevskaya's presence on the editorial board not only provided the journal with an entree into Russian mathematical circles but also helped to widen the journal's international circulation.[87]

It is important to note that this success story of international mathematical networking did require Kovalevskaya's institutional basis in Stockholm. In fact, when Sonya Kovalevskaya had returned to Russia in 1874 with her husband years before this last stage of her career,

> she discovered that the combination of her sex and her politics and her German degree made her unacceptable as a job candidate [in Russia]. Chebyshev and other Russian mathematicians were devotees of the French rather than the German school of analysis. They were in the

[87] [Barrow-Green 1994], p. 154.

> process of throwing off the domination of German scientists in their Academy of Sciences, were evolving their own approach to mathematical problems, and consequently looked with suspicion on those who had done their work entirely abroad.[88]

This reminder leads us to the focus of the next chapter. But before turning to it, we should at least mention in passing that Sofya Kovalevskaya was awarded the *Prix Bordin* of the Paris Academy of Sciences in 1888—she was 38 years old at the time—for her work on the spinning top. The prize money had been raised to 5,000 Francs on this occasion.

1.2 Nation Branding through Science

In this section we will speak of *Nation Branding* in a non-terminological manner in order to capture various ways of associating (often emphatically) nationality with accomplishments which *a priori* do not carry a national stamp.[89]

1.2.1 Chemical Elements

The following story is a staple for lecture courses on the history of chemistry. The very first International Congress devoted exclusively to chemistry was convened in September 1860 in Karlsruhe, Germany. One of its concerns was to settle once and for all divergent calibrations of atomic vs. molecular weights of gases. A commission of nine men alternately convened behind closed doors, and interacted with the plenary attendance of 127 chemists. By the end of the congress the problem was not settled. But the Italian chemist (and, by the way, also politician ...) Stanislao Cannizzaro (1826–1910) was there. He had published a paper on Avogrado's law relating volumes of gas to the number of molecules present.

> His paper, published in Italian, was at first ignored. Then he presented his argument and distributed his paper as a pamphlet ... in Karlsruhe. Some chemists were immediately persuaded, others read Cannizzaro's paper on the train going home and were persuaded by the time they got to their destination; and others, of course, missed the point. That conference was the turning point for the acceptance of Avogadro's hypothesis, almost a half-century after Avogadro first proposed it. Now that hypothesis could bring order to the whole of chemistry.[90]

One of the participants of the meeting in Karlsruhe, Dmitri Ivanovich Mendeleev (1834–1907) from St. Petersburg,

---

[88] [Hibner Koblitz 1985], p. 7.

[89] I am aware that today the term is mostly used to describe a typical phenomenon related to globalization—see e.g. [Kerr & Wiseman 2018], pp. 208–209. Contrary to this, the examples discussed in this section are from the turn of the nineteenth to the twentieth century, concern scientific rather than economic products, and form a rather casual bunch.

[90] [Levere 2001], p. 116.

> was able to benefit from the resulting consistency in determining atomic weights. He began to write down each element on its own card, together with its atomic weight . . . , its properties, and analogous elements. Then he looked for the best arrangement of the cards, the arrangement that would most fully bring out analogies in properties and relate them to atomic weights. He concluded that the properties of the elements were in periodic dependence upon their atomic weights. By periodic he meant regular and recurring. This was the origin of the periodic table of the elements, which has evolved and grown since Mendeleev's time, but which still appears in every classroom where chemistry is taught. And of course the periodic table is based on Mendeleev's periodic law: "The properties of the elements, as well as the forms and properties of their compounds, are in periodic dependence or (expressing ourselves algebraically) form a periodic function of the atomic weights of the elements."[91]

One of the exciting consequences of the periodic table of elements was that it had well-located gaps which called out to be filled.

> Where necessary, Mendeleev left blanks, so as to keep known elements in positions that corresponded to their chemical properties. Then, with remarkable confidence, he predicted that those blanks would later be filled by hitherto undiscovered elements, and he went on to predict the atomic weights and the chemical natures of the 'missing' elements. There was a blank after zinc, in the same group as boron and aluminum. Mendeleev predicted that this blank would be filled by an element with properties similar to those of aluminum and having an atomic weight of 68 and specific gravity of 6.0. In 1875 the missing element was discovered . . . , with atomic weight 69.9 and specific gravity 5.96.[92]

1.2.1.1 Gallium and Germanium

This discovery of Mendeleev's hypothetical 'eka-alumimum' was realized by the Frenchman Paul-Émile Lecoq de Boisbaudran (1838–1912), a son of wine merchants, in his own private laboratory: "On 27 August 1875, between 3 and 4 p.m., I noticed the first evidence of the existence of a new element, *which I called 'gallium' in honor of France* (or *Gallia*, in Latin)."[93] Apparently, he saw his discovery as a personal contribution to restoring the glory of French science in the face of competitors (cf. Section 1.1.4 above).

About a decade later, the Saxon Clemens Winkler (1838–1904) published a first little note announcing the discovery of another element: "After several weeks of arduous searching I can announce today with certainty that the 'argyrodite' [an ore found near Friedberg in Saxony] is a new element . . . , *to which the name 'germanium' be given*."[94] After months of thorough experiments Winkler could finally show that this new element was indeed one that Mendeleev had postulated and referred to as 'eka-silicium'. While he was working on this, colleagues started reacting to his

---

[91] [Levere 2001], pp. 118–119.

[92] [Levere 2001], p. 119.

[93] *Annales de chimie et de physique*, ser. 5, Vol. 10 (1877), p. 103; my translation and emphasis. We leave aside the question of whether or not Lecoq de Boisbaudran was aware of Mendeleev's prediction, or if his search for this new element was based on his own reasoning.

[94] *Berichte der deutschen chemischen Gesellschaft* 19 (1886), p. 210; my translation and emphasis.

first announcement. Mendeleev himself took a while to be convinced that the new element was indeed eka-silicium.[95] But the discussions in the chemical community also went beyond such technical issues:

> One should think that this accomplishment could only inspire the purest joy. Yet also in this respect dispute has not been lacking; for the name of the new element, the name 'germanium', has encountered the unlikely destiny of becoming a stumbling block. I would not mention this circumstance here, irrelevant in itself as it is, were it not for the editor of the *Moniteur scientifique*, Herr Quesnesville in Paris, who, flinging off all objectivity, publicly summoned me to abandon the name 'germanium', which according to him *a un goût de terroir trop prononcé*[96], and replace it by Mendeleev's provisional term 'eka-silicium.'
>
> One may certainly wonder if the international character of science can suffer such a patriotic nomenclature, and if it had not been more appropriate to choose the sought name from a myth or some other domain which passes for being neutral. ... In general the choice of a name has little importance insofar as petty criticism will always find something to complain about. For instance, Herr Quesnesville's suggestion is all the more incomprehensible as exactly the same sort of patriotic naming has occurred much earlier in France with the discovery of gallium. Should Lecoq de Boisbaudran now call his gallium 'eka-alumimum'; or should L.[ars] F.[redrik] Nilson [1840–1899] now call his scandium 'eka-boron'? ...[97]

So Winkler concedes that the patriotic naming of elements may be hard to reconcile with the international character of the scientific enterprise. But it is easy for him to hide behind the fact that a Frenchman had started this way of transferring one's scientific success to national honor, thereby elevating in a single flash both his personal fame and the glory of the fatherland.

Several chemical elements thus carry patriotic names. There is also 'europium' (no. 63), from the family of rare-earth elements, like scandium, and since transuranium elements produced in the process of constructing the atomic bomb during World War II were named following analogies with rare earths, the element no. 95 corresponding to europium was called ... americium.

1.2.1.2 Marie Curie

A special example of national patriotic naming occurred more than a decade after Winkler's germanium. It concerned the radioactive element no. 84. Pierre (1859–1906) and Marie Curie (1867–1934) in Paris claimed its name even before the discovery was fully established:

> We believe that the substance which we have extracted from the pitchblende contains a hitherto unknown metal, similar to bismuth as far as its analytic properties are concerned. If the existence of this new metal is confirmed, *we propose to call it 'polonium', from the name of the country of origin of one of us.*[98]

[95] *Journal für die praktische Chemie* 142 (1886), p. 182–183.

[96] I.e., 'the name smelled too much of the old sod.' Quoted in French in the German text.

[97] From Winkler's article in *Journal für die praktische Chemie* 142 (1886), here p. 183–184; my translation. The Swede Nilson had named the newly discovered 'scandium' (no. 21) in honor of Scandinavia.

[98] [Curie 1898], p. 177; my translation and emphasis.

Marie Curie had obtained French citizenship after her marriage. In 1903 she received the Nobel Prize in Physics jointly with her husband. After he died in a traffic accident in 1906 she was given his professorship at Paris University and became the first woman to be a University Professor in Paris. In 1911 she would receive her second Nobel Prize, this time in Chemistry for the discovery of polonium and radium, confirming her as the most distinguished woman in science on Earth. That same year Parisian newspapers had launched a scandal about her alleged affair with Paul Langevin (1872–1946), also spurning her as a foreigner to France. The Borel couple—Émile Borel (1871–1956) at the time was adjunct director of the École Normale Supérieure—sheltered Marie Curie from the mob and took an active part in the finally successful struggle to keep her in Paris.[99]

Fig. 1.4 Extract from a well-known group photo by Benjamin Couprie of the 1911 Solvay Conference in Brussels. Marie Curie in conversation with Henri Poincaré. Standing on the right Paul Langevin and Albert Einstein .

[99] Cf. the moving chapter IX in honor of Marie Curie: *Rue d'Ulm. Cabale contre Madame Curie*, which Émile Borel's wife Marguerite (1883–1969), alias Camille Marbo—herself daughter of the mathematician Paul Appell and a well-known writer—included in her autobiography [Marbo 1968], pp. 101–122.

1.2.2 Nation, Culture, Science

Analytical chemists are not supposed to be poets. If they 'go to encounter for the millionth time the reality of experience' in their laboratories, they do not as a rule 'forge in the smithy of their soul the uncreated conscience of their race.'[100] Indeed, there does not seem to be much poetry in labeling an element which one has laboriously succeeded to isolate with the Latinized version of the name of a nation. But the fact is that this act could succeed in transferring to a whole nation the pride and honor of a personal scientific achievement, and by the same token give heroic glamour to the scientific feat and the person(s) behind it. This highlights the emotional-political weight that the concept of nation had acquired since the beginning of the nineteenth century, especially through its connection with the Romantic movement. In fact, according to the German poet Novalis, "Romanticizing something is to raise its exponent to a higher power. The lower self becomes its higher self ... If I give the banal a higher sense, ... then I romanticize it; and conversely so, when I logarithmically bring down the higher, unknown, mystical and infinite into a common expression."[101]

> If there is such a thing as Romantic nationalism, we must conceive of it, not as a lump of facts or a cloud of semantics, but as a knot, a tight tangle, a node in the mycelium of intellectual and cultural developments. Romanticism and nationalism, each with their separate, far-flung root-systems and ramifications, engage in a tight mutual entanglement and *Wahlverwandschaft* in early-nineteenth-century Europe; and this entanglement constitutes a specific historical singularity. We can give this singularity a name: Romantic nationalism. And we may understand that to mean something like: the celebration of the nation (defined in its language, history, and cultural character) as an inspiring ideal for artistic expression; and the instrumentalization of that expression in political consciousness-raising.[102]

This broader cultural link of the chemical nation branding is particularly evident in the case of Marie Curie. With the appeal to the Polish Nation she placed herself in the tradition of artists like Adam Mickiewicz (1798–1855) and Frédéric Chopin (1810–1849), who had also spent important years of their lives in Paris and France.

Such ready 'exponentiations' of scientific accomplishments to the national scale can be observed in many guises. For instance,

> the death of a scientist is an occasion for national mourning even if the government does not order a state funeral. Here too, the scientific genius of one of the professors is put on a par with the national genius. When Poincaré died, the dean went as far as claiming that 'the death has been felt by the whole nation', and he quoted from [Paul] Painlevé's [1863–1933] funeral eulogy: "All Frenchmen knew that a man lived among us whose brain was in a way the scientific mastermind of humanity. There is no nation which has not envied

---

[100] Adapted from the famous, ironic ending of James Joyce's *A Portrait of the Artist as a Young Man*. Cf. [Leersen 2013], p. 18.

[101] Quoted from [Leersen 2013], note 14.

[102] [Leersen 2013], p. 283. Cf. more generally Leersen's *Encyclopedia of Romantic Nationalism*, which is accessible online at [URL 03].

> us our Poincaré, none which has not paid deference to the pre-eminence of his genius." This progression of the dithyramb goes beyond all commemorative speeches of professional scientific communities.[103]

In a multilayered analysis of the reception of Galois's ideas by Camille Jordan (1838–1922) and other authors towards the end of the nineteenth century, Frédéric Brechenmacher has pointed out similar nationalistic 'exponentiations' during the gradual apotheosis of Évariste Galois (1811–1832), for instance in the context of the centenary celebration of the *École Normale Supérieure* in 1894:

> ... in France, the mathematical sciences were mainly divided between analysis, geometry and applications. At the turn of the century, several authorities such as Jules Tannery [1848–1910], [Émile] Picard [1856–1941], Henri Poincaré, Jacques Hadamard [1865–1963] contrasted the 'richness' of the power of unification of analysis with the 'poverty' of considering algebra and/or arithmetic as autonomous disciplines. These official lines of discourse usually pointed to recent developments in Germany in the legacies of Kronecker or Richard Dedekind. Promptly following Picard ..., a review of Heinrich Weber's 1895 *Lehrbuch der Algebra* highlighted how Galois had introduced the 'fundamental ideas' of Algebra as it was practiced in Germany; had he lived longer, all 'French Science' would have had a different orientation .... The celebration of the centenary of the *École Normale Supérieure* had aggrandized Galois's reputation, to the level of one who merited entry into the pantheon of Science .... Like other *grands savants*, Galois became involved in the nationalistic anti-German discourse.[104]

National mindsets in science; Pierre Duhem. The preceding quote contains another way to give science a national or nationalistic twist: instead of appealing to great scientists as exceptional heroes whose adulation serves the pride of the fatherland, they were often presented as exponents of a national mindset, or 'style.' This trope of national mindsets influencing all intellectual activities was widespread towards the end of the nineteenth century. Remembering this cautions us to interpret the numerous creations of international organizations during that period as plain, honest to god expressions of a purely internationalist credence according to which the lofty universality of science was called upon to unite all of humanity.[105] To be sure, the rhetoric about national mindsets and their expressions in science would become much more aggressive during World War I when it was part of propaganda. Its further repercussions would mark the 1920s and 1930s, thus in particular the first decade of the IMU. However, the trope itself was endemic before the war.

To cite just one well-known example, the first edition of Pierre Duhem's (1861–1916) philosophy of physics[106] appeared already in 1906, and largely relied on earlier papers of his, published since 1893. In his book, Duhem describes—and criticizes, not to say ridicules—Michael Faraday's (1791–1867) and Maxwell's model- and equation-oriented approach to electrostatics and electromagnetism as the deplorable

---

[103] [Charle 1994], p. 185; my translation.

[104] See [Brechenmacher 2011], p. 281; cf. the copious references given there.

[105] We echo here the warning in [Schroeder-Gudehus 1978], p. 33, note 44.

[106] For reference, see the commented edition by Sophie Roux of the second edition [Duhem 1914].

consequence of the "ample" (as opposed to "profound") British mindset (*esprit anglais*).[107] We shall encounter other similar arguments, with respect to mathematic(ian)s, in some of the following paragraphs.

1.2.3 Nation and Mathematics

Mathematics does not seem to know the kind of nation branding we have seen in analytical chemistry, possibly because conceiving of a new mathematical object is not quite the same sort of thing as presenting evidence for a hitherto unknown physical object, like a star, a feature on the Moon, or indeed a chemical element. What about other forms of nation branding in mathematics? Googling 'Irish algebra' instantly produces a slew of URLs on the discovery of the quaternions by William Rowan Hamilton (1805–1865). And they rarely fail to mention that he inscribed the formulae on Dublin's Broom Bridge. I am not sure what this teaches us about current search algorithms.

Leaving the nineteenth century for a moment, I know of two examples in the twentieth century where the name of a nation has been transferred to mathematical objects: *Polish spaces* in general topology[108] and *Japanese rings* in commutative algebra.[109] Both terminologies are routinely used today, and both apparently originated in the Bourbaki environment. According to his own account, Roger Godement (1921–2016) first proposed the notion 'Polish space' jokingly to his Bourbaki confreres in 1949 after having learned the subject from Kuratowski's textbook, and realizing the contributions of Polish mathematicians to this theme,[110] such as Wacław Sierpiński (1882–1969), Kazimierz Kuratowski (1896–1980), and Alfred Tarski (1901–1983). The expression 'Japanese rings' in commutative algebra seems to be due to Alexander Grothendieck (1928–2014), probably inspired by numerous results (and counter-examples . . .) in this area due to Japanese mathematicians, in particular Masayoshi Nagata (1927–2008).[111] Neither case is an instance of nation branding in the sense we have encountered in the nineteenth century: the reference to a nation here only serves to acknowledge collectively a group, or 'school' of mathematicians.

Keeping with the historical focus of this chapter, let us go back and look at explicit examples of nation branding in mathematics in the nineteenth century.

---

[107] See [Duhem 1914], Chap. IV: *Les théories abstraites et les modèles mécaniques*.

[108] A Polish space is a separable topological space which can be metrized by a distance with respect to which it is complete.

[109] These are integral domains A whose integral closure in any finite field extension of their quotient field is a finitely generated A-module.

[110] See [Godement 2003], p. 67, footnote 33.

[111] See EGA IV, *Publications mathématiques de l'IHES* 20 (1964); §23, pp. 213–217.

1.2.3.1 Ernst Eduard Kummer

At the turn of the year 1841/1842 Ernst Eduard Kummer had reoriented his own work towards Higher Arithmetic in order to be eligible for a professorship in Prussia.[112] In his essays and speeches on patriotic or historical occasions he would portray pure mathematics as being germane to the period of mathematical research in Germany in which he was both an observer and a participant. The kind of mathematics he had in mind was strongly marked by the compound research field which, from the 1820s to the 1850s, brought together number theory (quadratic forms, cyclotomy, and their generalizations), the algebraic theory of equations, and the analytic theory of elliptic functions (and their generalizations). In [Goldstein & Schappacher 2007], Section 3, we have called this historically remarkable conglomerate research field 'Arithmetic Algebraic Analysis'.

Kummer, when looking at the history of science, tended to see a sequence of cycles in three periods: creative burst – consolidation – decline. In this perspective, he felt—as he explained in a book review in 1846—that pure mathematics in Germany, especially in Prussia, was in full creative expansion to new horizons. For him this also explained the dominance of research journals at the time, to the detriment of comprehensive book treatments. Influenced by Hegel's philosophy,[113] according to which history is an expression of the world's Spirit (*Geist*), he would conclude that, contrary to what was happening in France at the same time, "we, the Germans, are now still in the first period of the mathematical sciences, we are certain that the creative force of the Spirit is on our side and we hope that it will continue to bear fruit in the heroes of our science."[114] It was in this nationalistic perspective that Kummer, as well as a number of German authors after him, dressed up Dirichlet as a quintessential German mathematician, "since it is the German genius that pulled him back to his fatherland and gives his works their admirable depth"[115], even though there is probably no better example of a fusion of French (specifically Fourier's) and German (specifically Gauss's) influences in the whole history of nineteenth century mathematics than Dirichlet's applications of analytic methods to Gauss's Higher Arithmetic.[116]

On 6 July 1871, about half a year after the proclamation of the German Empire in Versailles, Kummer even tried to capture the essence of the German National Spirit at the Berlin Academy's yearly commemoration of its founder Gottfried Wilhelm Leibniz (1646–1716). "If we want to avoid the usual one-sidedness of interpreting all that is noble, beautiful, and grand as a very specific disposition and talent of our own nation, we have to look for a reliable measuring stick of our appraisal in the cultural history of the German Nation."[117] He found such a point of reference

[112] [Kummer 1975], Vol. I, p. 38. Cf. [Goldstein & Schappacher 2007], p. 42.

[113] For an entertaining account of this, cf. [Rowe 2013].

[114] [Kummer 1975], Vol. II, p. 696: my translation.

[115] [Kummer 1975], Vol. II, p. 695; [Goldstein & Schappacher 2007], p. 45.

[116] Cf. [Goldstein & Schappacher 2007], pp. 29–32.

[117] [Kummer 1975], Vol. II, p. 834; my translation.

in the Protestant Reformation, seen not merely as a religious protest movement, but as a typical expression of German inwardness (*Innerlichkeit*): the reflex of pursuing earnest personal self-scrutiny and engagement, instead of formal, outward rituals. According to Kummer, it was this tendency to inwardness which also distinguished German poetry, and which "gives German science its special character, and which is visible in Leibnitz [*sic*], especially in his philosophical, but also in his mathematical creations."[118]

1.2.3.2 Charles Hermite

The shelving of approaches according to national mindsets witnessed in physics (cf. Section 1.2.2 above) could also be found in mathematics, and was not a German privilege at the time. It could easily be associated with opinions about new, modernizing trends in the writing of mathematics. Hermite for example, in a letter written to Mittag-Leffler on 6 October 1884, criticized the set-theoretic setting which Mittag-Leffler, following Cantor, had chosen for the exposition of his theorem on the representation of meromorphic functions with prescribed poles, which had just appeared in *Acta Mathematica*:

> Two aspects have to be separated regarding your work: the results you have obtained, and the presentation which you chose for them. In other words, one has to distinguish between the substance and the form. As far as the substance is concerned, I am surely echoing all analysts in saying that your work, together with Weierstrass's famous theorems, lays the foundation of the theory of uniform functions, and Picard, with whom I was discussing your work, agrees completely with me on this. He is not opposed either to my opinion about the form of your exposition. We both agree that, in proceeding from abstract, completely novel concepts to the reality of analysis by a sequence of deductions, you have subjected yourself to the German tendency, and thus to a spirit which is not ours. Indeed, for the French spirit it is absolutely necessary to proceed the other way around, in showing as carefully as possible how a new concept arises from earlier notions, and to thus let the reader participate in the origin, the birth of more general propositions from known special cases without ever leaving, if I may say so, the objective reality.[119]

One of the reasons for quoting this passage here is that Hermite's son-in-law Émile Picard would essentially reproduce its main point 31 years later, in an anti-German propaganda brochure [Picard 1916], on behalf of the Paris Academy, which was part of the intellectual warfare on the academic front—cf. Section 3.2 below.

1.3 Felix Klein, a Sample of Projects he was Involved in

Any book about the history of the *International Commission on Mathematical Instruction* (ICMI), and *a fortiori* the present book homing in on the IMU, has to address the influential role that Felix Klein played on national and international

[118] [Kummer 1975], Vol. II, p. 835; my translation.

[119] Quoted from [Goldstein 2011], pp. 126–127; my translation.

mathematical scenes at the turn from the nineteenth to the twentieth century. The Proceedings volume of the 2008 Centennial Symposium of the ICMI splits the 100 year long history of this commission into two half-centuries: the 'Klein Era' from 1908 to World War II,[120] and the 'Freudenthal Era', named after Hans Freudenthal (1905–1990), post World War II until 2008. The 'Klein Era' is characterized there as having been "dominated largely by mathematicians with a substantial, but peripheral interest in education, of whom Felix Klein was by far the most notable example, plus some secondary school teachers."[121] At the 1912 ICM in Cambridge, England, the chair of the section on didactics, Charles Godfrey, called Klein "our natural leader."[122]

The present theme of mathematical nation branding leads us to other aspects of Klein's activities. Yet we are not primarily interested in Felix Klein as a research mathematician; neither his description of geometries by the groups which leave them invariant—the so-called *Erlangen Program*—nor his book on the icosahedron—i.e., the general quintic equation—nor even his rivalry with Poincaré about the 'fuchsian' functions need to hold our attention here. Instead we focus on the period of his life which Klein himself described as "social effectuality replacing the lost genius"[123], more precisely, we focus here on the period which began with his appointment at Göttingen University in 1886. In close collaboration with the increasingly powerful Friedrich Althoff (1839–1908), a high-ranking official of the Prussian Ministry of Education in Berlin, Klein subsequently built up Göttingen to become a leading mathematical center superseding Berlin after the deaths of Kummer, Kronecker, and Weierstrass. David Hilbert was appointed in 1895 and a chair for Hermann Minkowski was created in 1902. This was a tremendous achievement. But again, it must not overshadow Klein's quasi-omnipresent influence on local, national, and international scenes. It was thanks to this ubiquitousness that he emerged as a key figure of Science International in the making, especially for mathematics. We thus have to take into account several different arenas: from the Göttingen Academy to the Mathematical Encyclopedia; from Klein's attitude towards the applications of mathematics to his American connections; from IAA to ICMI.

Looking at this panoply of projects, a certain overall pattern emerges: all of Klein's numerous 'social' activities belong to the broad realm of science politics and every one of them can be seen as an attempt to *integrate* diverging domains, actors or institutions, behind a common scheme which could be presented as scientifically grounded. Note that putting it like this is neutral with respect to the distribution of power; in particular, it does not imply that his initiatives always met with consensus.

---

[120] As a matter of fact, Felix Klein died in June 1925. Cf. Chapter 9 for a sketch of the history of the ICMI.

[121] See the opening section of Hyman Bass's (b. 1932) chapter in [Menghini et al. 2009], pp. 9–10. Cf. [Hodgson 2009] and [Weigand et al. 2019]. Cf. Chap. 9 below.

[122] See Proceedings ICM 1912, Vol. 1, p. 54.

[123] My translation of Klein's biographical note quoted in [Gierl 2004], p. 44: *Soziale Wirksamkeit als Ersatz für das verlorene Genie.*

1.3.1 Attempts to Federate Pure and Applied Science

An instance of this was Klein's attempt to integrate engineering disciplines into German universities, with a view to bringing pure and applied science together. This idea, which was further encouraged by Klein's visits to the US, questioned the existing relationship between universities and polytechnical institutes in Germany at the time. It therefore met with strong resistance from both sides, and essentially failed on the national level.[124] Only in Göttingen did Klein manage to effectively promote applied science, which for him was a necessary corollary of mathematics in a highly industrialized world. Typically for Klein, his efforts in this direction created more than one institution in Göttingen.

In 1898, he managed to set up the 'Göttingen Association for the Promotion of Applied Physics and Mathematics' (*Göttinger Vereinigung zur Förderung der angewandten Physik und Mathematik*). This was a joint venture between Göttingen University and industry, and as such rather a novelty on the German academic scene. It was co-directed by Klein and the influential chemical industrialist Henry Theodor Böttinger (1848–1920).[125]

Furthermore, an offer from Yale University, which Klein turned down, persuaded Althoff to support Klein's plans. As a result, new departments for technical physics at Göttingen University were created around 1900. In 1905 they were grouped together under the name 'Institute for applied mathematics and mechanics' (*Institut für angewandte Mathematik und Mechanik*).[126] It would soon have two noteworthy directors: Carl David Tolmé Runge (1856–1927), the very first professor of applied mathematics in Germany, and for the 'applied mechanics' department, the expert in fluid dynamics Ludwig Prandtl (1875–1953), who conceived of the boundary layer theory and who would continue to acquire and expand various research facilities in Göttingen over the following decades, marking the pace of aviation technology, in particular also its military use.[127]

1.3.2 The Unlikely Resurrection of Scientific Academies as (Inter)National Agents of Science

Another development which would turn out to be significant for Science International also has one of its roots in Göttingen, in a project of Felix Klein's. The initial scene here was neither the university nor a joint venture with industry, but the Göttingen

[124] For details, see [Manegold 1970].

[125] Cf. [Manegold 1970], Chap. 5.

[126] See [Rammer 2004], pp. 446–451, who not only carefully explores the origins of this combined institute, but also mentions its early relationship with the collection of mathematical instruments and models, another favorite of Felix Klein's.

[127] Cf. Cordula Tollmien's chapter in [Becker et al. 1998], pp. 684–708; [Epple & Remmert 2000]; Florian Schmaltz's chapter in [Schumann, Schauz 2020], pp. 227–261; as well as the references given there.

Academy of Sciences, which at the time was still called a 'Society': *Gesellschaft der Wissenschaften zu Göttingen*. We have mentioned scientific academies at the beginning of this chapter as a hallmark of the *Republic of Letters*, i.e., the eighteenth century organization of science across state boundaries. Most of them were still around at the end of the nineteenth century, and still are today; others, like the academies in Leipzig and Vienna, had been founded alluding to this tradition in the middle of the nineteenth century. At the end of the nineteenth century, organizers of science were wondering how to put them to use for their projects in a new world dominated by Nation States.

It was in such a context that Felix Klein was invited in 1888 to react to a far-reaching reform project for the Göttingen Academy proposed by Paul de Lagarde (1827–1891).[128] Klein's counter-proposal was similar to de Lagarde's original scheme in that it foresaw a substantial enlargement of the Academy in terms of budget, number of its members, and circle of influence (Northern Germany). On the other hand, Klein added his favorite idea to integrate applied science and engineering, as well as a few other new specialities, into the scientific scope of the Academy.[129] The grand scheme stood no chance of being accepted or realized at the time. Yet Klein did manage to force a reform of the statutes of the Göttingen Academy, largely oriented on those of the Berlin Academy, by making this a condition of his refusing a job offer from Munich. This reorganization took place in 1893. It gave the Göttingen Academy legal capacity, independently of the university.

Klein's projects and activities for the Göttingen Academy would be a seed among others for a surprising combination of scientific institutions: the *Verband wissenschaftlicher Körperschaften*, usually referred to as the 'Cartel' (*Kartell*) of Scientific Academies, which was founded in 1893 at Leipzig. In the beginning it combined the Academies of Vienna, Munich, Leipzig and Göttingen. The Berlin Academy, although some of its members had played a role in the initiative, would finally join the Cartel only in 1906. The Heidelberg Academy followed suit in 1911; it had been founded in 1909 largely with this joining in mind.

The Cartel was in touch with the Royal Society of London—particularly with the physicist Arthur Schuster (1851–1934), of German origin—and other Academies abroad—special mention should be made of Gaston Darboux in Paris[130]—and mutual bonds were strengthened by welcoming key actors as corresponding members.[131] On 9 October 1899, the networking gave birth in Wiesbaden to the *International Association of Academies*, or IAA for short. The original members of IAA were the Academies of Berlin, Göttingen, Leipzig, London (*The Royal Society*), Munich, Paris (*Académie des sciences*), St. Petersburg, Rome (*Reale Accademia dei Lincei*), Vienna, and Washington (*National Academy of the USA*). Nine other academies

[128] See [Gierl 2004], pp. 30–60. The orientalist de Lagarde propagated a kind of religious German nationalism which included a hateful antisemitism; cf. [Sieg 2007].

[129] See [Gierl 2004], p. 51.

[130] Cf. [Rowe 2018].

[131] See for instance the chronological table of memberships of Theodor Mommsen, Felix Klein, Gaston Darboux, Eduard Suess, Hermann Diels, Arthur Schuster, and Wilhelm August von Hartel in the Cartel academies and the Royal Society, in [Gierl 2004], p. 429.

were invited to join: Amsterdam, Brussels, Budapest, Christiania (today's Oslo), Copenhagen, Madrid, Stockholm, as well as two other French academies: *Académie des Inscriptions et Belles Lettres* and *Académie des Sciences Morales et Politiques*. The IAA in turn was one of the ancestors of the International Research Council (IRC) created as a corollary of World War I. The IRC would transform in 1931 into ICSU, the *International Council of Scientific Unions*, which in 2018 merged into the *International Science Council*, or ISC for short: the global non-government organization which today houses the IMU as well as many other scientific unions.

This is the timeline. What are we to make of it in a historical perspective? Writing the history of (scientific) academies is a rather subtle exercise. The members of an academy come from different scientific specialities and, apart from their disciplinary education, ideals, and interests, they naturally tend to have their own cultural or political priorities. At the same time, being a member of the academy bestows on every one of them an elite distinction which, as they interact, serves to arrange bonds or coalitions. These can often not be reduced to the tenets of the individual actors, as scientists or citizens, but are genuine products of the context afforded by the learned society. As a result, decisions of an academy may be difficult to read in terms of scientific or political strategies. In any case, a detailed historical study involving academies is obliged to constantly change focus between individual members, groups of members (which may rearrange themselves depending on the subject under discussion), and the academy as an institutional whole. On top of this—unless we are talking about the Royal Society of London, which as a private body is only answerable to itself—to understand an academy one also has to take into account the politics of the government it depends on.

Since this is not the place to go into great detail about the Cartel of Academies and the IAA,[132] let me just mark a few points to put these peculiar international[133] institutions into perspective.

1.3.2.1 Awkward Structures

The Cartel of Academies and the IAA summoned venerable institutions to serve in a context that differed profoundly from that of the eighteenth century. Academies which had once acted in their own right across boundaries within the *Republic of Letters* now became local, and indeed national nodes in an international trust which was somehow situated at a higher level. The publications of academies—even if they continued to hold some importance, especially for concise research announcements—had lost their dominant position to the wealth of specialized scientific journals created during the nineteenth century. In response to this, the newly created trusts tried to transform the learned societies from "publication-oriented

[132] The most elaborate history of the early years of the Cartel and of the IAA which I am aware of is due to Martin Gierl; see [Gierl 2004], a book of 667 pages, as well as the concise account [Gierl 2014]. Cf. [Greenaway 1996] and [Schroeder-Gudehus 1966].

[133] Already the Cartel was international because it involved Austria-Hungary along with (at the outset) three German states: Bavaria, Saxony, and Prussia.

Fig. 1.5 First General Assembly of the IAA, cover of the invitation to the festive *Soirée du 20 avril 1901* (Courtesy Göttingen Academy of Sciences).

academies into project-oriented academies."[134] But apart from a few exceptions—such as the systematic cataloguing of the scientific literature initiated by the Royal Society[135]—the projects that were taken up typically belonged to one or few special disciplines, while all the Academies and thus also the framework of the Cartel and the IAA were of course supra-disciplinary. The Cartel and the IAA thus oversaw the collaboration of academies each of which had to form a special commission for every single project adopted, unless it chose not to participate in that project at all, as every member academy was free to do.

It may then be difficult to decide if an international project appearing on the agenda of the IAA was not simply dressed up in an inter-academic garb, being in fact an international *disciplinary* venture which would have been mounted anyway, organized by the experts—see for instance the data-driven creation of the International Statistical Institute mentioned above, Section 1.1.2. This observation was actually a major point raised by the influential pathologist and anthropologist Rudolf Virchow (1821–1902) during the discussions at the Berlin Academy which led to the initial refusal of the Berliners to join the Cartel: "Herr Virchow also pleaded against the *Verband*, arguing in particular that joint scientific projects have already been organized easily and successfully without recourse to an association of academies."[136]

This argument certainly glosses over the question of available resources. The new confederations would indeed make it easier to obtain public funding for certain projects.[137] On the other hand, both before and after the creation of the Cartel and the IAA, academies would help certain projects of international scope, without being able to guarantee their further financing. A case in point is the famous *Carte du Ciel* project initiated by the French astronomer Ernest Mouchez (1821–1892), i.e., the photographic cartography of all stars down to a certain magnitude. The initial launching of this project in Paris in 1887 was supported by the Paris Academy, but in the sequel the participating observatories had to fend for their own funds.[138]

Another awkward problem was that the leading experts one wanted to enlist for a given project might not be members of any of the participating academies. In other words, the international organizations that were created did not adequately represent all the participating nations. In the words of the physicist Arthur Schuster, who had been one of the architects of the IAA on behalf of the Royal Society,

> the constitution of the Royal Society has the great advantage of being truly representative of the [British] Empire. In France, on the other hand, no one can belong to the Academy of Sciences who is not domiciled in Paris. Similarly, although Germany possesses four Royal academies (Berlin, Gottingen, Leipzig, Munich), each of them is confined, as regards ordinary members, to its own locality, so that a professor of the Universities of Bonn or Heidelberg, however eminent he may be, could not become a[n ordinary] member of any of these academies. Neither in France nor in Germany can the academy therefore be called truly

[134] [Gierl 2014], p. 91; my translation.

[135] On this project, see for instance [Schuster 1906], pp. 233–234.

[136] [Gierl 2004], p. 249; my translation. Concerning Virchow in general see [Goschler 2002a]; for his scientific nation branding [Goschler 2002b].

[137] See for instance [Gierl 2004], pp. 388–394.

[138] Cf. [Lamy 2008].

> representative. The disadvantages which may arise from this defect have been minimised by adopting a rule that the International Association of Academies may appoint committees for the discussion of special questions, and that members of these committees need not be members of any of the constituent academies. This to a large degree obviates what would otherwise be a considerable difficulty.[139]

Furthermore, also the scientific structures of the various Academies were not always sufficiently compatible to make for a fruitful match. Quoting Schuster again:

> Most Continental academies contain both literary and scientific sections, and at the organising meeting held at Wiesbaden, marked attention was directed to the fact that there was no body in England that could be considered as representative of literary studies. If matters had been left as they stood then, this country would have been altogether unrepresented as regards half the activity of the association. Efforts were made in consequence to take a more liberal view of the branches of knowledge coming within the range of the Royal Society, and to include literary subjects. Very unfortunately, in my opinion, these efforts failed, and a charter was granted to the British Academy, which has now been included as a separate body among the list of academies forming part of the association.[140]

To illustrate the relevance of both the philological and the science branch of academies, we may mention that one of the projects which originally motivated the idea of the Cartel was the *Thesaurus Linguae Latinae*, i.e., a complete Latin vocabulary. On the other hand, the influential geologist Eduard Suess (1831–1914) in Vienna helped to establish the Cartel with a view to formatting international geological networks, for instance for gravitational measurements.[141]

1.3.2.2 Modernizing Research Politics around 1900

Regardless of their intrinsic structural problems the newly created Cartel and the IAA were naturally planned and seen as a contribution on the highest scientific level to the visible, largely industry-driven modernization of their time. World Fairs for instance were not only dazzling national, economic, and popular shows, but also inspired scientists and attracted international congresses.[142] To quote an example from mathematics, which is unrelated to scientific academies but clearly reflects the spirit of the period: When Hilbert was preparing his famous lecture proposing 23 mathematical problems for the new century—which he first presented at the ICM held in August 1900, as one among many venues, at the Paris World Fair—Hermann Minkowski congratulated Hilbert on his manuscript like this: "Now you really have taken a general lease for the mathematics of the twentieth century, and all mathematicians will readily accept you as the general manager."[143]

---

[139] [Schuster 1906], p. 259.

[140] [Schuster 1906], p. 259.

[141] See [His 1902], cf. [Gierl 2004], *passim*.

[142] [Gierl 2004], pp. 333–350.

[143] Letter of Minkowski to Hilbert dated Zürich 28 July 1900 in [Minkowski 1973], p. 130; *Nunmehr hast Du wirklich die Mathematik für das 20. Jh in Generalpacht genommen und wird man Dich allgemein gern als Generaldirektor anerkennen.*

After the massive creation of jobs and journals, especially for pure science, in the nineteenth century the world of science was looking for new structures of governance. Considering the daunting number of international organizations founded in the last decades of the nineteenth century,[144] the Cartel and the IAA thus appear as natural responses to the ambient civilization, and the word 'cartel' becomes less surprising. Still, with its recourse to academies and their history, this Cartel appears less up-to-date than the foundation—finally realized only in 1911—of the *Kaiser-Wilhelm-Gesellschaft* (today the Max Planck Society) with its specialized research institutes outside of universities.

The sort of compromise that the Cartel and the IAA tried to strike is well reflected in Hermann Diels's[145] (1848–1922) analysis of the scientific academies in his detailed 1906 account of the then current organization of science:

> Following the theoretical development of our university system in the nineteenth century, the purely 'academic' view of academies was developed fully, and maybe even exaggerated. In fact, the contempt for engineering, which has never existed like this, for instance, in the English and French institutes, could have turned to disaster for the German academies, if the situation had not been turned around at the right moment by exterior and interior influences.[146]

Another, more political category of the time which, not surprisingly, transpires in the work of the Cartel and the IAA is imperialism. Viewed from London, this is all the more natural as in British history the nineteenth century was one of Empire building rather than Nation building. Imperial temptations were swelling on the continent as well, and left their unmistakable traces in colonial research projects cultivated by the participating academies under the roof of the IAA, such as the geophysical observatory in Samoa of the Göttingen Academy which received a special allocation from the German Empire,[147] or a disastrous geophysical (gravitation measurement) expedition in Tanzania which involved several ministries and other imperial institutions.[148]

Worldwide imperial ambitions and national pride would of course at times generate tensions, and a lot of diplomacy was required in the administration of the international consortium.

> It is not my desire to disguise the difficulties which have sometimes been encountered in providing for joint undertakings on a large scale. Whether national or international, combined work between men of different temperaments always requires some suppression of personality. Even stronger feelings may be involved when a central office or bureau has to be selected which specially distinguishes one locality. The advantage gained by the locality is often one of appearance rather than of reality, for these central offices should be the servants rather than the masters of the undertaking. In order to prevent national feeling being aroused by any preference given to one nation, it has been customary in some cases to have a president who belongs to a different country from that of the director of the Central Bureau;

---

[144] See the decade-by-decade lists given in [Gierl 2004], pp. 348–350.

[145] The classical philologist renowned for his edition of pre-socratic Greek texts.

[146] [Diels 1906], p. 622; my translation.

[147] See [Gierl 2004], pp. 269; 391, and [Gierl 2014], pp. 104–106.

[148] [Gierl 2014], pp. 102–103.

there are also a vice-president and a secretary, all belonging to different nations. It is thought that such a distribution of office may assist in preserving harmony. I believe that this is the case, but sometimes at the risk of impaired efficiency. It cannot be denied, however, that the seat of the central office of an important undertaking confers a certain dignity, and it is quite natural that a country should feel some pride in the distinction.

England on the whole has not done so badly. We should not forget that in a great portion of the world all clocks strike the same minutes and seconds. Before long all civilised countries (except Ireland) will have adopted the Greenwich meridian for their standard of time, and we may rightly, therefore, call Greenwich the central bureau of universal time.

. . . . . .

I am afraid I have only given a very inadequate account of the serious interests which are already involved in international scientific investigations. But if I may point once more to Indian meteorology, and remind you of the vital importance of an effective study of the conditions which rule the monsoon, you will, I think, realise how impossible it is to separate scientific and national interests. The solution of this particular problem requires an intimate cooperation with Central Asia and Siberia—a cooperation which has been easily secured. I do not wish to exaggerate the civilising value of scientific investigation, but the great problems of creation link all humanity together, and it may yet come to pass that when diplomacy fails—and it often comes perilously near failure—it will fall to the men of science and learning to preserve the peace of the world.[149]

Fig. 1.6 The delegates of the IAA in Rome, May 1910.

[149] [Schuster 1906], p. 259.

1.3.2.3 Rhetoric and Reality

The preceding quote leads from a rather sober and candid description of basic difficulties of Science International to a final internationalist apotheosis ... whose predicament would be cruelly tested eight years later. Schuster's rhetorical turn is not atypical of public statements left behind by the IAA. Some of them are more emphatic and sound above all like vibrant echoes of the extraordinarily sumptuous receptions (see Fig. 1.5) that the hosts of the tri-annual meetings would prepare for the delegates—an impression that several historians have not tried to hide.

> To gauge the role of internationalism in practical international science, it would be necessary to carry out detailed studies of specific organizational initiatives and ventures. The rhetoric that surrounded these often masked the significance of both motives and achievements. The International Association of Academies (IAA), for example, was certainly steeped in internationalist rhetoric, yet it accomplished almost nothing from the time it was created in 1899 until it was disbanded during World War I.[150]

Not surprisingly, it is also easy to quote examples of the underlying nationalism accompanying all the glamorous and scientifically valuable gatherings and projects.[151] Therefore, when Émile Picard at the 1910 meeting in Rome praised the IAA as a great moral force, a veritable *Conseil supérieur de la Science*, this may seem a little over the top, especially when confronted with very concrete, factual national divergences which apparently had the potential to slow down the functioning of IAA commissions[152]:

> There are other deficiencies of uniformity which perhaps appear trivial, but which yet lead to the waste of a good deal of time. Such, for instance, is the position of the index in scientific books. The index is placed sometimes at the beginning, sometimes at the end, and sometimes neither at the beginning nor at the end. Some books have no index, some have two, one for the subject-matter and one for names of authors. The loss of time which arises from one's ignorance as to where to look for the index cannot be estimated simply by what is spent on the search, but must include the time necessary to regain the placidity of thought which is essential to scientific work.[153]

1.3.3 The Encyclopedia of the Mathematical Sciences Including Their Applications

Having heaped so much know-it-all historical scepticism on the Cartel and the IAA, let us now pass on to a successful and significant international project of Felix Klein's which the Cartel helped to realize, the "Encyclopedia of the Mathematical Sciences including their applications" (*Encyklopädie der Mathematischen Wissenschaften*

---

[150] [Crawford 1992], p. 41. Cf. [Greenaway 1996], pp. 13–16, for a brief overview of the IAA meetings.

[151] See for instance [Schroeder-Gudehus 1966], pp. 47–49.

[152] Cf. [Schroeder-Gudehus 1966], p. 46.

[153] [Schuster 1906], pp. 235–236.

mit Einschluß ihrer Anwendungen), the first volumes (1898–1904) of which were "edited on behalf of the Academies of Sciences of Göttingen, Leipzig, München, and Vienna", that is, the initial Cartel, "and with the collaboration of numerous expert colleagues." In the spirit of this chapter, we are particularly interested in the appreciation of various national contributions to the mathematical sciences in the context of the Encyclopedia, before World War I.

In Section 1.2.3.1, we have quoted Kummer who, in the middle of the nineteenth century, saw such a creative explosion of (pure) mathematics in Germany that comprehensive book treatments would have to wait until a later period after the new original ideas had settled in. Although the tremendous amount of new mathematics developed during the nineteenth century needed to be taken stock of, this kind of consideration was not a concern for Felix Klein and his colleagues at the launch of the encyclopedic project in the early 1890s. Indeed Klein himself recalled during a lecture course in the Winter of 1910–1911, of which Erich Hecke (1887–1947) took notes, that

> also with respect to the Encyclopedia the opinion has been voiced that its idea was a symptom of the fact that mathematical productivity was petering out, so that the only thing one could do was to collect what had been achieved. Well, that was not our idea when we were launching that plan. And the development of science since 1894 has in fact taken a different path. The Encyclopedia had more to do than to collect, it had to work through the extremely heterogeneous material in order to present it in a unified way.[154]

Klein's suggestion of a unified presentation meant that the scientific and technical applications of mathematics were certainly part and parcel of the project, albeit with a stress on the mathematical substance of the applications.[155]

1.3.3.1 The public Image of Mathematics

Another impetus of the project was the desire to strongly inscribe mathematics into the ambient culture.[156] That such a desire was felt in Germany at the time—in a country where no institution like the École Polytechnique in Paris consecrated mathematics as a key discipline (see Section 1.1.4 above)—can be seen for instance from the extremely defensive opening of Aurel Voss's (1845–1931) chapter about the mathematical sciences[157] in another encyclopedic collection of the time: *Die Kultur der Gegenwart* (The Culture of the Present). The general editor of *Die Kultur der Gegenwart* was Paul Hinneberg (1862–1934), former private secretary of the historian Leopold von Ranke (1795–1886),[158] had connections with the circle around Friedrich Althoff. Klein was actually the editor in charge of the volume in which

[154] My translation of the quote in [Tobies 1994], p. 11.

[155] Cf. the remarks on the importance of 'applied mathematics' in the Encyclopedia in [Gispert 2001], pp. 94–96.

[156] Cf. the first section of [Tobies 1994].

[157] [Voss 1914].

[158] Renate Tobies [Tobies 1994], pp. 7–8, points to traces of Ranke's thought in Klein and the Encyclopedia.

Fig. 1.7 Felix Klein among principal correspondents ([Rowe 1985], p. 75).

Voss's text appeared. The kind of culture required of the intended readers of this collection is indicated by the fact that Voss's German text contains untranslated quotes in Latin, French, and English.

In spite of the great attention Klein gave to physical and technical applications, and even though he fully accepted that mathematics and the experimental sciences were "natural allies" in high-school education, Klein also declared that "mathematics-in-itself is a pure branch of the humanities."[159] This twofold emphasis—which incidentally reminds us of the two classes that existed in each of the academies of the Cartel—sets Klein apart from mathematicians like Kronecker, Hermite, or even Poincaré for whom mathematics was a natural science and its foremost goal, the classification of given objects.

1.3.3.2 International Collaboration for the German Encyclopedia

From what was said so far, the Encyclopedia appears to be a typically German enterprise of the period 1894–1914, with the mild proviso that the Cartel also involved the Vienna Academy. However, just as the Cartel spawned the IAA, Klein

[159] My attempt to translate the quote in [Tobies 1994], p. 11: *die Mathematik ist an sich eine reine Geisteswissenschaft*. Cf. [Gispert 2001], p. 103: "Confronted with the need to legitimate the importance of their discipline and its teaching both to engineers and to intellectual elites, mathematicians insisted on this double character of their discipline. One of the principal artisans of the [French edition of the] Encyclopédie, Émile Borel (1881–1956), for example elevated mathematics to the level of the 'human sciences,' one which contributed to 'the formation of free men whose reason only yields to facts,' and insisted on 'making evident for all ... the points of contact between mathematics and modern life, the only means to prevent [them] from being suppressed one day as useless, as a financial saving'."

actually gave the Encyclopedia a truly international dimension, and in more than one way. Thus we read in the introductory report, signed 30 July 1904 by Walther von Dyck (1856–1934) as president of the editorial committee of the Encyclopedia:

> Even though we certainly want to claim the whole enterprise as a German one, by its principles and execution, it is of the utmost importance—lest it represent a one-sided point of view—that in the way in which individual domains are approached and expounded, all the voices are articulated which have shaped their specific development. The lasting achievements of each science are an international asset, distilled from the totality of the work of all scholars of all times and countries. But the various nations and ages have participated in various directions and put various emphases and preferences on the individual domains, with characteristic differences in method and type of presentation. This has to be reflected in the Encyclopedia in the way the material is presented according to its historical development, and in the choice of collaborators. Indeed the enterprise today counts, apart from the base team of German authors, collaborators from America, Belgium, England, France, Holland, Italy, Norway, Austria, Russia, and Sweden.[160]

Concretely, Felix Klein used his international contacts and trips—some of which were occasioned by Cartel or IAA affairs—to exchange about the Encyclopedia project and seek out potential collaborators. But to what extent does a foreign collaborator of the Encyclopedia represent a national approach to the subject he—or she[161]—was covering? The Russian mathematician Dmitry Fyodorovich Selivanov (1855–1932) from St. Petersburg, for example, wrote the short article on difference equations, and subsequently expanded it into a German textbook at the request of the publisher. He had not only studied with Hermite in Paris but had also been influenced by his Berlin years, which brought him into close contact with the circle of Kronecker's students, in particular with Jules Molk.[162] The sources of his subject—which belongs to the more applied sections of the Encyclopedia, before the chapters on analysis, and is meant to provide tools for making tables etc.—contain a strong mix of authors from different countries. Klein was nevertheless interested in it, and his orientations presumably played a role.

> When in the Summer of 1895 in the mathematical seminar of Prof. F. Klein and D. Hilbert questions of interpolation and difference calculus came up, a lack of related textbooks was noticed. The most easily accessible ones, for the German reader, by Lacroix and Schlöhmilch are partly outdated, whereas [George] Boole's [1815–1864] is less compatible with the German taste because of its insistence on symbolic methods.[163]

[160] *Encyklopädie der Mathematischen Wissenschaften mit Einschluß ihrer Anwendungen*, Vol. I, part 1, pp. XIII–XIV; my translation. The order of the nations at the end is alphabetical in the German original.

[161] As far as I can see, exactly one woman appears among the authors of Klein's German Encyclopedia: Tatiana Ehrenfest(-Afanaseva) (1876–1964), who co-signed the chapter on Statistical Mechanics in Vol. IV–4 with her husband. The couple had first met in Göttingen. In 1917, during the war, Klein's friend von Dyck would note with utter disgust that he could never talk to Paul Ehrenfest (1880–1933) and his friends without the wives also joining the conversation: *dass da immer auch die Frauen dabei sein müssen!*; see [Tollmien 1993], p. 193. – At least one other woman and wife was *translating* chapters of the Encyclopedia; see the end of the following Section 1.3.3.3.

[162] Cf. the obituary notice [Rothe 1934]. On Molk, see Section 1.3.3.3 below.

[163] From the translator's preface to [Markoff 1896], p. IV; my translation. The treatise by Boole alluded to here is [Boole 1860].

This motivated the translation of Markov's book into German. Andrey Andreyevich Markov (1856–1922) is remembered today for his 'chains', or more generally for the Markov Property.[164] However, he did teach difference calculus at St. Petersburg for years, and Selivanov's chapter in the Encyclopedia, while it lists both [Boole 1860] and [Markoff 1896] in the initial literature overview, does not have a single explicit reference to Boole in the text, but refers to specific pages of Markov's textbook four times. This is certainly compatible with the German reservation with respect to Boole's approach, but it is insufficient to establish a genuine Russian moment in the presentation of difference calculus, even if the textbook Selivanov used was written by a St. Petersburg colleague of his. It is thus not as easy to unravel national preferences at work in the German Encyclopedia. Still, certain major associations between nations and mathematical domains seem clear:

> In the end, already for the first three volumes on pure mathematics, there were 92 authors, among them 32 foreign mathematicians. According to Klein, the Encyclopedia should particularly reflect the important geometric school of the Italians, the analysis of the French, and British mechanics. ... The forewords to all the volumes of the Encyclopedia point out his part in the selection of authors, the arrangement of the material all the way to details of exposition.[165]

1.3.3.3 Translating the Encyclopedia

There is another remarkable international aspect to the Encyclopedia project: a French and an English edition of it were envisaged of which the former was partly realized and published, until it was stopped short by the death of the editor Jules Molk (1857–1914) and by World War I. Molk was first brought into this project, not by Klein or one of the mathematicians involved, but by the publisher Alfred Ackermann-Teubner (1857–1941) who participated in mathematicians' gatherings in Germany and in France and who would associate the Paris publisher Gauthier-Villars with the project. As a matter of fact, both the German and the French edition of the Encyclopedia were a commercial success.[166] Molk in turn managed to convince excellent French mathematicians such as Maurice Fréchet (1878–1973), Émile Borel, and Élie Cartan (1869–1951) to collaborate.

Jules Molk's life was itself marked by international influences. He was born and grew up in Strasbourg; when he was 13 years old, his hometown was annexed to Germany. He studied mathematics at the Zürich Polytechnicum, where one of his professors was Georg Frobenius (1849–1917), who would much later (1892) become Kronecker's successor in Berlin (from where he would subsequently try to fight Felix Klein's growing influence). From Zürich Molk went to Paris, and there he obtained a scholarship for advanced studies in Berlin from 1882 to 1884. This would bring him

---

[164] The inspiration of this theory may be linked to an internal Russian controversy about the compatibility between modern mathematics and Russian orthodox religion—see [Graham & Kantor 2009], pp. 67–71.

[165] See [Tobies 1994], p. 21; my translation.

[166] See [Tobies 1994], p. 31.

into contact with Leopold Kronecker, who directed his thesis in algebraic number theory. The Berlin connection is the reason why Molk's French version of Alfred Pringsheim's (1850–1941) Encyclopedia article on the theory of irrational numbers is one of the best extant sources about Kronecker's view on this matter.[167] As of 1898 Molk was professor of rational mechanics at Nancy University.[168]

Every installment of the French edition of the Encyclopedia contained the following short introductory note (*Avis*) to the reader:

> **Note.** In the French edition we have tried to reproduce the essential features of the articles of the German edition. As to the mode of exposition, however, we have mainly followed the French traditions and habits.
>
> This French edition has a very special character thanks to the collaboration of German and French mathematicians. Indeed, the author of each article of the German edition has indicated the modifications which he thought appropriate for his article, and on the other hand, the writing of the French version of each article has prompted exchanges in which all parties have participated. Additions which are specifically due to the French collaborators are placed between asterisks. The importance of such a collaboration, realized for the first time in this French edition of the Encyclopedia, will be obvious to everyone.[169]

Archival sources illustrate this very active exchange and the extent to which the preparation of the French edition offered new points of view and questions to the mathematicians around Felix Klein.[170] And on the French side, the "volume of the *Encyclopédie* dealing with mechanics, of which a third of the planned articles were published, offers a rather singular picture within the class of publications of mathematics and applied mathematics in France. Aside from a few long background pieces—of which certain had been the occasion of important 'theoretical additions in the French style' by French authors—, there are three articles dedicated to ballistics, a totally marginal field within the French mathematical community. The authors profited from the existence of a French edition to offer themselves a tribune for their research; indeed, one whole article is devoted exclusively to French work in this area, an exclusivity unparalleled in the rest of the *Encyclopédie*."[171]

Another confrontation of the German and the French edition has been suggested by Hélène Gispert by comparing two *different* articles in the two editions: on the one hand the German-Danish collaboration of Max Dehn (1878–1952) and Poul Heegaard (1871–1948) on *Analysis situs* (Vol. III–1–1 in the German edition), and on the other the article on set theory which was completely revised for the French edition by the Frenchman René Baire (1874–1932), who is remembered by mathematicians today essentially for the 'Baire Category Theorem.' Whereas Max Dehn turned his article with Heegaard into a "modern manifesto of topology"[172], Baire on the other hand "was extremely reticent to produce an exposition of set theory independent

[167] Cf. [Petri & Schappacher 2007], pp. 366–368.

[168] For Molk's biography, cf. the obituary [Vogt 1914].

[169] My translation.

[170] See for instance [Tobies 1994], Section 4.2, pp. 38–52, especially about new developments in analysis.

[171] See [Gispert & Tobies 1996], p. 418.

[172] See the title of Section 7.3 in [Epple 1999].

of other more classical domains to which it might be applied."[173] Baire's attitude should remind us of Hermite's and Picard's criticism quoted above, Section 1.2.3.2. The kind of 'modern mathematics' which is pushed by Dehn and Heegaard, and avoided by Baire, is that of the new formal axiomatics which Hilbert had introduced and executed in his *Foundations of Geometry* of 1899, thus wiping the slate clean now that the existence of non-Euclidean geometries had entered the mathematical mainstream. Just as Hilbert had studied the axioms of geometry, which use words like 'points', 'lines', 'surfaces' in a way that is indifferent to the possible nature of these things, Dehn and Heegaard develop topology "as a theory dealing with aggregates of uninterpreted elements, for which only combinatorial rules are specified."[174] They define 'point complexes' which can give rise to 'complexes of line segments' (*Streckenkomplexe*) and thus create a formal theoretical framework to study Poincaré's homologies on 'polyhedra.' This way they established the transformation of *Analysis situs* into the beginning of twentieth century combinatorial topology.[175] Speaking anachronistically, comparing Baire to Dehn and Heegaard puts the French edition of the Encyclopedia in a position opposite to what Bourbaki's (different sort of) encyclopedic project would put forward as of the mid 1930s.

For yet another example of an interesting comparison, one may open the monumental study by Bottazzini and Gray.[176] The authors contrast the treatment of complex numbers—and their (im-)possible generalizations to higher dimensions—, algebraic analysis—which Molk read as a sort of propaedeutics to Weierstrass—, and complex function theory in the German and the French versions of the Encyclopedia.

> The outcome, the image of complex function theory that [the French edition] offered, was nonetheless slightly different from the German one. Indeed, it was a combination of the prevailing images in France and Germany, respectively, a medley of the heritage of Cauchy's and Weierstrass's traditions, with Riemann's geometric approach left in the background. It was a provisory mixture, well represented by [William Ford] Osgood's [1864–1943] article [in Vol. 2–2 of the German Encyclopedia], and therefore destined to be outdated very soon.[177]

The differences in the treatment of complex function theory thus also reflect different appreciations of the nineteenth century history of mathematics, i.e., of the period for which the Encyclopedia tried to include succinct historical accounts. This phenomenon shows time and again when comparing the two editions.

> In some cases the versions of the history of the nineteenth century in the two editions do not agree. The French edition modifies the German one by offering [a] different or complementary version of developments in which certain actors do not have the same importance. The historical accounts in the two editions, by their very nature, seek to identify milestones in the development of mathematical ideas. Yet these accounts show that, for the nineteenth century—a century in which scientific communities began to be structures in

[173] [Gispert 2001], p. 102.

[174] [Epple 1995], p. 392.

[175] See [Epple 1999], pp. 229–233.

[176] See [Bottazzini & Gray 2013], Section 10.10, pp. 745–749.

[177] [Bottazzini & Gray 2013], p. 749.

> a national framework—these milestones were not universally agreed upon. This history of mathematics could therefore be read in a variety of ways according to the traditions in which the authors were enrolled.[178]

All this shows a mutually enriching, complicated negotiation of national approaches. Contrary to what we often tacitly tend to take for granted today when looking at the main mathematical literature, there was no generally, internationally accepted standard before World War I.

In general, and on a purely formal aspect, the new passages added by the French translator-authors and their style led to noticeably longer articles. As soon as the first issues of the French edition were out, this in turn induced some German authors to write longer pieces.[179]

Fig. 1.8 The Göttingen Mathematical Society in 1902. At the table is Felix Klein. Sitting on his left are the physicist Karl Schwarzschild (1873–1916) and Grace Emily Chisholm Young. Sitting on his right is David Hilbert. Source: [Arch. SUBG], Cod. Ms. K. Schwarzschild 23 : 1, 16.

---

[178] [Gispert 2001], p. 106.

[179] [Tobies 1994], p. 34, footnote 105.

Grace Chisholm Young, née Grace Emily Chisholm (1868–1944), was contacted as coordinator for this undertaking, together with her husband William Henry Young (1863–1942), and she actually started by translating several articles.[180] Disenchanted with mathematics at Cambridge she had come to Göttingen, studied with Felix Klein, who backed Prussian efforts to open universities to women, and obtained her PhD under his direction in 1895. This was not the first mathematical doctorate delivered to a woman at Göttingen; Sofya Kovalevskaya had also obtained hers, albeit *in absentia*, from the same university back in 1874. However, Grace Emily Chisholm (still unmarried at the time) was the first who was also invited to take the associated oral exams, which she did. As for the Encyclopedia in English, here is a brash account of what happened:

> The English edition never even reached the press, and only cost Grace and Will and a few other enthusiasts a great deal of time and effort. Klein had told them that it would take them ten years' work, and for a number of years they tried to make it a success; but in the end all that was to be shown for it was a few manuscript translations from the German made by Grace and certain others. When Will started trying to build up interest in it in 1900, he found A.[ndrew] R.[ussel] Forsyth [1858–1942] at Cambridge openly hostile and many other prominent mathematicians unwilling to spare the time and effort needed: the Cambridge attitude to mathematics, dominant in the English speaking countries, saw to the demise of the project.[181]

1.4 World Mathematics before World War I

In the last three decades before World War I, attention to national distinctions and feelings of national pride or imperial supremacy were extremely common, but by and large they *peacefully coexisted*—if we may put it like this—with increasing contact and collaboration among scientists from different empires or countries. This coexistence prevailed in scientific communities and in the minds of individual scholars. It was in fact a much more general phenomenon, as witnessed for instance by an article on "International Unions" published in 1887 in the very first volume of *Revue d'histoire diplomatique*, the French journal on foreign diplomatic history and international relations; the paper starts out by describing this very dualism of national and cosmopolitan views:

> Among the diplomatic realities which mark this *fin de siècle*, one of the most interesting is undoubtedly the rapid procreation of international unions ... Indeed, nobody will have any illusions: never have international treaties been less respected, never have the rivalries been more bitter, the racial hatred more alive, the instruments of destruction more terrible. At the same time, in a singular contradiction, never have there been more efforts to bring peoples closer together and unite them with respect to certain interests and for the common ideal of humanity.[182]

[180] [Tobies 1994], p. 22–23, footnote 74.

[181] [Grattan-Guinness 1972], p. 139.

[182] From the opening paragraph of [Lavollée 1887], p. 331; my translation. I owe this reference to the opening section of [Erdmann 2005].

Similarly, in 1901, the German polar explorer Georg Balthazar von Neumayer (1826–1909) candidly declared what reads like a paradox:

> While I have always upheld—as should be clear from what I said before—the national conviction that we have to promote the expansion of our maritime powers by large scale oceanographic and geographic research, it does give me the greatest satisfaction that the problem of the Antarctic will now be tackled, not in a unilateral, national way, but jointly with other nations.[183]

We are not interested here in exploring how the coexistence of these potentially conflicting tendencies was concretely negotiated by individual scholars or groups of scientists; the preceding Section 1.2 contains a few illustrations of this. The point we are making now is that this very coexistence is characteristic of the pre-WW I period of Science International; it is what distinguishes these decades from all earlier and later historical periods.[184] Indeed, the same sort of coexistence is *not* to be found earlier in the nineteenth century when international networking was in its infancy, i.e., when it was not a generally visible reality yet in terms of events and technical or industrial developments, and in particular lacked international structures. "International nongovernmental organizations [INGOs], though few and far between until about 1890, subsequently multiplied to reach a peak in 1910 (not exceeded until 1945), before falling back again in the run-up to the First World War."[185] It was also in 1910 that the *Union of International Associations* was created in Brussels with a view to federating some 400 associations. This was the proud result of years of preparatory work by the Belgian visionary lawyers Henri La Fontaine and Paul Otlet. But World War I brutally brought this 'peaceful coexistence' of the national and cosmopolitan attitudes to an end (see Chapter 3 below) and seared its lasting mark into subsequent enterprises towards Science International.

Scientific international unions lagged a little behind the general trend of INGOs, but some of them were indeed founded before World War I. For example, the *International Seismological Association* (ISA) was officially created during the Second International Seismological Congress in July 1903 in Strasbourg. Georg Gerland (1833–1919) had worked towards this since the 1880s, and by 1900 he had negotiated the setting up of a seismological measuring station at the (German) University of Strasbourg. Given the subject matter, it is not surprising to see that the ISA had a number of members from outside of Europe and the US: Argentina, Chile, Congo State, Japan, Mexico. This Association would be enlisted in activities of the IAA as of 1904.[186] But most of the international scientific unions bore the stamp of the war when they were founded. For example the International Astronomical Union (IAU),

---

[183] Quoted from [Soutschek & Nickelsen 2019], p. 235; my translation.

[184] Cf. [Crawford 1992], p. 43, where the author takes her time interval of reference to be 1880–1914.

[185] See [Osterhammel 2014], p. 505, with reference to the detailed statistical study [Boli & Thomas 1999], which not only analyzes extensive data about creations of INGOs but also studies their correlation with indicators of administrative, economic, and technological development. Cf. the lists in [Gierl 2004], pp. 348–350, which were already cited in Section 1.3.2.2. For a general historical reflection about internationalism before World War I, see the concise text [Rasmussen 2004], with reference in particular to her 1995 PhD Thesis.

[186] See [Odenbach 1911], [Schweitzer & Lay 2019], as well as the historical literature cited there.

the International Union of Biological Sciences (IUBS), the International Union of Pure and Applied Chemistry (IUPAC), and the International Union of Radio Science (URSI) were first instituted in 1919, and the International Mathematical Union (IMU) yet a year later, in 1920—see Sections 4.1.1 and 4.3.2 below.

In the absence of an international union for a given branch of science, one can still try and monitor the buildup towards the respective INGO before World War I by following international contacts and projects—this is what I have tried to do for mathematics in previous sections of this chapter—and by looking at:

1.4.1 International Congresses

There are various sorts of international conferences. We have mentioned in Section 1.2.1 the outstanding early case of the Karlsruhe Chemical Conference of 1860 which helped trigger a major advance for the subject, even though it failed to reach an immediate result. Other meetings served practical needs. In November 1885 representatives from various European countries and professions (musicians, musicologists, critics, ...) got together in a ministerial building in Vienna to fix an international standard for concert pitch. Already around 1830, the core musical repertoire had stabilized across Europe, and the European market of musical performances involved an increasing number of musicians and ensembles touring various countries. In 1858, the *Académie française* had set the national *diapason normal* at 435 Hz for the A above middle C, and French wind instrument makers had built their produce in a way compatible with this rule in principle. This attempt to limit the strain on the singers' vocal chords was repeated at the 1885 international conference in Vienna. In practice though, the pitch used for instance at the *Scala* in Milan, in the middle of the nineteenth century, tended to exceed 440 and often 450 Hz. In spite of this intractable reality, the norm of 435 Hz was subsequently even inscribed into the Versailles treaty after World War I (Article 282, item 22)! The pitch was finally notched up to 440 Hz only at the last meeting of the International Federation of the National Standardizing Association—founded in 1926, and reconstituted under another name after World War II—in 1939 in London.[187]

Compared to such a concrete agenda, the five pre-war International Congresses of Mathematicians (ICM) appear in the first place as social gatherings. They were organized at Zürich in 1897, Paris in 1900 (cf. Section 1.1.5.4 above), Heidelberg in 1904, Rome in 1908 (where the Guccia Medal was awarded, see Section 1.1.5.5), and Cambridge, UK in 1912.[188] These events set the rhythm of regular ICMs which

[187] Cf. [Finscher 1998], Vol. 8, column 1828.

[188] For a brief overview of these congresses, cf. [Barrow-Green 1994].

we still know today, and they were visible tokens of Mathematics International in those years.[189] At the same time, as far as each of them was organized *ad hoc*, they also show mathematics still in search of an international organization.

1.4.1.1 History International before World War I

It is a widely echoed idea in texts about Mathematics International that "mathematicians proudly considered themselves and their discipline to be peculiarly international, perhaps because the language of mathematics sits on top of vernacular tongues."[190] The idea is hardly compelling in itself because the same formalism can be used to express diverse mathematical practice; recall for instance the differences between the German and the French edition of the *Encyclopedia* discussed in Section 1.3.3.3 above. More importantly, however, the ostensive beginning of mathematical internationalism in the late nineteenth century as shown by the first International Congresses of Mathematicians (ICMs) may not have needed the help of any specific predisposition of mathematics towards uniting different nations; it can be seen first of all as a symptom of the general trend which is the main theme of the present Section 1.4. To get a feeling both for the importance of this general trend, and for what may indeed be special to the case of mathematics, let us briefly compare the early ICMs to the first International Congresses of Historians.

Historians are in the habit of discussing various methodological approaches, but clearly history neither was nor is a formalized exact science, and its research domains have a much more direct, non-metaphorical affinity to the realm of politics than mathematics, even including statistics. And yet, the early international congresses of historians follow a chronological and geographical pattern which is very similar to that of the first ICMs[191]: An initial *Congrès international d'histoire diplomatique* took place in 1898 in The Hague [just one year later than the first ICM in Zürich]. Then followed a *Congrès international d'histoire comparée* which [just like the second ICM] was one of the many events embedded in the Paris Universal Exhibition of 1900; the historians gathered there two weeks before the mathematicians. This second congress was called by the French government because the organizing committee appointed at The Hague meeting had disintegrated over a personal rift. These tensions incidentally opened up the historians' international congresses beyond diplomatic history, giving them a more academic flair. The following International Congress was already officially announced for 1902 in Rome [where the mathematicians would get together in 1908] when a leading local organizer, the classical historian and archaeologist Ettore Pais (1856–1939), a former student of Theodor Mommsen's (1817–1903) in Berlin, was attacked by Italian colleagues for having used an approach of source criticism which was deemed to be too German. As a result the Rome

[189] For the historical significance of regular international congresses, cf. the special volume "International Cooperation and the Making of Science. Scientific Congresses from 1865 to 1945" of the *Revue germanique internationale*, Vol. 12, 2010, edited by P. Rabault-Feuerhahn & W. Feuerhahn.

[190] See [Riehm & Hoffman 2011], p. 83.

[191] As a general reference for this section, see chapters 1 through 6 of [Erdmann 2005].

Congresso Internationale di Scienze Storiche had to be postponed to April 1903. The historians gathered again in 1908 in Berlin [whereas the pre-war ICM which was held in Germany took place in 1904 in Heidelberg], and their last, surprisingly uncontroversial get-together before World War I took place in London in 1913 [one year after the Cambridge ICM]. The Berlin congress had drawn criticism from several Berlin historians, which reminds us of the reluctance with which the Berliners finally joined the Cartel of Academies—see Section 1.3.2 above. Just as the ICMs, the five pre-war congresses of historians were organized *ad hoc*, without recourse to an international body. Various ideas of such an institution were proposed, but only on 13 May 1926 would the *Comité international des Sciences Historiques* be officially created.

The historians' greater proximity with the world of politics may show in the organizational hurdles encountered, for instance in preparing the congresses in Paris and Rome, and the way they were overcome: essentially by government intervention. Accordingly, the fact that these five International congresses of historical sciences took place at all impressively vindicates the general trend alluded to above. Comparing with the mathematicians, already the regularity of the ICMs at four-year intervals indicates that mathematicians in Switzerland, France, Germany, Italy, and the UK at the time were not only gathering efficiently behind these projects, but they were also well placed to successfully solicit support from local and national administrations as well as private donors (such as scientific publishers). The extent to which the international, formal language of mathematics may have eased their endeavors seems difficult to gauge.

1.4.1.2 The Pre-war ICMs

For their first congress the mathematicians were careful to select a host country known for its attention to international causes, the most emblematic example being the International Committee of the Red Cross in Geneva founded in 1863. Hurwitz's welcome speech at the 1897 ICM in Zürich avoided any allusion to national motives, insisting merely on the contrast between lonely work in the study and the exchange of ideas with colleagues.[192]

In 1900, Hilbert's very much abridged presentation of problems for the new century (cf. Section 1.3.2.2 above)—he only had time for 10 of his 23 problems—was oddly placed in the session under the double heading *Bibliography & History, Teaching & Method*. (Hilbert's problems would acquire their formidable reputation and influence on future research only once the complete text was published; this happened within a year in German, French, and English.) During that session at the Congress, after Hilbert's talk, the proposal was launched that the mathematicians take an active part in the movement for adopting a universal language. Even though this finally led to nothing, the idea had mobilized quite a few colleagues for years,

[192] See Proceedings ICM 1897, pp. 22–23.

for instance Giuseppe Peano (1858–1932).[193] A French-German & German-French mathematical dictionary was also published on the occasion of the Paris ICM.[194] By the way, language questions also popped up at the historians' international congresses, for instance in London in 1913 when Russian was proposed as an additional official language of the congress besides English, French, German, and Italian.[195]

A recurring theme at the ICMs was that of a unified mathematics classification scheme. In the Proceedings of the earlier International Mathematical Congress held in Chicago 21–26 August 1893,[196] for instance, the various contributions are labeled in the table of contents "according to the notation proposed by *la commission permanente du répertoire bibliographique des sciences mathématiques*, Paris, 1893." This refers to a classification published as a 100 page book, which was in fact the result of an International Congress on mathematical bibliography held 16–19 July 1889 in Paris, the *Congrès international de bibliographie des sciences mathématiques*, whose president was Henri Poincaré. The initiative of this international meeting came from the French Mathematical Society.[197]

> The history of the *répertoire bibliographique* is in fact part of the more general history of internationalizing science and cultural productions. During the second half of the nineteenth century, the idea of an international space conceived as a unit in its own right was slowly making its way: the issue was no longer just the opening of national scientific practices to foreign countries . . . , it was also about rendering scientific and intellectual productions accessible everywhere.[198]

Following this trend, committees were formed in 1893 in Chicago and at the subsequent ICMs, to pursue this issue further. The subject continued to be on the agenda of the IMU and the ICMs afterwards. But it was neither the ICMs nor the IMU which would initiate the two still existing mathematical review journals—*Zentralblatt* and *Mathematical Reviews*—and the corresponding Mathematics Subject Classification scheme MSC.[199]

---

[193] For mathematicians in touch with the various movements in favor of universal languages like Esperanto, Volapük, Peano's Latino sine Flexione, Ido, etc., see [Gray 2008], Section 6.1.2, pp. 376–379.

[194] See [Müller 1900].

[195] See [Erdmann 2005], p. 61.

[196] The series of ICM proceedings downloadable from the IMU webpage is preceded by those of the International Mathematical Congress held in Chicago 21–26 August 1893 (in the context of the World's Columbian Exposition), and of the subsequent mathematical colloquium organized upon Felix Klein's initiative in Evanston 28 August – 9 September 1893. See [URL 04].

[197] See [Commission 1893]; a summary account of the International Congress is printed on pp. VI–IX, preceding the classification index itself. For more information and further questions about the *répertoire*, see [Rollet 2007].

[198] See [Rollet 2007], p. 263; my translation.

[199] See Section 6.3 below. Cf. [Fraser 2017], where mathematics is embedded in a broader spectrum. Cf. also the first pages of [Schuster 1906].

Few slots were given to applied mathematics at the first two ICMs; this changed as of 1904. The Encyclopedia project (which did include important applied sections) was presented by Klein at the ICM in 1904, and by von Dyck (in Klein's absence) in 1908. In 1912, Klein (who again could not participate himself) recommended by letter that the encyclopedic project *Die Kultur der Gegenwart* (cf. Section 1.3.3.1 above) be presented in the subsection on didactics. However, it was not this but another one of Felix Klein's numerous projects which would foster the institution we have already mentioned and which was created at the 1908 ICM in Rome: a *Comitato internazionale* for questions concerning the teaching of mathematics in various countries, i.e., today's ICMI.[200] This can be understood as a repercussion of mathematics teaching reforms in the main European countries:

> In France after 1902 the study of Euclid's Elements was dropped in favor of treating geometry as a physical science based on the study of rigid body motion, and concrete experience was stressed; Latin was no longer essential for higher education and the importance of science was promoted. Mathematicians supported these moves. Borel, for example, argued that mathematics must be made theoretical and practical or one day it will not be taught. In Germany, Klein's energetic participation in the so-called Meran reform movement, which had begun in 1905, promoted functional thinking and geometric intuition in German schools, Klein's books entitled *Elementary Mathematics from an Advanced Standpoint* date from this period. The International Commission on Mathematics Instruction (ICMI) was established in 1908 at the International Congress of Mathematicians in Rome, when it was resolved that "the Congress, recognizing the importance of a comparative study on the methods and plans of teaching mathematics at secondary schools, charges Professors F. Klein, G[eorge] Greenhill [1847–1927], and Henri Fehr [1870–1954] to constitute an International Commission to study these questions and to present a report to the next Congress." Klein was its first president and served until 1920, when the ICMI dissolved in the acrimony following the First World War.[201] In 1924 these reforms were reversed in France: the former emphasis on the humanities was restored, mathematics and science were cut back, and the earlier reform was denounced as too German and utilitarian.[202]

ICMI would remain the only international organization created at a pre-war ICM.

Beyond these official international initiatives discussed at the first ICMs, the concrete impact of the congresses on the international mathematical community, however strongly it may have been felt by participants, is not easy to evaluate. We do have the breakdown of the participants according to nationality.[203] Looking at them globally and asking which continents were represented, the only non-European[204] "country"—the Proceedings actually refer to it in German as a *Land*—from which participants to the first ICM in 1897 are listed, was *Nordamerika*. These North

[200] See Proceedings ICM 1908, Vol. I, p. 33. Cf. the beginning of Section 1.3 above; see Chapter 9 below for a brief overview of the history of ICMI.

[201] Fehr contested that ICMI was actually dissolved; cf. Chapter 9.

[202] [Gray 2008], p. 38; see also [Tobies 2019]. Cf. [Menghini et al. 2009], in particular Jeremy Kilpatrick's chapter, pp. 25–39.

[203] See the numerical breakdown in Proceedings ICM1897, p. 78; Proceedings ICM 1904, p. 23; Proceedings ICM 1908, p. 20; Proceedings ICM 1912, p. 28. Only for the 1900 ICM in Paris, one has to extract these numbers from the total list of participants by hand.

[204] We are counting Russia among the European countries.

Americans accounted for less than 3% of the participants in Zürich. In Paris in 1900, the non-European countries represented were Argentina (1 participant), Canada (1), Japan (1), Mexico (1), Peru (2), Turkey (1), and the USA (19). Since we are looking for continents, we should add Algeria (1), even though it was at the time a French *département*. Altogether these non-European participants in 1900 accounted for roughly 10% of the gathering, the US mathematicians by themselves for about 7.6%. In 1904 in Heidelberg the only non-European countries represented were Argentina (1), Canada (1), Japan (2), and the USA (15), altogether less than 5.7%. Rome in 1908 was similar, but with different countries: Canada (1), Egypt (1), Mexico (1), Tunisia (2), and the USA (16). Since the total participation had risen to 535 participants, this makes altogether less than 4% from outside of Europe. Finally, in 1912 at Cambridge, England, we encounter among 574 participants for the first time three colleagues from a previously unrepresented subcontinent: Argentina (5), Canada (5), Chili (1), Egypt (2), India (3), Japan (3), Mexico (2), and the USA (60). This adds up to more than 14%, the US alone representing more than 10% of the congress, a notable notch-up with respect to the previous ICMs.

Counting the happy few who were able to participate in a congress, especially when it took place far from home, can give at best a spotty image of how mathematics spread around the world. Still, individual cases behind these figures are very instructive, and some of them also remind us vividly of the importance of national feelings. Here is one remarkable example:

1.4.2 Japan Goes West

On the morning of 8 August 1900 in Paris, after Hilbert's abridged problem lecture, and before the subject of a universal language was proposed, the President of the Section for History and Didactics, the German historian of mathematics Moritz Cantor (1829–1920), gave the floor to Rikitaro Fujisawa (1861–1933) from the Imperial University of Tokyo, for a "Note on the Mathematics of the old Japanese school" presented in English. Rikitaro Fujisawa was the first mathematician from Japan to attend an ICM. Here the word 'mathematician' has to be stressed, and it has to be understood in the Western sense, because Fujisawa was neither a historian of mathematics nor an expert of the traditional *wasan* mathematics which he chose to talk about.[205] Instead, by dismissing *wasan* mathematics, whose "nomenclature and the notations are as clumsy as they are awkward"[206], he obviously wished in the first place to present himself as a staunch fighter for the superiority of Western mathematics. Even as he tried to explain the peculiarities of the old Japanese school to his audience, he constantly resorted to comparisons with the history of European mathematics, especially from the seventeenth century. In his conclusion he alluded to the Japanese educational reform of the early Meiji period (1868–1877), which

[205] See [Horiuchi 1996] for a concise historical treatment of that subject.

[206] See Proceedings ICM 1900, p. 379.

had been devised in accordance with the fifth principle of the Imperial Oath of 1868: "Intellect and learning shall be sought throughout the world, in order to establish the foundations of the Empire."[207]

> Before concluding this brief discourse, may I be permitted to repeat once more what I have said in the beginning. I have been speaking of things which are now entirely obsolete and which can have at most historical interest, leaving however the chance that some really valuable things might still be found in those regions of this mathematics through which I have not happened to pass. It was surely a wise policy on the part of the educational authorities that they, in organising the new system of education, put this mathematics of the old Japanese school entirely out of sight, and were anxious to introduce free and unmolested the mathematics which has no schools and whose universal language is intelligible to all the civilized nations.[208]

This non-expert lecture on the history of Japanese mathematics, dressed up for European ears, was the unlikely prelude to the enduring success that mathematicians from Japan would achieve on the world mathematical scene in the twentieth century. It was at Tokyo University—founded in 1877, and transformed into the 'Imperial University' in 1886—that Fujisawa managed to implant a research tradition for mathematics. He expanded the department's special library and acquainted students with modern research early on.[209] Fujisawa had previously profited from the opening of the Japanese educational system; he was initially sent to London for his studies. From there he soon went to Berlin, and on to Strasbourg, where he obtained his PhD under Elwin Christoffel (1829–1900) in 1886. When the Germans had been building a new university in Strasbourg after the annexation of Alsace-Lorraine following the 1870 war, proposals for the mathematical chairs were solicited from Leopold Kronecker in Berlin. Apart from Christoffel, Kronecker had also led Theodor Reye (1838–1919) to Strasbourg. Fujisawa's German education faithfully reflected the overall "Germanization of the political system and of learning in Japan 1881–1945."[210]

> Fujisawa's period in Strasbourg ultimately played an extremely important role both in the history of mathematics in modern Japan and in his own career. Awarded a doctorate in July 1886 for a thesis on Fourier series, Fujisawa returned to Tokyo in 1887 and became the second professor of mathematics at his alma mater, ushering in a drastic change in the mathematical curriculum. Fujisawa seemed to have acquired both the mathematical acumen and the political characteristics of his teacher at Strasbourg. It is reported that when a student of his at the University of Tokyo asked why he studied mathematics, Fujisawa's answer was "For the nation!" No words seem more appropriate to characterize not only why Fujisawa but also why Japanese mathematicians as a whole studied mathematics before 1945.[211]

---

[207] We refer to and quote from [Sasaki 2002], p. 235. For more background and details, see [Sasaki 2001] as well as the study of the Japanese reform of the educational system and what it meant for mathematics in [Kümmerle 2021].

[208] End of Fujisawa's lecture in Proceedings ICM 1900, p. 393.

[209] See [Kümmerle 2021]; I heartily thank the author for having shared parts of his text with me.

[210] See [Sasaki 2002], pp. 236–238.

[211] [Sasaki 2002], p. 239.

Fig. 1.9 Rikitaro Fujisawa in 1922.

Apart from the Paris ICM, Fujisawa would also attend the Cambridge ICM in 1912. After the war, in another world, Fujisawa's brilliant student Teiji Takagi (1875–1960) would address the Strasbourg ICM in 1920, reporting on his completion of class field theory and correction of Hilbert's 12th problem, which Heinrich Weber and a number of Hilbert's European PhD students had failed to achieve. Takagi himself had profited from the Japanese educational policy, and visited Göttingen in 1900–1901, where he obtained some advice from Hilbert for his own thesis.[212]

[212] See [Schappacher 1998].

1.4.3 Mathematical Associations

There are other ways, besides focussing on the early ICMs, to gather insights into the global mathematics community before the First World War. We take our clue from the following passage describing the preparation of the 1904 ICM in Heidelberg:

> Already in June 1903 a first invitation to the congress was sent to 2000 mathematicians of all countries. The list of invitees was compiled on the principle that the members of the big mathematical societies should be invited first: *Deutsche Mathematiker-Vereinigung*, *Société mathématique de France*, the *London Mathematical Society*, the *Wiskundig Genootschap te Amsterdam*, the *Circolo Matematico di Palermo*, the mathematical societies of Moscow and Kazan, the American Mathematical Society. For other countries like Hungary, Sweden, Norway, Spain, Portugal etc., mathematicians of these countries sent us lists of addresses. Beyond these personal invitations, our invitation was also distributed in the form of more than 25,000 flyers inserted in leading mathematical journals. The company B. G. Teubner also printed for free short announcements in all its mathematical journals.[213]

In contrast to scientific academies, which we have discussed before (Section 1.3.2), local or national mathematical associations assemble a disciplinary community. They are generally not elitist but grounded at the grassroots level. In the mathematically most visible European nations the formation of mathematical associations was an outgrowth of the professionalization of mathematics which had started in the early nineteenth century. In other parts of Europe and of the world, creating a mathematical society could express the craving for mathematical networking. The Union of Czech Mathematicians for example—which was created in 1869, and by 1912 could boast more than 1,000 members—grew out of a student union founded in 1862 with the goal to organize more advanced lectures on mathematics and physics.[214] In all cases, these societies or associations are locally grounded mathematical institutions so that their formation naturally reflects the peculiar constellation of mathematicians in the country at hand.

These institutions share some features with academies. For example, they usually come with a regular journal and sometimes other occasional publications,[215] and their membership lists often feature foreign members, reflecting both local conditions and advancements of internationalization. Here are some examples of the approximate percentage of foreign members in the last years before World War I: American Mathematical Society 8%, Circolo Matematico di Palermo 67%, Deutsche Mathematiker-Vereinigung 38%, Edinburgh Mathematical Society 9%, London Mathematical Society 19%, Société mathématique de France 37%.[216] On the other hand, unlike what happens in academies, among the members we may find mathematics teachers, or publishing houses. A mathematical association may serve the advancement of mathematics in its region or nation. The upshot is that an adequate treatment of the

---

[213] See Proceedings ICM 1904, pp. 6–7; my translation.

[214] See [Nový 1996], p. 508.

[215] An example of the latter are the reports on the development of mathematical subdisciplines commissioned by the German Mathematical Society DMV, one of which is Hilbert's well-known *Zahlbericht* on the theory of algebraic number fields, of 1897.

[216] See [Nový 1996], pp. 508–509, footnote 15. See also [Schappacher & Kneser 1990], pp. 8–9.

history of mathematical associations typically requires reviewing the corresponding regional or national histories of mathematics. This is highlighted for instance in the comparative study of the French and the German mathematical societies by Hélène Gispert and Renate Tobies.[217]

The following list, which is certainly still incomplete and would profit from additional search for more literature about the various mathematical societies mentioned, is put here to give at least a first impression of the rich variety of such associations that either existed before World War I or were created soon after the Great War. The latter ones are *printed in slanted type*.

Mathematical Societies founded before World War I
or in the wake of the war

Austria – Hungary

Union of Czech Mathematicians (1869)[218]
Jednota českých matematiků
Hungarian Mathematical and Physical Society (1891)
Matematikai és Fizikai Társulat
Austrian Mathematical Society (1903)

Belgium

Belgian Mathematical Society (1921)
Société Mathématique de Belgique; Belgisch Wiskundig Genootschap

Bulgaria

Union of Bulgarian Mathematicians (1898)

Denmark

Danish Mathematical Society (1873)
Dansk Matematisk Forening

---

[217] See [Gispert & Tobies 1996]. — Apart from the national cases, there were also occasional international mathematical societies, such as the ephemeral *International Association for Promoting the Study of Quaternions and Allied Systems of Mathematics* which was founded in 1899 and disintegrated shortly before World War I. The history of the Quaternion Association can be followed through the successive volumes of its bulletin.

[218] Originally founded as a student union in 1862 under the name Union for free lectures about mathematics and physics / *Spolek pro volné přednášky z matematiky a fysiky / Verein für freie Vorträge aus der Mathematik und Physik*. See [Nový 1996], p. 508.

Finland

Finnish Mathematical Society (1868)
Suomen Matemaattinen Yhdistys

France

Société mathématique de France (1872)[219]

Georgia

Georgian Mathematical Union (1923)

Germany

Hamburg Mathematical Society (1690)
German Mathematical Society (1890)
Deutsche Mathematiker-Vereinigung
Berlin Mathematical Society (1901)
Berliner Mathematische Gesellschaft

India

Scientific Society of Aligarh (1864)
Indian Mathematical Society (1907)[220]
Calcutta Mathematical Society (1908)

Italy

Circolo Matematico di Palermo (1884)
Italian Mathematical Union (1922)[221]
Unione matematica italiana

Japan

Tokyo Mathematical Society (1877)[222]
Tokyo Sugaku Kaisha

---

219 See [Gispert 2015].

220 Cf. [URL 05].

221 See [Giacardi & Tazzioli 2021].

222 As of 1886: Tokyo Mathematico-Physical Society. See [Kümmerle 2021] for a detailed discussion of Cambridge-trained Dairoku Kikuchi's role in steering this society towards Western mathematics.

The Netherlands

[Royal] Dutch Mathematical Society (1778)
Wiskundig Genootschap

Norway

Norwegian Mathematical Society (1918)[223]
Norsk matematisk forening
Norwegian Statistical Association (1919)
Norsk statistisk forening, NSF

Poland

Polish Mathematical Society (1917)
Polskie Towarzystwo Matematyczne

Romania

Romanian Mathematical Society (1911)
Societatea de Stiinte Matematice din Romania

Russia

Kazan Physico-Mathematical Society (1863)
Moscow Mathematical Society (1864)
Kharkov Mathematical Society (1879)
St. Petersburg Mathematical Society (1890)

Spain

Royal Spanish Mathematical Society (1911)[224]
Real Sociedad Matemática Española

Switzerland

Swiss Mathematical Society (1910)[225]
Schweizerische Mathematische Gesellschaft, Société Mathématique Suisse

---

[223] See Thomas Kalleberg's chapter in [Siegmund-Schultze & Sørensen 2006], pp. 133–146.
[224] See [González 2011].
[225] See [Colbois et al. 2010].

UK

London Mathematical Society (1865)
Edinburgh Mathematical Society (1883)
Trinity Mathematical Society (1918)

USA

American Mathematical Society (1888)
Pi Mu Epsilon (1914)
Mathematical Association of America (1915)

1.4.4 India's Entry onto the World's Mathematical Stage

Centres of learning on the Indian subcontinent have played a distinguished role in the history of astronomy and mathematics, and for several millennia exchange has occurred with other lettered civilizations. These local sanskrit or vernacular (e.g., Malayalam) scientific traditions are an important part of India's tremendous cultural heritage and as such somehow influenced all the actors of science politics in India at the end of the nineteenth century. But the stage that India was about to enter at the time was the science of that period. This science was measured according to standards set in Europe or North America. Furthermore, different in this respect from Japan (cf. Section 1.4.2), India was a British colony. Therefore any serious history of mathematics in India around the turn of the century has to address the phenomenon of colonial science. The diffusion model proposed for the spread of Western science in colonial constellations in an old paper of Basalla's[226] has been contested in many detailed studies.[227] V.V. Krishna has instead analyzed the emergence of a national science in India by proposing "that the most innovative scientists in colonial India were neither the 'gate keepers' nor the 'soldiers' of colonial science, but a third category of science personnel who were nationalists in the cultivation of science, and promoted technology, engineering, and the use of vernaculars, against the declared policies of the colonial power."[228]

> The third category consisted of scientists who struggled to create support structures for the cultivation of modern science and its advancement in the framework of emerging nationalism.
>
> ......

---

[226] See [Basalla 1967].

[227] Cf. for instance the volume [Petitjean et al. 1992] which contains several such studies.

[228] The quote is from a round table discussion recorded in [Petitjean et al. 1992], p. 34.

> Emerging nationalism after the 1870s and the ideological role of scientists in it is in no small measure [...] connected to the struggle of Indian scientists to achieve international recognition. Limited to their sphere of influence, they believed that advancing the frontiers of knowledge also meant giving a distinct national identity to their intellectual production.[229]

Even though Krishna's analysis can remind us of the dualism between national and cosmopolitan agenda which underlies the present Section 1.4, his model is largely based on what happened in chemistry and physics, and may not quite match the case of mathematics. Leaving this open here, let us simply mention two outstanding South Indian scholars, whose difficult early careers were helped by the existence of national scientific associations.

The first one is Chandrasekhara Venkata Raman (1888–1970) who in 1930 would be awarded the Nobel prize in physics. He is one of the major protagonists in Krishna's third, nationalist group of Indian scientists. After a brilliant education up to college level, ill health prevented Raman "from pursuing higher studies in physics at one of the great British universities, and India offered him no possibility of a further career in science." Continuing as a civil servant in the Indian finance department, he nonetheless conducted research in his spare time, mostly "at the laboratory of the Indian Association for the Cultivation of Science, which had been founded in Calcutta by Mahendralal Sircar in 1876."[230]

The other example is the mathematical genius Srinivasa Ramanujan (1887–1920), who was likewise getting nowhere professionally in India in spite of the fact that his prodigious talent had been noticed at school.

> In late 1906, several dozen professors at colleges in Madras, Mysore, Coimbatore, and elsewhere in South India received a letter from V. Ramaswami Iyer [1879–1966], in which he proposed the formation of a mathematical society. Behind the idea lay simple want. Just as Ramanujan had so depended on whatever few mathematical books had come his way, so did Indian mathematicians generally suffer a lack of books and journals from Europe and America. The society, in Ramaswami's conception, would subscribe to journals and buy books, then circulate them to members. Twenty-five rupees per year from even half a dozen members would be enough to get the society off the ground. He wound up with 20 founding members, all hungry for mathematical fellowship, and what was known first as the Analytical Club, then the Indian Mathematical Society, was born. Soon it was publishing a journal of its own. Just a dozen years later, at its second conference in Bombay, it would claim 197 members and be circulating 35 European and American journals.
>
> ......
>
> It was into this nascent new world that Ramanujan 'came out,' as it were, as a mathematician in 1911. He had met Ramaswami Iyer, the society's founder, the previous year when, in search of a job, he had traveled to Tirukoilur. Now Ramanujan's work was appearing in volume 3 of Ramaswami Iyer's new Journal—which, like most mathematics publications, opened its pages to provocative or entertaining problems from its readers.
>
> ......

---

[229] The quote is from V.V. Krishna's chapter in [Petitjean et al. 1992], pp. 57–72; here pp. 68, 69. I have taken the liberty to delete what I think is an erroneous double negation.

[230] Both quotes are from [Gillispie 1970–1980], Vol. XI, p. 264.

> Appearing in the Journal of the Indian Mathematical Society, Ramanujan was on the world's mathematical map at last, if tucked into an obscure corner of it. He was starting to be noticed.[231]

As is well known, Ramanujan finally agreed—after initial hesitations for religious reasons—to travel to Cambridge, England in the Spring of 1914. He worked with Godfrey Harold Hardy (1877–1947) for about four years and became a Fellow of the Royal Society as well as of Trinity College, Cambridge. But that was already during the Great War.[232]

[231] [Kanigel 1991], pp. 85, 86, 92.

[232] Cf. [Rice 2015]. This article explores the role of the London Mathematical Society (LMS) in the exchange between Ramanujan and Hardy.

Chapter 2
Unfirm Foundations

> *... mais il est arrivé qu'on s'est heurté à certains paradoxes,*
> *à certaines contradictions apparentes, qui auraient*
> *comblé de joie ZENON d'Elée et l'école de Mégare.*
> *Et alors chacun de chercher le remède.*
> *Je pense pour mon compte, et je ne suis pas seul,*
> *que l'important c'est de ne jamais introduire que des êtres*
> *que l'on puisse définir complètement en un nombre fini de mots.*
> *Quel que soit le remède adopté, nous pouvons nous promettre*
> *la joie du médecin appelé à suivre un beau cas pathologique.*
>
> But then it happened that one encountered certain paradoxes,
> contradictions which would have made the day of Zenon of Elea
> and the Megara school. And everybody was looking for the remedy.
> I personally think—and I am not the only one—that the key thing is
> to always only introduce objects which can be completely defined in
> finitely many words. But whatever the remedy adopted will be,
> we can be sure to experience the joy of a physician called
> to look after a beautiful pathological case.
>
> Henri Poincaré, The future of mathematics, Rome 1908[1]

In each of the three parts of this book we present a characteristic feature that highlights the state of mathematics in the corresponding period. The period considered in this first part, i.e., the nineteenth century extended to 1920, saw fundamental changes:[2] Non-euclidean geometries entered the stage and finally triggered a radical turn towards formal axiomatic mathematics, especially in David Hilbert's *Foundations of Geometry* (1899). Set theory was invented and first developed by Georg Cantor as a fundamental discipline; but it was soon ridden with paradoxes. Formal logic was remolded. Algebraic structures, such as groups and fields, began to be treated not as

[1] See Proceedings ICM 1908, Vol. 1, p. 182; my translation.

[2] Parts of the following paragraphs are loosely based on [Schappacher 2012].

N. Schappacher, *Framing Global Mathematics*,
https://doi.org/10.1007/978-3-030-95683-7_2

notions gathering concrete examples but as formal objects of a new kind. Also the concept of space was being rendered more malleable, both for the benefit of the new discipline of topology (*analysis situs*, as it was called then) and as a fitting home for general spaces of functions which would encompass oceans of functions in a single structure, where one had previously studied individual specimens.

Mehrtens's book. These historical phenomena call for a coherent historical narrative. Herbert Mehrtens (1946–2021) has been the first to propose such an account in his book [Mehrtens 1990]. He deliberately limited himself to pure mathematics—leaving the technological modernization of the nineteenth century outside of the scope of the book—and focussed mainly on developments in Germany. His key concept was that of *modernity* or *modernism*, which he used both as an overarching concept describing the epoch, and to designate the camps in opposition at the time: *modern* versus counter-*modern*, the latter sometimes radicalized to an anti-*modern* stance, for instance in the racist Nazi ideology of mathematical creativity. Mehrtens used the term *modern* to describe mathematical concepts that were conceived of as being logically independent of extra-mathematical data; *modern* mathematical theories work autonomously on a formal system. David Hilbert's paradisiacal freedom for the modern mathematician and his postulate that all mathematical problems can be solved in principle (possibly after a suitable context shift) were based on abandoning the traditional referential meaning of basic mathematical concepts.

In fact, let us recall *verbatim* what may well be the most provocative kick-off in the whole history of mathematics: the first sentences of Hilbert's landmark text of 1899, in which he would downgrade the classical axioms of geometry to implicit definitions.

> *Explanation.* We think three different systems of things: the things of the *first* system we call *points* and denote them by $A, B, C, \ldots$; the things of the *second* system we call *lines* and denote them by $a, b, c, \ldots$; the things of the *third* system we call *planes* and denote them by $\alpha, \beta, \gamma, \ldots$ ...
>
> We think of the points, lines and planes as having certain mutual relations, and denote these relations by words such as *'to be situated'*, *'between'*, *'parallel'*, *'congruent'*, *'continuous'*; the exact and complete description of these relations is given by the *axioms of geometry*.[3]

A year later, in 1900, in his sixth problem for the new century, Hilbert challenged mathematicians to axiomatize physics in a similar way, specifically the theories of probability and of mechanics.

Mehrtens also associated his mathematically specific concept of modernity with cultural and political orientations. This worked fine when confronting the liberal Hilbert with the reactionary Gottlob Frege (1848–1925), but on the whole it seems overstretched even when limited only to the situation in Germany.

Be that as it may, the momentous upheaval of the inner fabric of mathematics in the nineteenth century resulted in what would soon be known as the *foundational crisis of mathematics*. Mehrtens places this crisis essentially in the 1920s, after World War

[3] See [Hilbert 1899], p. 4; my translation.

I.[4] It is true that the war and the immediate postwar period singularly exacerbated the sense of crisis, also with respect to the foundations of mathematics—see for example the abundant political metaphors in Hermann Weyl's (1885–1955) article [Weyl 1921] on the "new foundational crisis of mathematics."

2.1 Mathematics Meets Literature

2.1.1 Mathematics and Name Worshipping

Yet, the peculiarities and paradoxes of set theory had sunk in during the first decade of the twentieth century, and they triggered noticeably different reactions in various national and cultural contexts. We have briefly indicated an example of this in Section 1.3.3.3 above, with Baire's article on set theory for the French Encyclopedia. Another intriguing case has been studied by Graham and Kantor; they contrast the French reactions—in particular of Baire, Borel, and Henri Lebesgue (1875–1941)—with those of their Russian counterparts, specifically Dmitri Fyodorovich Egorov (1869–1931) and Nikolai Nikolaevich Luzin (1883–1950), taking into account the influence of Pavel Alexandrovich Florenskii (1882–1937) and of the religious movement of Name Worshipping (*imyaslavie*), which is based on the idea that the name of God is God Himself.[5]

> Although Luzin was very close to a number of leading French mathematicians and cited his debt to them, his worldview was different. In their study of set theory, the French sought to distinguish the philosophical, mathematical, and psychological components and keep them separate. Luzin and some of his friends, on the other hand, believed that mathematics was linked to religion, but they could not be explicit about these links because of the hostile Soviet environment after the 1917 revolution.[6]

The Russian thinkers studied by Graham and Kantor had no problem accepting infinite hierarchies and a 'continuum' made up of uncountably many individual points, whereas French reactions—as echoed also in Poincaré's words at the 1908 ICM quoted at the beginning of this chapter—tended to stick in the last resort to objects which are completely definable in a finite number of words.

---

[4] See [Mehrtens 1990], section 4.1, pp. 289–307.

[5] The article [Graham & Kantor 2006] addresses this cultural/national comparison directly. The book [Graham & Kantor 2009] develops the full story as the authors see it.

[6] [Graham & Kantor 2006], p. 73.

2.1.2 Robert Musil

The Austrian engineer and writer Robert Musil (1880–1942)—who is best known today for his late, unfinished novel "The Man Without Qualities" (*Der Mann ohne Eignschaften*), the hero of which is a mathematician—published as early as 1913, i.e., before the war, a short essay "The Mathematical Man" (*Der mathematische Mensch*) in which we read in particular:

> Mathematics is the bold luxury of pure reason, one of the few that remain today. ...... We may say that we live almost entirely from the results of mathematics, although these themselves have become a matter of indifference to mathematics. Thanks to mathematics we bake our bread, build our houses, and drive our vehicles. With the exception of a few handmade pieces of furniture, of clothing, shoes, and children, everything comes to us through the intervention of mathematical calculations. All the life that whirls about us, runs, and stops is not only dependent on mathematics for its comprehensibility, but has effectively come into being through it and depends on it for its existence, defined in such and such a way. For the pioneers of mathematics formulated usable notions of certain principles that yielded conclusions, methods of calculation, and results, and these were applied by the physicists to obtain new results; and finally came the technicians, who often took only the results and added new calculations to them, and thus the machines arose. And suddenly, after everything had been brought into the most beautiful kind of existence, the mathematicians—the ones who brood entirely within themselves—came upon something wrong in the fundamentals of the whole thing that absolutely could not be put right. They actually looked all the way to the bottom and found that the whole building was standing in midair. But the machines worked! We must assume from this that our existence is a pale ghost; we live it, but actually only on the basis of an error without which it would not have arisen. Today there is no other possibility of having such fantastic, visionary feelings as mathematicians do.
>
> The mathematician endures this intellectual scandal in exemplary fashion, that is with confidence and pride in the devilish riskiness of his intellect.[7]

According to the literary scholar Andrea Albrecht: "Musil uses the foundational crisis of mathematics, unfolding during the first decades of the twentieth century, to draw an analogy between modern mathematics and modern poetry." She sets out to investigate "what exactly ... makes mathematics poetically so relevant for Musil."[8] Without going into details, let us simply note here that Musil—with all the mocking, ironic style of his essay—does convey a positive image of the 'modern' mathematician who manages to blissfully live with foundations the solidity of which he has himself eroded. This image is undoubtedly meant by Musil to serve as an example for the 'modern' poet of his time.

---

[7] Quoted from the English edition [Musil 1990], pp. 41–42.

[8] From the English summary of [Albrecht 2008], p. 218. Cf. [Albrecht & Bomski 2016], [Albrecht 2018].

2.2 Hermann Weyl's Changing Attitudes to the Foundational Crisis

One of the most influential, and internationally active mathematicians of the twentieth century, Hermann Weyl will naturally be mentioned more than once in this book. The reason to discuss him here, in connection with the foundational crisis, is that the sequence of markedly different positions he held within a few years makes him the most intriguing case among all influential mathematicians, as far as foundational problems are concerned.

Musil's ironic admiration of the mathematician who is confidently working away in spite of the shaky base below captures Hermann Weyl's first reaction to the foundational crisis quite well, which we now turn to.

2.2.1 The First Phase: Before the War

Weyl had followed discussions about logic and set theory since the first decade of the twentieth century, especially by Ernst Zermelo (1871–1953) and Émile Borel. Around the same time and in connection with his philosophical interests, Weyl met both his future wife Helene Joseph (1893–1948) and the philosopher Edmund Husserl (1859–1938) at Göttingen. In April 1909 he listened to Henri Poincaré's Göttingen lecture on Richard's Paradox.[9] On the occasion of his second thesis (*Habilitation*) in 1910, he presented a lecture "On the Definitions of the Fundamental Concepts of Mathematics" [Weyl 1910] which "provided a means to replace the vague idea of 'definite property' in Zermelo's *Aussonderungsaxiom* in [the latter's] theory of sets by a more precisely defined notion."[10] Already at that time he was not entirely happy with the contemporary, Hilbertian modern mathematical formalism in general, and set theory in particular, but he accepted them in the interest of formal rigor. Thus, before the war, his mixed feelings typically expressed themselves in the form of solemn, general appeals to the reader such as this one:

> May we say—as is suggested by what we have developed—that mathematics is the science of ε[11] and of those relations which can be defined from this concept via the principles discussed? Maybe such an explanation does actually determine mathematics correctly as for its logical substance. However, I see the proper value and the meaning proper of the system of concepts of logicised mathematics thus constructed in that its concepts may

[9] See [Poincaré 1910]. Jules Antoine Richard's (1862–1956) paradox of 1905 observed that the set of real numbers definable in finitely many words is countable; but one can obviously perform Cantor's diagonal argument on this set.—Contradiction.

[10] See [Feferman 1998], p. 250; see also pp. 258–259.

[11] Weyl refers here to the relation of 'being an element of' a set; his "science of ε" is set theory, formalized along the lines presented in the lecture.

> also be interpreted intuitionwise without affecting the truth of the statements about them. Furthermore, I believe that the human spirit has no other way to ascend to mathematical concepts but by digesting the given reality.[12]

2.2.2 The Second Phase: 1917–1918

By 1917, however, the world had changed, and so had Weyl's attitude to foundational problems:

> The goal of this treatise is not to drape the 'solid rock' on which the house of analysis is founded with a wooden scaffolding in the sense of formalism, and to then pretend to the reader, and finally even to oneself, that this is the real foundation. Here, we rather defend the opinion that an essential part of the said house is grounded on sand. I believe I can replace this faltering ground by props of reliable stability; but they do not sustain everything one generally takes today to be certain; the rest I sacrifice, since I do not see any other possibility.[13]

Indeed, in the book "The Continuum" which came out in 1918, Hermann Weyl was prepared, in order to save the logical solidity of the theory of real numbers, to sacrifice certain results which are generally regarded as crucial for analysis:

> ... that every bounded set of real numbers has an upper bound has then to be abandoned; such sacrifices do not make us waver on the path we have chosen.[14]

Similarly, Weyl was led to abandon the idea that every infinite set of real numbers has to contain a countable subset, in the specific sense of 'countable' afforded by his theory of definitions:

> If we adopt the concept of denumerability suggested by this proof[15], then naturally there is no reason at all to assume that every infinite set must contain a denumerable subset—a consequence from which I certainly do not shrink.[16]

Weyl's radical refusal to compromise on these matters is further confirmed by a remarkable wager he engaged in with his Zürich colleague George Pólya (1887–1985) on 9 February 1918:

---

[12] From the conclusion of [Weyl 1910]; my translation. A similar but more elaborate exhortation of the reader to see beyond the formalized presentation of intricate mathematics can be found in the foreword to Weyl's famous textbook on the *Concept of Riemann Surface*, [Weyl 1913].

[13] From the introduction to [Weyl 1918], p. iv; my translation.

[14] See [Weyl 1918], p. 23/24; my translation. The translation in [Weyl 1987], p. 32: "But such sacrifices should keep the path ahead clear of confusion", fails to adequately reflect the element of personal resolve in Weyl's sentence: ... *wir lassen uns durch solche Opfer an dem Wege, den wir eingeschlagen, nicht irre machen.*

[15] I.e., Cantor's diagonal argument, reconstructed in Weyl's setting.

[16] See [Weyl 1987], p. 28; translated from [Weyl 1918], p. 19.

As for the following two propositions:

1) *A bounded set of numbers has a precise upper bound.*

2) *Every infinite set of numbers contains a countable subset.*

Weyl predicts:

A. Within 20 years, i.e., by the end of the year 1937, Pólya himself, or the majority of the relevant mathematicians, will admit that the concepts of 'number', 'set', 'countable' which intervene in those propositions and on which we generally rely today, are totally vague; asking whether 1) or 2) is true or false makes as little sense as the same question addressed to the main claims of Hegel's philosophy of nature.

B. Pólya himself, or the majority of the relevant mathematicians, will have realized that the propositions 1) and 2) are in fact wrong according to a clear, possible and reasonable reading of their words (either several such interpretations are still being discussed, or a single one will have already been agreed upon). Or else, in the event that until then a clear interpretation has been found which renders at least one of the propositions true, then this will necessarily be the fruit of a creative achievement which will have given a new and original twist to the foundation of mathematics, and the concepts of number and set will have gained a meaning which we do not in the least suspect today.[17]

Purely from the point of view of mathematics and logic, the stubborn path that Weyl followed in those war years was an elaboration of the *predicative analysis* postulated by Poincaré, which Weyl worked out along the lines of the logic of definitions he had first laid down in the lecture mentioned above [Weyl 1910]. Poincaré had pointed out what he called vicious circles in definitions as the origin of such difficulties as Richard's paradox. To avoid them, only first-order quantifiers, extended to variables for the initial category of objects, must be used in forming judgments which can then serve to characterize sets. For Weyl, as for Poincaré, the initial category of objects were the natural numbers. However, neither Poincaré nor anyone else, only Hermann Weyl followed this idea through far enough to see what a truly predicative analysis would look like, and what could still be achieved in spite of the drastic sacrifices.[18]

Let us briefly review the biographical timeline which preceded Weyl's radical shift. In the Fall of 1913, the Weyl couple and the Hecke couple celebrated their double wedding in Göttingen, and Hermann Weyl was appointed professor at ETH Zürich. The war thus found the German couple in neutral Switzerland, where Weyl could at first continue working and teaching; their first son was born in February 1915. In spite of this splendid isolation, a letter from Weyl to Hilbert of March 1915 already indicates his apprehensions about the world events which distracted him from the "bloodless realm of numbers": "... one is actually a little ashamed to be occupied with such—I am by no means saying worthless, but complicated and distant

[17] See [Arch. ETH], Hs 91a:87 (Weyl papers), for the original handwritten document; my translation. A copy of these two pages is posted in the Mathematics Department's Lecture Hall at ETH Zürich. Cf. [Schappacher 2003], p. 15.

[18] For an explanation of Poincaré's idea and Weyl's formal execution of it in more current terminology of mathematical logic, see [Feferman 1998]. Here Feferman also notes, p. 265, that the iterative generation of judgments which is finally adopted in [Weyl 1918] is also not completely immune to non-predicative twists—a fact that Weyl seems to have overlooked—and he offers additional precisions. The final message of [Feferman 1998] is that an amazingly big chunk of analysis and functional analysis can actually be vindicated within a strictly predicative setup.

things—while [what is] utterly primitive and bitterly necessary, the soil and the earth on which we stand, is questioned and millions risk their lives for its protection." In May 1915, the German authorities called on him for military duty, which he would serve in a garrison near Saarbrücken. The service was rather unspectacular, no frontline fighting or immediate danger, but it completely disrupted the continuity of his research. A year later, in May 1916, the Swiss education authorities managed to obtain his release from German military duty. Upon his return to Zürich he would start looking for new orientations to guide his work, and he would renew his interest in philosophy.[19]

His philosophical readings were influenced by his colleague from the Philosophy Department of ETH, Fritz Medicus (1876–1956), a German who had been appointed there two years before Weyl. Originally an expert on Immanuel Kant (1724–1804), Medicus was working on Johann G. Fichte (1762–1814) at the time. And thus Weyl came to read Fichte. Even though he would keep a critical distance from what he called Fichte's zealotlike attitude, the book [Weyl 1918] does reflect a certain influence of Fichte's theoretical philosophy, and possibly also of other parts of Fichte's work: "We cannot go for a final clarification of the true nature of states of affairs, judgments, objects, properties; this task leads into metaphysical depths; about those one has to seek advice from men whose names one cannot mention among mathematicians without getting a pitiful smile, Fichte for instance."[20] Weyl's frequent appeals to his readers to check things against what is immediately evident for them may also be influenced by Fichte's style of writing. Besides Fichte, Weyl was also studying the German homilies of the medieval mystical thinker Meister Eckhart. This apparently added an element of personal quasi-religious search for a 'transcendental reality' to his scientific endeavors.[21]

Furthermore, in the Winter term 1916/17, Hermann Weyl was teaching a course on the Mathematical Theory of the Electromagnetic Field. In the next term, Summer 1917, he lectured on *Raum, Zeit, Materie*, i.e., Space, Time, and Matter. Only in the following Winter term 1917/18 did he turn to the Logical Foundations of Mathematics. These last two lecture courses correspond to the two books he would publish in 1918: the first edition of *Raum, Zeit, Materie*—probably his most famous book, which would go through numerous, different editions, as a mathematician's guide to the theory of relativity—and *Das Kontinuum* [Weyl 1918], the focus of this section. Coming back to this book, the full extent to which Weyl was consciously resigning himself to that poor but sure predicative analysis comes to the fore in the

---

[19] See [Sigurdsson 1991], Chap. II, in particular pp. 60–61 for the quotes from the letter to Hilbert.

[20] [Weyl 1918], p. 2. Note in passing the similarity of this quote with the problems underlying Wittgenstein's *Tractatus*; but Wittgenstein would of course not recur to Fichte in looking for answers.

[21] For the last point, cf. [Scholz 2001], p. 80. The most thorough and up-to-date exploration of the Zürich philosophical context of Hermann Weyl's and Fritz Medicus's thought is the book [Sieroka 2010]. Weyl's critical remarks about Fichte, for example Fichte the zealot, are to be found in Weyl's stenographic notes of his readings, of which Sieroka has managed to obtain transcriptions which he kindly shared with me.

Fig. 2.1 Hermann Weyl on a teeter-totter ([Pólya 1987], p. 109).

last part—as of [Weyl 1918], p. 65—which deals with possible applications of the mathematical continuum to physics (but independently of relativity), specifically the modeling of space and time. The basic message here is devastating:

> If we make precise the notion of set in the way here proposed then the claim that to every point on the line . . . correspond a real number, and vice versa, acquires a profound content. It establishes a peculiar link between what is given in our intuition of space and what is construed in a logical-conceptual manner. But this claim obviously leaves entirely the scope of what intuition teaches us or may teach us about the continuum; it is no longer

> a morphological description of what intuition offers us (which, most importantly, is not a set of discrete elements but a flowing totality); instead exact notions are planted on the immediately given reality, which by its nature is not exact.[22]

The inadequacy of Weyl's poor continuum becomes glaringly obvious when he tentatively tries to apply it to time. At the end of a longer passage describing this attempt he concludes:

> Well, I think everything we ask here is evidently nonsense: to all these questions, our intuition of time gives us no answer (from which we would expect conceptual clarification of the nature of its flow); just as someone gives no answer to questions that were obviously addressed to him only because of a mix-up, and that he therefore cannot understand. ... It is the great merit of [Henri] Bergson's [1859–1941] philosophy to have underscored this profound alienation of the world of mathematical concepts from the immediately experienced continuity of the phenomenon of time (*la durée*).[23]

The security of sound foundations is thus obtained in *Das Kontinuum* at the high price of violating our intuition of space and time. This *Kontinuum* is all but satisfying for the mathematician-physicist Hermann Weyl. But since he sees no way out of this dilemma during the war, he does sketch the principles of physical applications of the poor continuum. The price one has to pay is the almost complete incompatibility of these modern formal mathematical techniques with our human apperception of the world which resurfaces only occasionally; to determine a point in space-time, one has to fix a coordinate system:

> The coordinate system is the inevitable residue of the annihilation of the ego in that geometrical-physical world which reason carves out of what is given under the norm of 'objectivity'—the last meagre symbol even in this objective sphere for the fact that existence is only given and can only be given as intentional content of the conscious experience of a pure, sense-creating ego.[24]

Such was the situation in 1917, according to Weyl, of the most basic theory for mathematical physics: Threatened by abyssal paradoxes if we advance without strict control, stubborn courage and discipline can lead us down a path full of tremendous sacrifices towards safe grounds, from where we can apply our solid tools to reality—never mind that we fully know how woefully inept they are. In this way, the logical stance of analysis emerges as a parable for life in times of war.

We have come a long way from Musil's uplifting irony. But modern poetry had also come to respond to the senseless cataclysm of the Great War: In 1916, the Club *Cabaret Voltaire* was founded in Zürich where Hugo Ball initiated the international anti-poetic protest movement of *Dadaism*, or DADA for short.

---

[22] See [Weyl 1918], p. 37; my translation.

[23] See [Weyl 1918], p. 68; my translation.

[24] See [Weyl 1918], p. 72; my translation.

2.2.3 The Third Phase: After the War

The fact that Hermann Weyl did assimilate the background of war times in his book *Das Kontinuum* is confirmed *ex post* by the dramatic turn in his view on the foundations of mathematics which he took right after the war, when he fell in for Luitzen Egbertus Jan Brouwer's (1881–1966) intuitionist conception of the real numbers, via (a modification of) Brouwer's notion of choice sequences (*Wahlfolgen*). As mentioned at the end of the introduction to this chapter, Weyl's immediate post-war papers abound with openly political allusions—in slogans like "Brouwer is the revolution"—to the political situation in Germany, the aborted revolution as well as the first wave of inflation.

Weyl's intuitionist period did not last very long, for reasons which have still not been completely understood historically. We leave this to another occasion.[25]

[25] But cf. [Schappacher 2010], pp. 3271–3276.

Chapter 3
World War I

Cher Stravinski, vous êtes un grand artiste !
Soyez de toutes vos forces, un artiste russe !
C'est si beau, d'être de son pays, d'être attaché
à sa terre comme le plus humble des paysans !
Et quand l'étranger met ses pieds sur elle,
comme les blagues internationalistes sont amères !

Dear Stravinsky, you are a great artist.
Be a Russian artist with all your power.
It is so good to belong to one's own country, to
be attached to its soil like the humblest peasant.
And when the stranger sets foot on it,
how sour do the internationalist jokes turn.

Claude Debussy in 1915[1]

The delicate dualism of nationalist and internationalist orientations that characterized the pre-war period, as described in Section 1.4 above, came to an abrupt end once war was declared, at the beginning of August 1914. The pride and interest of the nation left no room for 'internationalist jokes', as Debussy put it. We will first, in Sections 3.1 and 3.2, document this new state of mind with respect to science, and mathematics in particular. World War I was not only marked by this surge of exclusive nationalism; it was also a modern, technical war which mobilized scientists, including mathematicians; this is the subject of Section 3.3. A short final section considers the (im)possibility of organizing international congresses during the war.

[1] Letter to Igor Stravinsky 24 October 1915, see [Debussy 1993], p. 361; my translation.

N. Schappacher, *Framing Global Mathematics*,
https://doi.org/10.1007/978-3-030-95683-7_3

3.1 To the Civilized World!

In Germany, members of the academic elite—to name but one example, the philosopher Rudolf Eucken (1846–1926), who had been awarded the Nobel prize in literature in 1908—were vying with one another from the very beginning of mobilization in giving heroic speeches that not only justified the war but exalted it into a kind of moral obligation in the name of genuinely German culture. Much of this enormous amount of quasi-academic literature prompted by the war strikes us today as being "more remote than texts of medieval philosophy."[2]

By early October 1914, a presumptuous *Aufruf*, i.e., appeal "To the Civilized World" (*An die Kulturwelt*) signed by 93 German artists, writers, and scientists was circulated and published in ten languages. It squarely rejected foreign claims about German war crimes in six paragraphs, each of which started with the words: "It is not true that . . . ": Germany had not started the war, was not guilty of having "wantonly violated" the neutrality of Belgium; no atrocities were committed by the German troops during their advance through Belgium, only inevitable acts of self-defence. Towards the end, the text insisted that any campaign against German militarism was *ipso facto* a campaign against German culture itself. "No other manifesto has ever damaged the foreign reputation of German science as much as this *Aufruf*."[3]

This assessment is supported by the fact that the occasional voices that begged to differ were hardly noticed at the time. The Berlin professor of medicine Georg Friedrich Nicolai (1874–1964), for instance, drafted a counter-appeal "To all Europeans", which was signed in particular by Albert Einstein, but would not be published before 1917.[4]

There was precisely one mathematician among the 93 men listed as signers of the *Aufruf*: Felix Klein. When confronted with his signature right after the war in an affectionate but determined letter from his former student Grace Chisholm Young, Klein refused to publicly distance himself from the *Aufruf*, even though he confirmed that he had actually agreed to sign on the sole basis of a short telegram, and was never shown the text before publication. He wrote that he himself would certainly not have formulated the text as it was published. His wish was that, by adding his signature, emotions would be calmed and international collaboration could continue, but obviously the *Aufruf* did just the opposite. Yet, "everybody will stick to his country in bright and in dreary days." When Grace Chisholm Young wrote to him again, suggesting an English decantation to be sent to The Times, Klein

---

[2] Comment by the historian of philosophy, especially medieval philosophy, Kurt Flasch in his book on World War I, [Flasch 2000], p. 63; my translation. For more examples and the beginning of an analysis of this ominous corpus of texts, see the first part of his book, [Flasch 2000], pp. 15–99.

[3] [Tollmien 1993], p. 139; my translation.

[4] Cf. for example [Ungern-Sternberg 2013], pp. 71–83.

took this badly and regretted in return what the IAA was turning into under allied rule—cf. below, Chapter 4. In this post-war correspondence with her former thesis advisor, Grace Chisholm Young never mentioned that she had lost a son in the war.[5]

While Klein was the only mathematician, his name under the 1914 *Aufruf* did appear in illustrious company. There was the German impressionist Max Liebermann (1847–1935) for example, who had once painted Klein's portrait, and next to several Nobel laureates one also finds the physicist Max Planck (1858–1947).[6] Even recent and otherwise careful historical studies claim that David Hilbert was also asked to sign the *Aufruf*, but refused.[7] The source of this legend is Constance Reid's biography of Hilbert.[8] One may note that in 1919 Klein was convinced he had been the only person in Göttingen to be invited to sign.[9] Reid's story also has to be confronted with the fact that Hilbert did sign, along with Klein and more than 80% of the full professors of Göttingen University, another "Declaration of academic teachers of the German Reich" of 14 October 1914 which solemnly picked up the last point of the earlier *Aufruf*:

> We, teachers at Germany's universities and institutes of higher learning serve science and pursue a work of peace. But we see with indignation that the enemies of Germany, the first of them England, try to construe a contradiction between the spirit of German science and what they call Prussian militarism. There is no different spirit in the German army than in the German people, because both are one, and we are also part of them.
>
> ......
>
> We do believe that the salvation of the whole European civilization will depend on the victory which German "militarism" will achieve, through the discipline of manhood, fidelity and the spirit of sacrifice of the united free German people.[10]

As the war wore on in the trenches, however, we do find indications of differing attitudes between Klein and Hilbert. Felix Klein was pleased to observe how the war demonstrated the technical relevance of mathematics, which had always been one of his chief concerns: "It is precisely those directions of physical and mathematical research promoted by the *Vereinigung* in Göttingen which now prove their immediate relevance in the war."[11] David Hilbert on the other hand signed a petition in July 1915

[5] See [Tollmien 1993], in particular pp. 182–185. Cf. [Grattan-Guinness 1972], pp. 159–160. See also [Aubin & Goldstein 2014], p. 16–17, for a sample of various ways to react to the loss of sons in the war.

[6] Cf. [Tollmien 1993], pp. 186–190, about the exchange between Klein and Planck after the War.

[7] See for instance the otherwise excellent book [Ungern-Sternberg 2013], pp. 27 and 72, where Hilbert is explicitly mentioned as the unique 'known' case of a consulted scientist who refused to sign.

[8] See [Reid 1970], p. 137. According to Reid's unlikely account, Hilbert would have even been able to read the final text of the *Aufruf* before making up his mind!

[9] [Tollmien 1993], p. 146, note 38, referring to a letter from Klein to Planck of 21 September 1919.

[10] Quoted from [Tollmien 1993], pp. 143–146; my translation. This declaration had been drafted by the well-known Berlin classical scholar Ulrich von Wilamowitz-Moellendorff (1848–1931).

[11] Quoted from [Tollmien 1993], p. 206; my translation. For Klein's *Göttinger Vereinigung*, see Section 1.3.1 above.

against possible German annexations of foreign territory.[12] And in 1917, Hilbert seized the occasion to deliver to the Göttingen Academy an obituary speech for Gaston Darboux, who had been Klein's major French partner on the international scene, especially in the IAA, and had been elected corresponding member of the Göttingen Academy in 1883—cf. Section 1.3.2 above. There was no tradition, and hence no obligation to present an obituary for a deceased corresponding member. Thus Hilbert's praise of Darboux's mathematics and his role in the Paris Academy as well as in Mathematics International, delivered in 1917, was nothing less than a political provocation in front of his distinguished colleagues at the Göttingen Academy.[13] In fact, at a secret meeting in March 1915, Darboux had voted with the majority of his colleague Academicians—but against Jaques Hadamard—in favor of stripping signers of the 1914 *Aufruf* of their membership in the Paris Academy, including Felix Klein.[14]

3.2 Intellectual Warfare

Hilbert's commemoration of Darboux reminded us of the pre-war period when scientific academies were rolled in as chief actors of Science International (Section 1.3.2). Since that period had never forgotten the national side of things, it is not actually surprising that during the war academies quickly mobilized themselves for national academic propaganda.[15] The five French academies thus published a series of studies "in the name of truth", *Pour la vérité 1914–1915*, whose program translates as: "The Germans have chosen as their motto 'Germany above all.' We do not reply 'France above all.' There is only one motto which is worthy of France: 'Above all, the truth.'" One of these texts was the essay about the "pretensions of German science" [Picard 1916] written by Hermite's son-in-law (cf. Section 1.2.3.2 above). Here Picard proceeds in four steps: First he presents a kaleidoscope of major events from the history of science since the fourteenth century, insisting on the predominant role of Celto-Latin and French savants; a few German scientists such as Johannes Kepler (1571–1630), Gauss, and Hermann von Helmholtz (1821–1894) are mentioned, and so is the tendentious forgetfulness in German historical surveys when it comes, for instance, to acknowledging the modernizing role of Cauchy for analysis. The second part addresses the question:

[12] The so-called 'Delbrück-Dernburg petition', which in turn was a reaction to 'Seeberg's pamphlet.' The latter had focussed on a total German victory including annexations. See [Tollmien 1993], p. 151. In Prochasson's contribution on Intellectuals in [Audoin-Rouzeau & Becker 2004], p. 671, Hilbert is mentioned with Delbrück and Einstein as one of the rare "dissidents" against the spirit of 1914.

[13] See [Rowe 2018], pp. 4091–4093.

[14] See [Aubin & Goldstein 2014], p. 143.

[15] Cf. the volume [Debru 2019] for various explorations of academies' involvements in war.

> It is thus a strange aberration that the German race claims to be the only one in the world capable of working for the scientific development of humanity. Is it possible to find reasons for this belief in their own superiority of so many German brains?[16]

Picard's answer starts from the observation that German scientific treatises often get lost in minor details, unable to single out the essential. He thinks that this default also explains the frequent recourse to purely formal issues (p. 23). The third section probes the philosophical penchant of German scientist, as opposed to French and Anglo-Saxon scientists who tend to rely on their common sense, and have no patience with talks about an alleged 'crisis of science' (p. 31). There follows a critical presentation of Kant's theoretical philosophy, in particular his hapless philosophy of geometry, the alleged certainty *a priori* of which was rendered worthless by Nikolai Lobachevsky's (1792–1856) non-euclidean geometry.

> There is a tendency of German science to lay down *a priori* certain ideas and concepts, and to then follow their consequences without concern for their accordance with reality, and even to enjoy this departure from common sense.[17]

In mathematics, Picard opposes to this attitude Hermite's conviction of the reality of mathematical truths, discusses similar problems in the physical and biological sciences, and following Émile Boutroux (1845–1921) finds the same dangerous formal approach also in Kant's ethics, the categorical imperative, and its potentially disastrous application in a context of war. The fourth and final section of Picard's essay deals with applied science and tries to console the relative inferiority of German science with the superior strength of German industry.

Predictable though it is, Emile Picard's essay does mobilize a certain knowledge of episodes from the history of science; it belongs to the educated ordnances rolled into the intellectual battles of World War I, just as Pierre Duhem's very similar 1915 harangues in front of young Catholic students of Bordeaux University.[18] Following Blaise Pascal (1623–1662), Duhem distinguishes between German scientists obsessed by the *esprit géométrique*—drawing logical conclusions from first principles even when these defy common sense—as opposed to sensible researchers like Louis Pasteur (1822–1895), who are endowed with an *esprit de finesse*. Duhem also discusses the historical sciences, quoting in particular Fustel de Coulange's (1830–1889) confrontation of the French historian's diligent search for truth, with the German colleague's blind spot: the assumed German superiority.

Scientific Academies were willing to tolerate contributions to intellectual warfare which were quite a bit more extreme. The psychiatrist and expert on hypnotism Edgar Bérillon (1859–1948) for instance went as far as trying to develop an ethnochemistry (*ethnochimie*) which, in particular, postulated fundamental differences of metabolism

---

[16] See [Picard 1916], p. 21; my translation.

[17] See [Picard 1916], p. 36; my translation.

[18] See [Duhem 1915]. This volume includes (pp. 101–143) Duhem's article: Quelques réflexions sur la science allemande, *Revue des deux mondes* Fév. 1915, which explicitly takes up his pre-war work mentioned in Section 1.2.2 above.

between Germans and Frenchmen in order to explain the bad smell of the former.[19] The fact that he could present such papers to the Academy is a sobering reminder of how deeply the Great War affected leading scientific institutions.

3.3 Mathematic(ian)s during World War I

The three authors mentioned in the previous subsection: Picard, Duhem, and Bérillon, were in their fifties when the war began. Even though certain members of this generation would play an important role in shaping Science International after the war, focussing on them partly obscures the tremendous, broad impact of the Great War on scientific biographies and projects. This is spelled out most impressively in the rich volume [Aubin & Goldstein 2014]. Its various chapters highlight the loss of talent in battle or through indirect consequences of the war; Ramanujan's death in 1920, for instance, may be associated with his living conditions at Cambridge during the war, never mind that several of his contacts there were pacifists or conscientious objectors.[20] More generally they undertake "the effort to reconstitute mathematicians' wartime experiences and confer historical meaning to this scattered amount of evidence." This work leads the authors "to revise our definitions of what constitutes a mathematician and what properly belongs to mathematics."[21] The mathematics of sound-ranging (used to locate enemy canons), new computational methods for heavy artillery ballistics, communication devices in the trenches, the heyday of fluid mechanics induced by the war: all these and other topics modified mathematics, not only its appearance and its context, but also its substance. The book also shows how the internal research organization in all allied countries was renovated in the course of the war, with some Academies taking an instrumental part.[22] This is not the place to delve fully into the enormous material encompassed by the volume [Aubin & Goldstein 2014]. Here are only a few aspects which may highlight the variety of developments triggered by the war.

[19] See [Lefrère & Berche 2010].

[20] See [Aubin & Goldstein 2014], pp. 20 and 76. More generally on pacifism among Cambridge mathematicians, see Section 2 of June Barrow-Green's chapter in [Aubin & Goldstein 2014], pp. 65–78.

[21] See [Aubin & Goldstein 2014], p. 17.

[22] For the example of France, see [Aubin & Goldstein 2014], pp. 149–155.

3.3.1 Vito Volterra and Mauro Picone

Among the countries which entered the war only in 1915, the most interesting one from the point of view of the history of mathematics is Italy.[23] Although *a priori* part of the Alliance with Germany and Austria, Italy was neutral until May 1915, and many intellectuals took part in the nationwide debate about whether or when to go to war against Austria, on the side of the *Entente*, in order to obtain regions considered as belonging to Italy. For Vito Volterra, his political engagement against the central powers (Germany and Austria-Hungary) and in favor of military intervention reoriented the traditionally strong scientific bonds between Italy and Germany (cf. Section 1.1.5 above) towards a resolute alliance with French, and also with US scientists, in particular the influential solar astronomer George Ellery Hale (1868–1938).[24] Even though Volterra's attitude was shared by a majority of Italian intellectuals, opinions did of course vary across the scientific community. At Padova for instance, Severi changed from socialist pacifism to 'revolutionary interventionism', whereas Tullio Levi-Civita (1873–1941) remained a pacifist throughout the war, and his stance with respect to Mathematics International after the war would accordingly differ from Volterra's, as we shall see anon.

Born in 1860, Volterra was only 4 years younger than Emile Picard, but he played a more active part in the mobilization of scientific resources for the Italian war effort than Picard in France. In fact, Volterra was in touch with Émile Borel, who had been placed as of November 1915 at the head of the Directorate of Inventions for the National Defence by the mathematician Paul Painlevé when the latter became Minister of Public Instruction, Fine Arts and Inventions concerning National Defence. Volterra was involved early on with the aeronautic program for military dirigibles and tried to promote collaboration on this with the British. In 1917 he was instrumental in creating the UIR, the Italian Office for Inventions and Research (*Ufficio Invenzioni e Ricerche*) for the war effort. In 1918, after the end of the war,

> the UIR was dissolved. But this experience was not lost on Volterra, who endeavoured to establish a new institution with the aim of perpetuating the collaboration of pure and applied scientists among themselves and with the military and industry. This was an institution that postwar Italy needed in order to reemerge on more solid ground and with a stronger national industry. In this sense the UIR sowed the seed of the [Italian National Research Council] CNR.[25]

An even stronger effect of the war on a scientific career is illustrated by the case of Mauro Picone (1885–1977), a relatively young mathematician who was ordered to deal with the major problem of artillery in the Dolomites caused by the difference of elevation between canon and target. He managed to calculate new firing tables adapted to the situation.

---

[23] For details, see the chapter by Pietro Nastasi & Rossana Tazzioli in [Aubin & Goldstein 2014], pp. 181–227.

[24] See [Guerraggio & Paoloni 2013]; cf. [Mazliak & Tazzioli 2009] for Volterra's war correspondence.

[25] See Nastasi & Tazzioli's chapter in [Aubin & Goldstein 2014], p. 203.

Fig. 3.1 Mauro Picone in 1903.

> Picone ... recalled his work for the Sixth Army as a life-changing experience. ... He concluded from this that mathematics was 'not only beautiful' but 'also useful.' This was a hard-won conviction that was forged during decisive years for the determination of his career path and scientific character. To be truly useful, moreover, mathematics had to be backed up by the proper organization of research resources. For Picone, therefore, the realization that mathematics may be useful would not lead to the bracketing of his wartime experience. On the contrary, Picone would pursue, even as a civilian, the military experience of appropriating the right resources to make mathematics useful. ... The establishment of the [Italian Institute for the Application of Computing] IAC was the concrete expression of an intuition born in the bunkers of WWI.[26]

[26] See Nastasi & Tazzioli's chapter in [Aubin & Goldstein 2014], p. 209.

In all the nations at war one finds such examples of scientific biographies reoriented by World War I. Ralph Howard Fowler (1889–1944), for instance, worked on British anti-aircraft gunnery, and after the war his research would turn to statistical mechanics.[27] Richard von Mises (1883–1953) was actively involved with aircraft development in Austria and would occupy the first chair of applied mathematics created in Germany at Berlin after the war.[28] And the German Theodor Vahlen (1869–1945) turned from pure to applied mathematics, in particular ballistics, as a result of his war experience, and for the rest of his mathematical career. Besides, he became a militant Nazi in the early twenties, and in 1933, he would become an influential Nazi official in the Ministry of Education, backed the creation of the ideologically tainted journal *Deutsche Mathematik*, and in 1938 he became President of the Berlin Academy of Sciences.[29]

3.4 International Congresses during the Great War

In neutral Switzerland, it was easier to continue established routines during the war. The Swiss Mathematical Society (SMG/SMS) had been established in 1910 as a separate section within the society of Swiss natural scientists, motivated both by the existing mathematical activity in various parts of Switzerland and the editorial project of Euler's Collected Papers (*Opera omnia*). The only meeting of this young society that would not take place, because World War I had just begun, was the general assembly in September 1914. In 1917, the meetings (which rotated regularly) took place in Zürich; on 30 May 1917 Hadamard came from Paris for a talk at the Spring meeting; and at the general assembly on 11 September 1917, two of the 15 lectures were given by foreign mathematicians who also had no position in Switzerland: Arnold Emch (1871–1959) from Urbana, Illinois, and David Hilbert from Göttingen.[30]

It was not altogether impossible to organize a truly international congress during World War I in a neutral country. An extraordinary reminder of this is the International Women's Conference which was held in The Hague from 18 April to 1 May 1915, with Jane Laura Addams (1860–1935; USA) and Aletta Henriëtta Jacobs (1854–1929; The Netherlands) as its main organizers. About 1500 women from Sweden, Norway, Germany, Hungary, Italy, Belgium, Austria, Canada, the United States of America, Denmark, Great Britain, and the Netherlands managed to attend this pacifist meeting and voted a resolution calling in particular for an immediate truce. But even here it was not possible to represent all major nations involved in

---

[27] See the various mentions of Fowler in J. Barrow-Green's chapter in [Aubin & Goldstein 2014].

[28] Cf. [Aubin & Goldstein 2014], p. 12, and Siegmund-Schultze's chapter in [Booß-Bavnbeck & Høyrup 2003], pp. 23–82.

[29] See [Aubin & Goldstein 2014], p. 43. Cf. the different aspects of Vahlen's life highlighted in [Siegmund-Schultze 1984] and [Inachin 2001].

[30] See [Colbois et al. 2010], pp. 70–75.

the war. The invited feminist organizations from France declined to attend the heavily contested meeting; in 1919, they would organize an international suffragettist congress exclusively reserved for delegations from the allied countries.[31]

World War I disrupted the quadrennial rhythm of ICMs (see Section 1.4.1.2). In 1912 in Cambridge, Gösta Mittag-Leffler reiterated his invitation "in the name of the members of the first class of Royal Academy of Sweden, the Swedish editorial board of the journal *Acta Mathematica*, and on behalf of all Swedish mathematicians"[32] to hold the ICM in 1916 in Stockholm, as he had already proposed in 1908 in Rome. The invitation was accepted *nem. con.* The Hungarian mathematician Emanuel Beke (Beke Manó, 1862–1946) "presented an invitation to the Congress to hold its meeting of 1920 in Budapest."[33] The decision about this offer was put off until the Stockholm meeting. Also Cyparissos Stephanos (1857–1917) "expressed the hope that the Congress would meet in Athens in 1920 or 1924."[34] But no ICM would be organized during World War I, and its aftermath would not come back to any of these old proposals.

"As a reaction to the changed political landscape in Scandinavia following the dissolution of the union between Norway and Sweden in 1905, the prominent Swedish mathematician Gösta Mittag-Leffler extended 'a brotherly hand,' calling for Scandinavian colleagues to meet for a congress of mathematicians in Stockholm in 1909. This event became the first in a series of biannual meetings which proved to be an important institution for Scandinavian mathematics."[35] The biannual rhythm was broken, however, during the second decade of the twentieth century, even before 1914:

> By 1914, prior to the outbreak of the war, [Mittag-Leffler] had proposed to his Scandinavian colleagues that the SCM[36], scheduled for 1915, instead should form part of the ICM [planned for 1916 in Stockholm]. By 1915, however, with no end to the war in sight, he realised that a full-blown international meeting would not be feasible the following year. Instead, he settled on a compromise by organising a meeting of the SCM in Stockholm and inviting a smaller number of foreign participants from both sides of the conflict. Thus, he envisioned utilising Scandinavian neutrality in mitigating international scientific relations. . . . One of the main reasons for which Mittag-Leffler hoped to host a meeting in 1916 instead of waiting for a declaration of peace was to ensure the future of the SCM. He believed that the existence of these meetings was fragile, and that further delaying a Scandinavian congress—which had already been postponed from 1915—might seem natural in light of the political climate, but might compromise the future of the meetings. . . . Mittag-Leffler did not want to abandon the aim of giving the next meeting of the SCM an international character, and his attempt at a 'hybrid' SCM was met with approval from his good friend and colleague [Niels Erik] Nørlund [1885–1981], who wrote to him: "It is surely the best way in which one can invite foreigners at the moment, when they don't really need to fear the public opinion in their own country because regardless of how it goes, the meeting will maintain the character

[31] See [Audoin-Rouzeau & Becker 2004], p. 623.

[32] See Proceedings ICM 1912, p. 42; my translation.

[33] See Proceedings ICM 1912, p. 42.

[34] See Proceedings ICM 1912, p. 43; my translation.

[35] From the abstract of [Turner et al. 2013], p. 385.

[36] This abbreviation for 'Scandinavian Congress of Mathematicians' is used in [Turner et al. 2013].

of a Nordic Congress. It isn't unlikely that such a congress can be at least as fruitful for science as a big international congress, and it is surely a much less risky undertaking." Thus, Nørlund pointed to the two things that distinguished a meeting held under the auspices of the SCM from a typical ICM, namely the neutrality and the smaller size offered by the former. Utilising the more flexible format of a formally Scandinavian meeting, Mittag-Leffler could pursue objectives on the international scene that were otherwise out of his reach. The 1916 SCM thus distinguishes itself from the earlier meetings by virtue of its international outreach. However, the event actually turned into a predominantly Swedish affair. Due to hindrances caused by the war and some opposition against staging a congress in wartime, many non-Swedish Scandinavians did not attend.[37]

World War I produced unheard of quantities of armament, both material and intellectual, and inflicted about 40 million casualties. When it finally ended, "the age of European power had come to an end. . . . The one nation that emerged apparently unscathed and vastly more powerful from the war was the United States."[38] It would take a long time and another war before the mathematicians in their international undertakings would come to grips with the new world order. The second part of the book will describe this bumpy transition.

[37] See [Turner et al. 2013], pp. 396–397.

[38] [Tooze 2014], p. 6.

Part II
Mathematical Consolidation in Times of Tempest: 1919–1949

> This pacific war of liberation from German hegemony,
> that is being prepared even during this bloody war and
> that must be continued more vigorously than ever
> after peace comes, must also be carried
> into the scientific domain.

Eugenio Rignano[39]

> Moved by insane delusion and reckless self-regard, the German people overturned the foundations on which we all lived and built. But the spokesmen of the French and British peoples have run the risk of completing the ruin which Germany began, by a peace which, if it is carried into effect, must impair yet further, when it might have restored, the delicate, complicated organisation, already shaken and broken by war, through which alone the European peoples can employ themselves and live.

John Maynard Keynes[40]

> After 'the war to end war' they seem to have been
> pretty successful in Paris at making a 'Peace to end Peace.'

Archibald Wavell[41]

This part of the book starts out with the first decade after World War I. It was marked by the exclusion of the central powers from Science International. This is what characterizes the first IMU, and precipitates its irrelevance as of 1928. We then focus on the short but rich period of new mathematical projects, international contacts and congresses, especially during the 1930s—not only ICMs but also the famous Topology Conference in Moscow in 1935. This decade, however, also saw the emigration or displacement of scientists on a global scale and led into the Second World War.

[39] Letter of the editor of the Italian journal *Scientia* to the editors of *Nature*, published in *Nature* 98 (Jan. 1917), p. 408. Quoted from [Schroeder-Gudehus 1966], p. 88.

[40] See [Keynes 1920], Introductory to Chapter 1.

[41] Archibald Wavell (later Field Marshal Earl Wavell), an officer who served under Allenby in the Palestine campaign, commenting on the treaties bringing the First World War to an end. Quoted from the opening page of [Fromkin 1989].

Chapter 4
The First IMU: Triumph and Demise

The foundation of the IAA was discussed in Section 1.3.2. Its structural peculiarities were highlighted by remarks made by one of its architects, the physicist Arthur Schuster, Fellow of the Royal Society. The upshot was that, as an international science consortium, the IAA was somehow suspended in mid-air between nations and disciplines because the member Academies each had their own international network of corresponding fellows, independently of their being part of the IAA, and each one of them tried to represent all disciplines as well as possible. And yet, taken together, these Academies did not count among their individual members all the relevant researchers, even for some of the most prestigious international projects.

World War I stifled IAA and thus opened the way for a radical overhaul of international science organizations. This reinvention of *Science International* as of 1918 was guided by war-inspired politics. Several of the international organizations created in the wake of the war did not survive for long, but certain underlying structures which were designed at that very special historical juncture are still in place today.

Let us look a little more closely at how the new structures came into being. The fifth general assembly of the IAA took place in 1913 in St. Petersburg. On this occasion,

> the chair—and the treasury—of the IAA passed to the Royal Prussian Academy, and with it the mandate to prepare the next regular meeting for 1916 in Berlin. Soon after the outbreak of war, Hermann Diels suggested to temporarily entrust the chair of the association to a neutral Academy, namely the Dutch one. The attempt failed, chiefly, it seems, due to resistance from the Paris academies; they would agree to the transfer only on the condition that the seat of the association be permanently moved to Amsterdam. This, however, was unacceptable to the German Academy. Among others, the American astronomer and Foreign Secretary of the National Academy of Sciences in Washington, George E. Hale, citing the approval of the Royal Society and appealing to the traditional French sense of chivalry, tried to move the Parisian colleagues to accept a temporary solution. When this failed, the fate of the association was more unclear than ever, all the more so as . . . most scientists on either side

N. Schappacher, *Framing Global Mathematics*,
https://doi.org/10.1007/978-3-030-95683-7_4

> tended to have serious doubts and apprehensions about resuming the former international cooperation after the war. Also the prestige and the readiness of the US Academy of Sciences to intervene in favor of the neutral members did not outlast the year 1917.[1]

Indeed, the entry of the USA into the war gave an additional impetus to the cooperation among scientists of the allied countries, naturally excluding the enemy, i.e., Germany and the other Central Powers. It was the war which enabled people like Hale in the USA[2] and Volterra in Italy (see Section 3.3.1 above) to set up national research councils in the interest of national preparedness. Even after the war, their goal was still to secure public and private support for both applied and fundamental research: "The problems of peace are inextricably entangled with those of war, and if scientific methods and the aid of scientific research were needed in overcoming the menace of the enemy they will be no less urgently needed during the turmoil of reconstruction and the future competitions of peace."[3] This conscious continuity across 1918 of a new type of research politics should be borne in mind when trying to explain the practice of excluding the Central Powers from all scientific meetings and all international organizations created after the war, and why it was initially backed by men like Hale.

> Taking as a model his own National Research Council, Hale suggested . . . the creation of an organisation that would go beyond the experiences of international collaboration undertaken up to then, and could stimulate and coordinate the great scientific undertakings that required the combined efforts of scientists from different countries. Up to then the most significant experiments in this sense were those carried out by the community of astronomers. Hale had praised their international dimension from the time of the congress of astrophysics held in Chicago in 1893 (which took place in parallel with that of the mathematicians).
>
> The new international agency was supposed to be constituted under the aegis of the most important national academies: the Royal Society of London, the Académie des Sciences of Paris, the Accademia dei Lincei of Rome and the National Academy of Sciences of Washington. Hale then proposed the constitution, within this Inter-Allied Research Council, of international unions for the various disciplinary sectors, taking as an example the International Solar Union which had been in operation since 1905. His proposal was formalised in April 1918 and unanimously approved by the National Academy. This opened the way for a series of inter-ally conferences for programming and organisation.[4]

4.1 The Framers of the Council, the IRC

The first of these inter-Allied meetings took place from 9 to 11 October 1918, when the end of the war was in sight. The Royal Society and the Belgian Academy of Sciences invited 35 representatives of the principal national academies of the

[1] See [Schroeder-Gudehus 1966], pp. 89–90; my translation.

[2] See [Kevles 1968]. For a rich, yet concise account of the reshaping of US research politics during and after World War I, in which Hale was one of the main actors, see also [Dupree 1957], starting on p. 307.

[3] From Hale's introduction to [Yerkes 1920], p. viii.

[4] [Guerraggio & Paoloni 2013], p. 100.

allied powers to London in order to discuss the new international order of science. Twelve of the participants were from the UK, seven came from France, six from the United States, three from Belgium; Japan and Serbia sent two each, and Brazil, Italy, and Portugal one representative. The Italian participant, both as delegate of the *Accademia dei Lincei* and as director of the Italian Office of Inventions and Research, was Vito Volterra.

As explained above, the fundamental principle underlying the inter-Allied conferences was that the Central Powers, in particular Germany, had to be absolutely excluded from this international cooperation. As the official reason for this the London declarations cited war atrocities. However, it was not, so they wrote, the mere fact that atrocities had been committed by the German troops. Atrocities were acknowledged to be a regrettable but inevitable reality of war. Rather the claim was that the atrocities committed by the Germans were actually the concrete manifestation of a systematic, uncivilized political mindset: the German *méthodes politiques*.[5] The infamous German *Aufruf* "To the civilized world" of 1914, which we have discussed in Section 3.1, could always be cited as a proof.

In practical terms, the outstanding feature of these inter-Allied meetings was their amazing efficiency (never mind that the minutes record tenacious discussions):

> Although the delegates had very different, or no mandates from their home academies and were supposed to limit themselves to clarifying discussions, the resulting *Résolutions de Londres* would gain crucial importance. The program contained therein not only slid undiminished into the statutes of the IRC, but these resolutions were, in spirit and letter, the real lifeblood of the new Council. The reservation of Article 1(2) and Article 3 of these statutes—the notorious 'exclusionary clauses'[6]—consisted of nothing but a dry reference to the first article of the London *Résolutions*. Eight years later, when the neutral academies finally won their withdrawal, the whole organization began to tumble, even though ... only the references to those clauses of the London Resolutions had been deleted, and not even the cumbersome procedures to propose and accept new members. The subsequent General Assembly (1928) then already devoted itself entirely to creating a new organization: the *Conseil international des unions scientifiques* (I.C.S.U.).[7]

Only six weeks after the London meeting, the next inter-Allied meeting was held in Paris from 26 to 29 November 1918. There it was decided

> to invite the following neutral nations to join the Council, which was to be called the International Research Council: Denmark, Spain, Monaco, Norway, Netherlands, Sweden, Switzerland, Czechoslovakia, Finland. A provisional Executive Committee was appointed to undertake the preparatory work for the Constitutive Assembly. M. Émile Picard, Secrétaire perpétuel de l'Académie des Sciences, Paris, was nominated President of this Committee, with Sir Arthur Schuster as Secretary.[8]

We have met Schuster (Section 1.3.2) and Picard (Section 3.2) before. Apart from them, the Executive Committee consisted of George Hale, the Belgian astronomer and explorer Georges Lecointe (1869–1929), and Vito Volterra. These "Big Five",

[5] Cf. [Schroeder-Gudehus 1966], p. 91, especially footnote 6.

[6] *Note by N.Sch.:* We quote these and a few other clauses *verbatim* below.

[7] See [Schroeder-Gudehus 1966], pp. 91–92; my translation.

[8] [Jones 1961], p. 4.

as *The Times* called them[9], jointly controlled the new framework of international scientific cooperation after World War I. Their individual framing of the situation probably differed, though. Whereas we have seen that Hale and Volterra worked for an international analogue of their National Research Councils, Picard and Lecointe fought above all for the French and Belgian cause to overthrow all German supremacy in scientific matters. From this arose their principle to exclude scientists from Germany or the Central Powers, not only from the new international associations, but from all scientific congresses. Picard had even warned that the admission of neutral states might allow the Germans to sneak into the new institutions.[10]

Fig. 4.1 Vito Volterra and Émile Picard at the Strasbourg ICM in 1920, drawn by Eugène-Michel Maeckler.

The five of them thus seized the historic moment to try and gain control of all aspects of scientific cooperation where German colleagues or institutions had traditionally played an important part. For example, across all the sciences, Germany was perceived as having had almost a monopoly for scientific review journals, which had to be broken. Projects to this effect did not make good progress, though.[11]

> The provisional Executive Committee met in Paris, 20–24 May 1919, when draft statutes for the Council, for International Unions of Astronomy, of Geodesy and Geophysics, and of a Union of physical societies were adopted. A committee for international co-operation

[9] *The Times*, 8 March 1921, p. 10; quoted in [Schroeder-Gudehus 1966], p. 106.

[10] Cf. [Schroeder-Gudehus 1966], pp. 106–108.

[11] Cf. Section IV.C.1.d: *Brechung des deutschen Referatenmonopols* in [Schroeder-Gudehus 1966], pp. 117–120. The case of mathematical reviews will be discussed below in Section 6.3.

> in chemistry had met in Paris in November 1918, which was followed by an inter-allied conference of associations of pure and applied chemistry in Paris in April 1919, at which a project of confederation between these associations was adopted.[12]

Thus, in contrast to what had happened twenty years before, when the IAA was born, the new international world of science after World War I was constructed top-down. First they put into place the International Research Council. Such a lofty, independent umbrella organization for the various international scientific unions had never existed before. The IAA had been a gathering of Academies each of which had both a regional and an international standing of their own. The IRC on the other hand was conceived to centrally control the political and diplomatic framework for projects involving scientists from various countries. Indeed, Article 1 of the IRC Statutes reads:

> 1. The purpose of the International Research Council is:—
>
> (1) To co-ordinate international efforts in the different branches of science and its applications;
>
> (2) To initiate the formation of international Associations or Unions deemed to be useful to the progress of science in accordance with Article I of the resolutions adopted at the Conference of London, October, 1918[13];
>
> (3) To direct international scientific activity in subjects which do not fall within the purview of any existing international associations;
>
> (4) To enter through the proper channels into relation with the Governments of the countries adhering to the International Research Council in order to promote investigations falling within the competence of the Council.[14]

Given the intended dominance of the IRC, the question of whether and in which way a given country may adhere to it acquires some importance:

> 3. The countries in the following list may participate in the foundation of the International Research Council, and of any scientific Union connected with it, or join such Union at a subsequent period:— Belgium, Brazil, United States, France, the United Kingdom of Great Britain and Ireland, [the Dominions of] Australia, Canada, New Zealand, South Africa[; and] Greece, Italy, Japan, Poland, Portugal, Romania, Serbia.
>
> After a Union has been formed, nations not included in the above list, but fulfilling the conditions of Article 1 of the resolutions of the Conference of London, may be admitted, either at their own request, or on the proposal of one of the countries already belonging to

---

[12] [Jones 1961], p. 4.

[13] To wit: "That it is desirable that the nations at war with the Central Powers withdraw from the existing conventions relating to International Scientific Associations in accordance with the Statutes or Regulations of such Conventions respectively, as soon as circumstances permit"; and "That new associations, deemed to be useful to the progress of science and its applications, be established without delay by the nations at war with the Central Powers with the eventual co-operation of neutral nations."

[14] Quoted from the original statutes of the IRC, [Schuster 1920], p. 222.

> the Union. Such requests or proposals shall be submitted to a vote of the Union concerned. A favourable majority of not less than three-quarters of the countries already forming part of the Union shall be required for admission.[15]
>
> 4. A country may join the International Research Council, or any Union connected with it, either through its principal Academy, its National Research Council, some other national institution or association of institutions, or through its Government.[16]

Article 4 pays its debt to the past organization of Science International under the reign of the IAA. Some countries, unlike the USA and Italy, did not yet have a National Research Council, so Academies were still allowed as national stakeholders. However, it was also possible for governments to step in directly.

One juridical detail of the statutes of the IRC bears mentioning because it spilled over into the statutes of all the scientific unions depending on the IRC. In the case of the first International Mathematical Union it may have been an additional factor precipitating its early death. Indeed, Article 23 of the statutes of the IRC reads:

> 23. The present Convention shall come into force on the lst of January, 1920, provided that at least three of the countries mentioned in Article 3 have signified their adhesion. It shall remain in force until the 31st of December, 1931, and shall then, with the assent of the adhering countries, be continued for a further period of 12 years.[17]

The standard history of the IRC/ICSU comments dryly: "This is an example of a 'sunset clause' so frequently included in the terms of reference of scientific projects and programmes. In this case it was to trigger the emergence of ICSU."[18] Frequent or not, one may wonder whether there were any particular reasons for writing such a clause into the statutes of the IRC. Lacking explicit sources about how the clause was proposed and accepted, one may guess that the framers of the IRC were conscious of doing things under pressure of time, at a turning point of history, and therefore invited a reassessment of the situation after a reasonable period had elapsed.[19]

Thus prepared by inter-Allied conferences and by the Executive Committee, the IRC was officially created at a Constitutive Assembly held in the *Palais des Académies* in Brussels, 18–28 July 1919. There were altogether 225 delegates representing 12 out of the 16 eligible nations listed in Article 3 quoted above. More than two thirds of the delegates were from Belgium (107) or France (48). 27 came from the US, 19 from the UK, and 15 from Italy; Canada, and Romania as well as the new

---

[15] Two years later, the second General Assembly of the IRC on 26 July 1922 introduced the following additional rule concerning membership in scientific unions: "That only countries which have adhered to the International Research Council are entitled to be members of the Unions connected with it."—See [Schuster 1923], p. 111, and [Jones 1961], p. 6.

[16] Quoted from the statutes of the IRC, in the version adopted at its official creation; see [Schuster 1920], pp. 222–223.

[17] See [Schuster 1920], p. 226.

[18] See [Greenaway 1996], p. 18.

[19] I thank Helge Holden for insisting on this question, and Danielle Fauque for a helpful exchange on the matter.

Serbo-Croat-Slovene state each sent a delegation of 2; Japan, New Zealand, Poland, and Portugal were each represented by one man. Hale himself was not able to come to Brussels.

Nations that had been neutral during World War I were thus excluded from the creation of the IRC. When they were invited to join the established Council later on, they had to take it or leave it. Scandinavian and Dutch scientists, for example, who had their own ideas about further collaboration with German colleagues, were very uncomfortable with this situation and would continue to lobby for a reversal of the basic IRC policy of exclusion.[20]

4.1.1 The First Scientific Unions within the IRC; Preparing for the IMU

Not only was the IRC officially created in July 1919, but the new structure was also immediately fleshed out by several international scientific unions whose formation was already sufficiently advanced:

> The Statutes of the *International Research Council*, of the *International Astronomical Union*, of the *International Union of Geodesy and Geophysics*, and of the *International Union of Pure and Applied Chemistry* should he considered as having been definitely adopted. The Statutes of the other proposed Unions are at present under the consideration of the Academies, and are therefore subject to modification.[21]

Altogether five international unions came into being as members of the IRC during the year 1919. Each of them had its executive committee (*bureau*, in French); all presidents of these five unions at their creation happened to be Frenchmen: *International Astronomical Union* IAU, Pres. Benjamin Baillaud (1848–1934); *International Union of Geodesy and Geophysics* IUGG, Pres. Charles Lallemand (1857–1938); *International Union of Pure and Applied Chemistry* IUPAC, Pres. Charles Moureu (1863–1929); *International Union of Radio Science* URSI, Pres. Général Gustave Ferrié (1868–1932); *International Union of Biological Sciences* IUBS, Pres. Yves Delage (1854–1920).[22]

Other union projects were also discussed in Brussels, but could not be finalized in 1919. This was the case in particular for physics—the physicists were not sufficiently advanced in their plans to even propose a provisional executive committee

[20] See [Fauque & Fox 2021], pp. 46–48, as well as the rich volume [Letteval et al. 2012].

[21] From Schuster's preface to [Schuster 1920].

[22] Cf. the interesting material—especially the slides of Danielle Fauque's talk—from the commemoration of these foundations held at *Institut de France* in Paris on 3 December 2019: [URL 06]. In the general discussion at the end of this session, the quest was voiced to restore today a more prominent role for the French language in international scientific exchange.

in Brussels—and for mathematics. The following account of the mathematicians' discussion in Brussels can be found in the official report about the Constitutive Assembly of the IRC.[23]

D. International Union of Mathematicians.

The sessions are chaired by Mr. de la Vallée Poussin.[24]

Mr. De Donder[25] acts as secretary.

The assembly accepts the proposed statutes unanimously...

The assembly expresses the wish to see an International Congress of Mathematicians organized in September 1920; M. Kœnigs[26] hopes that it will be possible to hold this Congress in Strasbourg. This proposal meets with unanimous consent.

... the assembly expresses the wish that the authors of mathematical papers or treatises send, right after the publication of their works, abstracts of these to an organism whose task it will be to centralize and coordinate all bibliographical abstracts; this organism will be lodged in Paris, or in another scientific centre, and will be linked as far as possible to an existing similar agency.

The delegates present [at the sessions discussing the project of an IMU] form the *Provisional Committee* of the International Union of Mathematicians. Its Executive Committee consists of:

Honorary Presidents: Mr. H. Lamb[27], E. Picard and V. Volterra;

President: Mr. de la Vallée Poussin;

Vice-President: Mr. W.H. Young[28];

Secretaries: Mr. de Donder, Kœnigs, Petrovich[29], and Reina[30].

[23] See [Schuster 1920], p. 26; my translation. For a while there were several different French names for this union. Whereas today it is generally written as *Union mathématique internationale*, or UMI, the 1919 project was called *Union Internationale de Mathématiciens* in [Schuster 1920]. The typed version of the proposed statutes prepared for the foundational meeting of the IMU in Strasbourg on 20 September 1920 uses instead *Union Internationale de Mathématiques*, both in the title and the text (Article 8)—a copy of this typescript has survived for instance in folder M 119 323 at Bibliothèque Nationale et Universitaire, Strasbourg. Afterwards both *Union mathématique internationale* and *Union internationale mathématique* are used; see for instance Proceedings ICM 1920, p. XXXIV vs. p. XXXV; [Schuster 1923], p. 52. The statutes of the IMU proposed in 1919 and mentioned in the following quote are reproduced in [Schuster 1920], pp. 185–189 for the French version—the only authoritative judicial reference—, and pp. 247–250 in English. They erroneously stipulate (Article 9) triannual—instead of quadrennial—meetings of the IMU General Assembly.

[24] The Belgian mathematician (as of 1928: Baron) Charles-Jean de la Vallée Poussin (1866–1962). In some of the sources we quote his name is hyphenated: de la Vallée-Poussin.

[25] The Belgian physicist, mathematician, and chemist Théophile de Donder (1872–1957).

[26] The French mathematician Gabriel Xavier Paul Kœnigs (1858–1931). I use the spelling of his name with 'œ' which dominates in Proceedings ICM 1920.

[27] The British mathematician Sir Horace Lamb (1849–1934).

[28] I.e., the husband of Grace Chisholm Young.

[29] The Serbian mathematician Mihailo Petrović (1868–1943). Since he was an avid fisherman on the Sava and Danube rivers, he acquired the Serbian name of Mihailo Petrović Alas. For more information on his life and mathematics, see [Dragović & Goryuchkina 2020].

[30] The Italian geologist Vincenzo Reina (1862–1919), Secretary of the Italian Society for the Advancement of Science.

> The remaining delegates in the Provisional Committee are: Mr. Demoulin[31], de Ruyts[32], Glaisher[33], Parenty[34], Stuyvaert[35].

Picard and Volterra were centrally involved in constituting the International Research Council. Both were famous and influential mathematicians, de la Vallée Poussin now being remembered above all for his proof of the Prime Number Theorem in 1896. However, in spite of their calibre, they were unable to raise the critical mass to officially establish an International Mathematical Union under the auspices of the IRC in Brussels in July 1919. Only five nations were represented there for the IMU project: Belgium, France, Italy, the UK, and Serbia. In his report delivered in Strasbourg a year later, Gabriel Kœnigs would also cite the more formal reason that National Committees for mathematics had not been created in a single country yet; such committees were, however, part of the IRC format for scientific unions.[36]

The minutes we have quoted above remind us of the interests that distinguish mathematics from other sciences whose international unions were further advanced: In astronomy or geology there is an immediate need to coordinate with partners all around the planet for the collection of measures, photographs, etc. The international plans of the mathematicians in 1919, on the other hand, focus on an ICM and on the international coordination of bibliography, through authors' abstracts.[37] This narrow scope would continue to characterize the first IMU during its short life, after its creation in Strasbourg in 1920.

As to the ICM, recall (Section 3.4) that Mittag-Leffler—in keeping with the four year rhythm which had been established before the war in the absence of any international union of mathematicians stipulating it—was to host an ICM at Stockholm in 1916, which was foiled by the war. Sweden was not represented at the constitutive assembly at Brussels in 1919; it had been a neutral country in World War I. Mittag-Leffler would never accept the decision to hold an ICM in Strasbourg in 1920. More precisely, he would continue to refuse—as did many German mathematicians—to count the Strasbourg event as a fully fledged ICM.[38]

---

[31] The Belgian geometer Alphonse Demoulin (1869–1947).

[32] The Belgian mathematician Jacques Deruyts (1862–1945).

[33] The British mathematician James Whitbread Lee Glaisher (1848–1928). There is an incorrect spelling in [Lehto 1998], pp. 310, 380.

[34] The French engineer Henry (Louis Joseph) Parenty, known for his book *Les tourbillons de Descartes et la science moderne*, Clermont-Ferrand 1903.

[35] The Belgian mathematician Modeste Stuyvaert (1866–1932).

[36] See Kœnigs's report in Proceedings ICM 1920, p. XXXIV. For the national committees in the IMU statutes, see [Schuster 1920], p. 247, Articles 3 and 4. The same provisions can be found in most of the other statutes of scientific unions voted in Brussels.

[37] Recall from Section 1.4.1.2 that questions of bibliographic classification had already been discussed at pre-war ICMs.

[38] See for instance the exchange between Mittag-Leffler and Edmund Landau published in [Siegmund-Schultze 2011].

It seems more difficult to estimate how quickly resistance against the IRC politics executed by the IMU accumulated in the US.[39] Indeed, some published histories of what happened seem to be written with hindsight in this respect; for example this one:

> The four secretaries of this provisional [mathematical] Union, representing Belgium, France, Roumania[40], and Italy, failed in their duty. Without consulting the United States or Great Britain, at least, they decided to hold at Strasbourg an International Congress of Mathematicians from which the Central Powers of Europe were excluded.[41]

It is true that no American was among the mathematicians who met in Brussels when the provisional executive committee of the IMU was instituted. However, the UK was certainly represented. Also, it was the assembly's, not the secretaries' decision to hold the exclusive Congress in 1920, if possible in Strasbourg. More generally, US scientists who cared could not miss the fact that the whole politics of the IRC and its scientific unions systematically excluded Germany and the Central Powers. So it seems that this surly account of the Brussels meeting projects back onto the year 1919 the critical attitude which would indeed grow in the US over the following years, and which will be discussed below—see especially Section 4.4.2.

4.2 The UAI; the League of Nations; the ICIC

Before moving on to Strasbourg, let us stop for a moment to get a fuller picture of international organizations created in 1919–1920.

First of all, whereas the old IAA had tried to unite both sciences and the humanities under a common roof, the scope of the IRC extended only to the exact sciences, in accordance with the interests of the framers of this council. Indeed, we have pointed out Hales's and Volterra's background in war-oriented research. As to Émile Picard—whose wartime writings on the history of science we have discussed in Section 3.2—it is true that he would finally succeed in 1924 in obtaining one of the famous 40 seats of the literary *Académie française*. However, his international activities in 1919–1920 were more in keeping with his position of permanent secretary of the Paris *Académie des sciences*, in which he had succeeded Darboux in 1917. As for the UK, let us recall that the Royal Society also did not represent the humanities. It is therefore not surprising that, formally on a par with the IRC, an organism was created for the international cooperation of the humanities:

> The natural sciences were followed by the humanities. Initiated by the humanities departments of the academies of the Allied countries, but including the neutral states from the beginning, an International Academic Union (*Union Académique Internationale*) [UAI] was

[39] For a collection of passionate early reactions opposing the exclusion policy, see for instance [Riehm & Hoffman 2011], pp. 121–125.

[40] In fact, as we have noted above, Petrović Alas was from Serbia.

[41] See [Archibald 1938], p. 19. This remark is quoted in [Lehto 1998], p. 24, and Archibald's account is accepted there as evidence that "resentment against the IMU arose in the American Mathematical Society even before the Union had been officially founded."

> founded in 1919. Like the *Conseil International de Recherches* [IRC], the Union also had its headquarters and place of assembly in Brussels. [The Belgian medievalist historian] Henri Pirenne [1862–1935] became its first president. The statutes of the Union did not decree the explicit exclusion of Germany or the Central Powers, but there was left no doubt that their scholars would not be admitted. Only the neutral states were urged to join, and a three-quarters majority was prescribed for all applications. The report on the Union's founding conference, co-authored by Pirenne, stated in emotional and ambitious terms that a new republic of scholars would be created in the spirit of friendly, trusting, freedom-loving brotherhood. Whereas the former Association of Academies [IAA] had been in danger of becoming the tool of nationalistic ambitions striving for scientific hegemony over the world, the new Union was to become an intellectual League of Nations, with the goal of "making researchers accustomed to thinking like human beings and seeking nothing else but the truth."[42]

The further evolution of international unions and congresses in the humanities during the 1920s, and their politics of exclusion of (participants from) the Central Powers followed by and large the same pattern that we will see in the case of mathematics. This is despite the fact that Pirenne himself—who had been a prisoner of war in Germany during World War I—saw historians in an especially complicated situation, with war experiences challenging their very methodology: "One must also take into consideration that a Historical Congress is something quite different from a congress of physicists or mathematicians."[43]

Secondly, even if the new setup of Science International starting with the IRC was top-down, as we pointed out, this does not mean that the IRC and the UAI were at the top of the pyramid of international organizations. At the crest of the pyramid outlined by the Versailles treaty, the *League of Nations* was to oversee a new international order. This novel institution was President Woodrow Wilson's (1856–1924) brainchild, based on the idea of general peace and international security, conceived along with a new quality of international law enforced by the *Permanent Court of International Justice*, which would actually function in The Hague from 1922 to the beginning of World War II.[44] Also on this highest political level, the devastating experience of World War I motivated the search for international structures that could, ideally, confine violent nationalism. The League of Nations got off to a "faltering start"[45] though, to say the least, and the American lawmakers prevented the US from joining it. The story of the League of Nations, from its actual foundation in 1920 to its limbo in World War II and its official end in 1946, shows it was all too often dwarfed by the nation states it was supposed to control. Yet the League did survive longer than the IRC, and with hindsight it emerges as a blueprint for the *United Nations*. The UN would carry this torch into the age of newly independent decolonized states and of the Universal Declaration of Human Rights.

[42] See [Erdmann 2005], p. 71, and the references given there.

[43] From a letter of Pirenne to the Danish historian Aage Friis (1870–1949), dated 17 September 1922; I quote the translation from [Erdmann 2005], p. 78. Cf. the whole chapter 8 of [Erdmann 2005].

[44] See Chapter 8 below for remarks on international justice after World War II.

[45] This is the title of Chapter 3 of [Henig 2019].

During the second half of the twentieth century, the UN would interact with the world of science especially through UNESCO, the Paris-based *United Nations Educational, Scientific and Cultural Organization*—see Section 8.1.2 below. UNESCO was founded in 1945 as the immediate successor of the *International Committee on Intellectual Cooperation*, or ICIC for short.

This committee of the League of Nations had functioned from 1922 to 1939. It brought together twelve outstanding scholars who were supposed to be chosen not to represent a nation, but for their scientific and cultural excellence and charisma. Alas, no mathematician ever sat on this committee. Its first President—from 1922 until his withdrawal for substantially political reasons in 1925—was the philosopher Henri Bergson. (Bergson, by the way, would receive the Nobel Prize for Literature in 1928.) After considerable back and forth, Albert Einstein (1879–1955; German born, Swiss citizen and pacifist; 1921 Nobel Prize of Physics, awarded in 1922) finally accepted in 1924 the renewed offer of a seat on the committee. There was exactly one woman among the members of this committee: Marie Curie (see Section 1.2.1.2), whom Einstein considered his "sister in defiance."[46] Here is how Einstein described the work of the committee in an article for a German newspaper:

> I have just returned from the session of the League Committee on Intellectual Co-operation in Geneva and should like to convey some of my impressions to the German public. The object of the committee is to initiate or foster efforts which may promote international co-operation between the scientific and intellectual communities of various countries in the hope that national cultures, heretofore separated by language and tradition, may thereby be brought into closer communication. Rather than entertaining utopian schemes, the committee has initiated several modest but fruitful projects on a small scale, such as the international organization of scientific reporting, the exchange of publications, the protection of literary property, the exchange of professors and students among various countries, etc. Thus far, the greatest progress has been achieved in the sphere of international reporting.
>
> While the specific projects just mentioned may be of little interest to the general public, much consideration should be given to the question of what attitude the German people and the German Government ought to adopt in principle toward the League of Nations. . . . [47]

Then follows an appeal that Germany join the League of Nations. This actually happened in September 1926. Seven years later, under the Nazis, Germany would leave the League again.

We see from this clipping that the committee intersected with activities about scientific reviewing under the guidance of the IRC. This point of intersection, or competition, was officially discussed between the two international bodies. In the end, neither of them achieved substantial progress in this direction.

In spite of the resources which the French government provided for the ICIC by establishing in 1926 the *International Institute for Intellectual Cooperation* in Paris, the overall record of the ICIC is meagre. This is hardly surprising given the ambient political situation. Augustus Trowbridge (1870–1934), the physicist and head of the

---

[46] See the translation from Einstein's (German) letter to Marie Curie of 25 December 1923 in [Rowe & Schumann 2007], p. 196.

[47] Albert Einstein in *Frankfurter Zeitung* of 29 August 1924; translation quoted from [Rowe & Schumann 2007], pp. 196–197.

Paris office of the International Education Board (see Chapter 5 below), commented on a January 1926 meeting of the ICIC: "It is perhaps inevitable, but it seemed ... that in spite of the general protestations of internationalism the nationalistic prejudices are very close to the surface all the time ... On this very minor stage one could see played what was played on the world stage in 1918–1919."[48] As Brigitte Schroeder-Gudehus points out at the end of her detailed and merciless analysis, the story of ICIC does shed a sobering light on the potential of intellectual luminaries to promote international understanding. The history of the ICIC offers "nothing more than a dim reflection of the conflicts which were fought by the politicians, both on and off the stage of the League of Nations."[49]

4.3 Strasbourg

It was Louis XIV who in 1681 integrated Strasbourg into the Kingdom of France and gave the Cathedral—which had been reformed into a protestant church—back to the Catholics. As a token of commemoration of this, a statue of the Sun King on a horse was added to the front of the Strasbourg Cathedral in 1823. Around the same time, at the dawn of the age of nation states, fighters for German national unity began to claim the integration of Strasbourg and its region: Alsace-Lorraine, with its local family of Alemannic dialects, into Germany.

During the German-French war of 1870–71, the center of Strasbourg was seriously damaged by Prussian artillery; the rich library at *Temple Neuf* was destroyed. After the end of that war, Alsace-Lorraine was annexed to the freshly created German Empire and Strasbourg became the capital of this newly added *Reichsland*. Part of the Alsatian elite—including for instance a number of professors of medicine and chemistry, but also artisans of *art nouveau* design—opted for France and left Strasbourg and Alsace. Some settled in Nancy (in the French, non-annexed part of Lorraine) where a new French Medical Faculty was set up as a visible counterpart to the German University planted in Strasbourg[50], others moved to Paris.[51]

In Strasbourg the Germans constructed in the 1870s and 1880s a whole new quarter, called *Neustadt*, North-East of the historic centre of town. It included a generous campus for the new German university—except for the Faculty of Medicine which was accommodated on the opposite side of town. At the time, the new structures were internationally considered to be a showcase of up-to-date university architecture, with separate institute buildings not only for the observatory but also for chemistry,

[48] Quoted from [Siegmund-Schultze 2001], p. 61.

[49] See [Schroeder-Gudehus 1966], pp. 135–179, for the story of the ICIC; p. 179 for the quote; my translation. See also [Voges 2019].

[50] See [Bonah 2000].

[51] See [Carneiro & Pigeard 1997]. Cf. Pierre Laszlo's chapter in [Simões et al. 2015], pp. 89–105.

physics and branches of biology. An essential part of the resources needed for these large-scale constructions came from reparations that France had to pay having lost the war.[52]

The mathematicians, it is true, did not get their own institute building. Their library and limited office space was lodged within the main building (*Kollegiengebäude*), next to the quarters for the humanities and various lecture halls, and it would remain there, with moderate later extensions, through some thirty-five years of the German and more than forty years of the French university, until the 1960s. In a letter to David Hilbert dated 19 June 1899, after passing through Strasbourg, Hermann Minkowski remarked that the rooms allocated to mathematics there "are really opulently equipped."[53] And in contrast to most other German universities—notably Göttingen, where the joint faculty of sciences and the humanities would even outlast World War I and create friction, for instance for Emmy Noether's (1882–1935) career[54]—mathematics in Strasbourg belonged to an independent *Faculty of Sciences*, right from the beginning of the German university.

Fig. 4.2 The new buildings of Strasbourg University from the 1870s / 1880s. Source: *L'Univers illustré*, 29 November 1884. The main building, where the mathematics library was lodged, along with other disciplines and lecture halls, is shown in the center. This is also the building where the 1920 ICM would be held. The campus behind it can be seen in the central picture above; its institute buildings, for chemistry, physics and botany, are depicted in the lower row; astronomy is in the upper left corner. The remaining three institutes shown belong to the Faculty of Medicine.

[52] See [Nohlen 1982]. Cf. Part 1 of [Schappacher & Wirbelauer 2010], especially pp. 56–57.

[53] See [Minkowski 1973], p. 36: *wahrhaft opulent eingerichtet.*

[54] Cf. [Tollmien 1990]. See Section 6.1 below.

The new German university in Strasbourg adopted the name *Kaiser-Wilhelms-Universität* (KWU) in 1877. Notwithstanding its strong international connections,[55] it was an ostentatious monument of German nationalism. The frieze of 36 named statues on the roof of the main building offers the onlooker a collection of major protestant reformers around the Northern corner, and for the rest a parade of great German scholars (or considered as such, Copernicus for instance).[56] Most of the men displayed here who had been active in the nineteenth century were linked to the Prussian university system of the nineteenth century (cf. Section 1.1.1 above). The attitude of intellectual colonialism expressed in the architecture also transpired in everyday reality and led to a lasting estrangement between the new institution and the Alsatian population.[57]

From the very beginning of World War I, one of the obvious goals of France was to recover the territories lost in 1870. Concrete plans for the French future of Alsace-Lorraine in case of victory began in February 1915 when the French Ministry of Foreign Affairs at Quai d'Orsay installed a *Conférence d'Alsace Lorraine*. Already in March of the same year it envisaged the replacement of KWU by a French university. Thus Christian Pfister (1857–1933), professor of history at the Sorbonne and of Alsatian origin, elaborated an almost 100-page-long detailed report about the existing KWU and how it should evolve into a French institution. The *Conférence d'Alsace Lorraine* adopted this roadmap in the early Summer of 1918, and it was put into practice as of 22 November 1918, i.e., right after the crushing of the short-lived Strasbourg soviet republic (*Räterepublik*). The French university started to function on 15 January 1919.[58] And on 22 November 1919, the *Université de Strasbourg* could officially celebrate its first *rentrée*, i.e., the opening of a new academic year. It was on this occasion that Pfister would pronounce the oft-quoted words:

> Today we do not celebrate the inauguration of a new university, but after 49 years, launch the new academic year of 1870.[59]

Of course, in spite of this victorious appeal to ideal continuity, both the facilities and the personnel were essentially new. A few days earlier, on 17 November 1919, Maurice Fréchet—whom we have mentioned before (Section 1.3.3.3) as a collaborator of the French edition of the Mathematical Encyclopedia—had given his inauguration lecture about "Mathematics at *Université de Strasbourg*" which Émile Borel would subsequently publish in his *Revue du mois*.[60] Fréchet was only one of a number of distinguished French scholars who accepted to move from Paris

---

[55] Cf. Christian Bonah's chapter: *Une université internationale malgré elle*, in [Crawford & Olff-Nathan 2005], pp. 29–35.

[56] See [Nohlen 1982], p. 181. Cf. [Denis 2005].

[57] See chapter 4, "The German University and Alsatian Society" in [Craig 1984], pp. 100–135.

[58] See [Schappacher & Wirbelauer 2010], pp. 60–63, for more details and additional literature.

[59] *Nous célébrons aujourd'hui, non pas l'inauguration d'une Université nouvelle, mais après quarante-neuf années, la séance de rentrée de 1870.* See for instance Françoise Olivier-Utard's chapter in [Crawford & Olff-Nathan 2005], p. 137.

[60] See [Fréchet 1920]; cf. the discussion below.

to Strasbourg in 1919 with the explicit goal to establish the renewed French university as a showcase of French civilization on the left bank of the Rhine. Other colleagues who moved to Strasbourg in 1919 include the mathematician Georges Valiron (1884–1955), the mathematical physicist (expert in fluid mechanics) Henri Villat (1879–1972)—who would chair the organizing committee of the Strasbourg ICM—, the physicist Pierre Ernest Weiss (1865–1940)—his talk at the ICM will be discussed below—; the astronomers Ernest Esclangon (1876–1954) and André Danjon (1890–1967), the seismologist Edmond Rothé (1873–1942); the historians Marc Bloch (1886-1944) and Lucien Febvre (1878–1956)—whose joint Strasbourg period would result in the foundation of the *Annales d'histoire économique et sociale* and the well-known school of historical method named after this journal—, the sociologist Maurice Halbwachs (1877–1945), and the anti-Freudian psychologist Charles Blondel (1876–1939).

In June 1919, the young mathematician Louis Antoine (1888–1971) also arrived in Strasbourg. He had been wounded three times in the war. The last time, in April 1917, it had cost him his eyesight. His Paris colleagues, especially Henri Lebesgue, oriented him towards topology, a young field with a still limited literature. His Strasbourg thesis was finished in 1921. In it he constructed a certain fractal set which would influence the development of topology at the time.[61]

4.3.1 Maurice Fréchet in Strasbourg

Fréchet's inauguration lecture of November 1919, more precisely its first section: *Comparaison des méthodes*, is interesting for us because in it Fréchet sets out to compare the German to the French way of teaching mathematics at university level. He motivates this comparison by the peculiar circumstances which brought the French teaching staff to Strasbourg.

> ... every one of the professors who arrived here last January had an open mind for all kinds of suggestions, ready to adopt the best methods regardless of their origin. We naturally had some ideas about the German teaching system. But we were about to truly discover its qualities and defaults on the spot. The results of our findings differ slightly, but actually depend very little on the course of studies under scrutiny.[62]

The upshot of Fréchet's comparison—developed over almost five pages, which only occasionally echo anti-German rhetoric from the war years—is basically this: The main advantage of the German system is that students are systematically initiated to research-type work, in advanced courses, and when they have to write a memoir as part of their graduating exam: "The student is put into contact with science in the making."[63] Consequently, in order to increase the number of specialized courses,

---

[61] For Louis Antoine's case, see [Aubin & Goldstein 2014], *passim*, and especially p. 165. For the mathematical influence of Antoine's wild set, cf. [Epple 1999], especially pp. 336–338.

[62] [Fréchet 1920], p. 338; my translation.

[63] [Fréchet 1920], p. 339; my translation.

a new chair of Higher Analysis was created in Strasbourg. The main advantage of the French system on the other hand is, according to Fréchet, that the basic math curriculum is much better organized and covers a variety of disciplines, insisting on their mutual interconnections. It thus gives the French students a sound and broad basis to rely on, whereas a student in the German system may get lost in the confusing variety of fairly specialized courses, often offered by *Privatdozenten* whose living depends on attracting students.

Fréchet's lecture reflects both the peculiar situation of Strasbourg University right after World War I and his own openness for international exchange. In those years, Fréchet tried to maintain as far as possible the number of foreign students at Strasbourg University. In his frequent talks abroad, particularly on the British Isles, he would persistently advertise Strasbourg University as being far superior to the typical provincial universities in France.[64]

Given his key role in the French renewal of Strasbourg University, it bears mentioning that Fréchet would keep a rather low profile at the *Congrès international de mathématiques*[65] when it was held at his new home town.

4.3.2 The IMU Founded in Strasbourg

From the inter-Allied point of view, the time was ripe in September 1920 to complete the creation of the International Mathematical Union, within the scheme laid out by the IRC. Indeed, by that time National Committees for Mathematics had been created in Belgium, France, Italy, the UK, and the USA. Apart from these five allies, delegates representing Czechoslovakia, Greece, Japan, Poland, Portugal, and Serbia were also present in Strasbourg. Altogether eleven countries thus launched this new format of Mathematics International:

> On 20 September 1920, in a [lecture] hall of the [main building of] Strasbourg University, the meeting of the delegates of allied or associated countries could be held as planned. The provisional statutes drafted in Brussels were definitively accepted, an official executive committee was formed which, apart from a few additions, confirms the provisional committee created in Brussels in the various functions. It is composed of:
>
> *Honorary Presidents*: Mr. Jordan[66], Lamb, Picard, Volterra.
>
> *President*: de la Vallée Poussin.

[64] Cf. [Siegmund-Schultze 2005], pp. 186–187.

[65] This label, rather than *Congrès international des mathématiciens*, was used by Gabriel Kœnigs in his report, Proceedings ICM 1920, pp. XXXIV–XXXIX. Cf. the remarks about the labeling of this congress in Mittag-Leffler's exchange of letters with Edmund Landau, reproduced and discussed in [Siegmund-Schultze 2011], p. 115.

[66] Eighty-two year old Camille Jordan was present in Strasbourg and would also act as Honorary President of the ICM.

Vice-President: Appell[67], Bianchi[68], Dickson[69], Larmor[70], Young.

Secretary General: Kœnigs.

Secretaries: de Donder, Hatzidakis[71], Petrovich, Pompeiu[72].

Treasurer: Demoulin.

> As soon as it was established, the International Mathematical Union took two decisions. The first concerns mathematical bibliography; the editors of mathematical journals will be invited to require the authors of articles they are printing to write *themselves* a short abstract of their paper.
>
> The second decision concerns the date and place of the coming Congresses. In fact, according to its constitution, it belongs to the International Union to take this initiative. It was decided that the Mathematical Congresses (*Congrès de mathématiques*) should take place once every four years, i.e., in 1924, 1928, etc. As for the place, two proposals were submitted simultaneously, by the Belgians and the Americans. The first propose Brussels, the others New York or its whereabouts. It was jointly agreed that the 1924 Congress would be held in New York, the following one in Belgium.
>
> On the following day, 21 September [1920], the definitive constitution of the International Union and its statutes were communicated to all [representatives of] neutral countries present. They were informed that they were free to join us and we have every reason to hope that the majority of the neutral nations will form National Committees, like we have done, which will have the necessary authority to legally represent them within the Union.
>
> In this way, the work is accomplished which we had envisioned in Brussels, and which delights all the more our French heart as this has been realized at Strasbourg University.[73]

The eleven Countries listed above had thus installed a novel framework for what they saw as the future of Mathematics International. The reality was going to be different. To this day, no ICM has been hosted in either New York or Brussels. Much more seriously, the war-inspired inter-Allied approach severed potential continuities and soon provoked more tensions than momentum. It would naturally come across as unilateral, not only in Germany and the other Central Powers, which were excluded, but also in places which, as we have seen, had been hot springs of mathematical internationalism in Europe before the war, such as Stockholm and Palermo. Concerning Stockholm, we have recorded (Section 4.1.1) Mittag-Leffler's contempt for the Strasbourg events. As to the *Circolo matematico di Palermo*, its founder Guccia had died in 1914, and it fell to Michele De Franchis (1875–1946) to steer the *Circolo* through the stormy waters of the post-war years. He refused to align himself with Picard's and Volterra's politics and kept German members:

---

[67] The French mathematician Paul Émile Appell (1855–1930). Born in Strasbourg, he left when the city had become German and began his university studies in Nancy, where his friendship with Poincaré began. Émile Borel's wife (the writer with the pen name Camille Marbo, whom we have mentioned in Section 1.2.1.2) was Paul Appell's daughter.

[68] The Italian mathematician Luigi Bianchi (1856–1928).

[69] The US mathematician Leonard Eugene Dickson (1874–1954).

[70] The Irish physicist and mathematician Sir Joseph Larmor (1857–1942).

[71] The Greek mathematician Nikolaos Hatzidakis (1872–1942).

[72] The Romanian mathematician Dimitrie Pompeiu (1873–1954).

[73] From Kœnig's report, Proceedings ICM 1920, p. XXXV–XXXVI; my translation.

> A Sicilian may be justifiably proud to observe that during the years immediately following the First World War, the Circolo was the only scientific organization in the world where one could see such names as Hilbert, Klein, [Max] Noether, Landau , Picard, de la Vallée Poussin, Hadamard, Borel, and Lebesgue side by side.[74]

At the same time De Franchis had to deal with dwindling material resources for the *Rendiconti*.[75]

We have already alluded to growing US opposition in the years following the Strasbourg ICM, see the end of Section 4.1.1 above. Norbert Wiener (1894–1964) was one of the Americans who came to Strasbourg on the occasion of the Congress, even though, according to his own account, his principal reason for the trip was the desire to meet Fréchet. When he looked back on his participation at this exclusive event later in his autobiography, mindful of how quickly the tide had turned against the IRC and the IMU, he almost felt he had to apologize:

> The last International Mathematical Congress before the war had taken place in England in 1912, at Cambridge. The congress which was to have taken place in 1916 was clearly impossible and was allowed to go by the board. The next one, in 1920, did not find any adequate machinery established for its organization. France decided to step into the gap and celebrate an international congress in the newly re-Gallicized city of Strasbourg and at its university, now French. This had become the second university of France and the only provincial university with a great tradition of its own.
>
> In many ways this was an unfortunate decision. It was one which later led me to regret my little share in sanctioning the meeting by my presence. The Germans were excluded as a sort of punitive measure. In my mature, considered opinion, punitive measures are out of place in international scientific relations. Perhaps it would have been impossible to hold a truly international meeting for another couple of years, but this delay would have been preferable to what actually did take place, the nationalization of a truly international institution. All that I can say for myself is that I was young and that I did not feel myself in a position of direct personal responsibility for the course taken by international science.[76]

4.3.3 The 1920 ICM in Strasbourg: "la grande manifestation patriotique et scientifique"

The *Congrès international des mathématiciens* at Strasbourg[77] began two days after the meeting which had given birth to the IMU. Horace Bryon Heywood (1883–1977), who had studied in Paris, was among the attendants. In 1912, when he was

[74] Quoted from Aldo Brigaglia's chapter on the *Circolo* in [Parshall & Rice 2002], p. 196.

[75] See the succinct discussion of this problem in the chapter by Nastasi and Tazzioli in [Aubin & Goldstein 2014], pp. 210–211. Cf. [Cerroni & Brigaglia 2021].

[76] See [Wiener 1956], pp. 49–50.

[77] The description of this congress quoted in the section title is from Kœnig's report at the closing session. However, the corresponding paragraph, towards the end of the speech, is not reproduced in Proceedings ICM 1920 because it principally dealt with the need for more donations to print the proceedings. It is preserved in print, in the original French, in the Portuguese brochure [da Costa Lobo 1921], p. 21.

Fig. 4.3 Gabriel Kœnigs in 1920(?).

working at the University of London, he would publish jointly with Maurice Fréchet a French textbook which Jacques Hadamard in his preface would call "the result of an international collaboration that cannot be applauded strongly enough."[78] Soon after the Congress he would publish an account in *Nature* which followed the model of an official report. It starts like this:

[78] See [Fréchet & Heywood 1912]. The quote from Hadamard's preface is on p. IV; my translation.

> This congress was opened at Strasbourg University on September 22 by the Rector, M. S. Claréty [sic][79]. The officers of the congress were then elected as follows :— *Honorary President*: M. Camille Jordan. *President*: M. Emile Picard. *Vice-Presidents*: Prof. Leonard Dickson, Sir Joseph Larmor, Prof. Nörlund[80], M. de la Vallée-Poussin, M. H. Villat and M. Volterra. *Secretary*: M. Koenigs.
>
> The delegates numbered 188 and represented 26 nations, amongst which may be mentioned Argentina (4), Australia (1), Brazil (1), Canada (1), Czecho-Slovakia (12), India (2), Japan (2), the Philippine Islands (1), Poland (4), Russia (1), and Serbia (2). The expenses of the congress, including the publishing of the proceedings, have been completely provided for. Of the sum required, 78,000 francs was contributed by public bodies, by industrial and commercial concerns, and by private persons. An interesting fact is that the French Government made its contribution of 10,000 francs through the Ministry of Foreign Affairs, thereby recognising, it would appear, that such a congress has a certain significance in international politics. The subscriptions of delegates produced a further sum of 12,000 francs.[81]

The count of participants given by Heywood differs slightly from what was published a bit later in the Proceedings of the 1920 ICM, p. XV. There 200 scientists from 27 countries are listed. 80 of them were from France, the other countries with a two digit number of participants were Switzerland (14), Czechoslovakia (12), USA (11), Belgium (10), and Spain (10). Compared to the pre-war ICMs (see Section 1.4.1.2 above), the size of the meeting did not reach 40% of the numbers attained since 1908, roughly falling back to the level of the first ICM in Zürich (1897). Both the exclusion of the Central Powers and the postwar situation in general certainly contributed to this. The 14% of non-European participants at Strasbourg almost match the share counted at Cambridge in 1912, but US participation dropped to 5.5%. Had German mathematicians been invited, these two percentages would of course have been even lower.

Heywood's report continues with summaries of the five plenary lectures that were presented at the Congress: by Larmor, Dickson, Volterra, de la Vallée Poussin, and Nørlund. Since Larmor tried to uphold "æther theory" against Einstein's relativity, Heywood mentions in passing one of the two section talks which also struggled with relativity. Finally, Heywood goes into some detail about a lecture which was given but would not be reproduced in the Proceedings:

> Prof. [Pierre Ernest] Weiss, the director of the Strasbourg Institute of Physics, gave an account of the methods of soundranging in use in the French Army during the war. The method normally employed was the same as that in use in the British Army. A useful

[79] The historian Sébastien Charléty (1867–1945), who had come to Strasbourg from Tunis after the war, actually held more than one high-level position in the regional educational and cultural administration, and he also spoke at the opening in the name of the governor of Alsace-Lorraine.

[80] The Danish mathematician Niels Erik Nørlund spoke for Denmark at the opening and closing ceremonies of the Congress—see Proceedings ICM 1920, pp. XXVI and XXX. But since he taught at Lund University, he could be mistaken for a delegate of Sweden. This is what Mittag-Leffler and Landau did in their 1922 correspondence—see [Siegmund-Schultze 2011], p. 112, note 20. No Swedish *délégation* to the Congress is listed in Proceedings ICM 1920, p. VIII; but exactly one participant of Swedish nationality is counted, p. XV. This was the well-known statistician Harald Cramér (1893–1985) who actually spoke on behalf of Sweden at the closing ceremony.

[81] See [Heywood 1920], p. 196.

> alternative was the *méthode à courtes bases*, in which six or more microphones were placed in pairs. The microphones of each pair were about a hundred metres apart, so that the gun locus became a straight line (asymptote), and at once gave the direction of the hostile gun. The installation was very simple, and could be made in an hour, while single sets of observations could be reduced and reported in a minute. This method was used, not for the accurate location of gun emplacements, but for determining quickly which one of the known hostile batteries was in action. Guns were also successfully located by observations of the *onde de choque*. The normals to this wave-surface determine a caustic which is nearly constant in form for high-velocity shells. To locate the gun emplacement, a standard caustic drawn on tracing-paper was fitted by trial to the normals determined by the instruments. This method was used when atmospheric conditions made the spherical wave imperceptible, and, although less accurate, it gave very good results. A case was quoted where 80 per cent. of the hostile emplacements were correctly located solely by *ondes de choque*.[82]

Before the war, Pierre Weiss had established his own institute on magnetic research at the Zürich Polytechnique; he was now about to do the same in Strasbourg.[83] Incidentally, since his days at ENS, Pierre Weiss was a close friend of Élie Cartan, who attended the ICM and gave a lecture. Cartan was accompanied by his wife and four children, among them his eldest son Henri Cartan (1904–2008), who would start teaching regularly at Strasbourg University in 1931, and marry Pierre Weiss's daughter Nicole in 1935.

Heywood's account of Weiss's lecture is an appropriate reminder of the mobilized mathematical environment of the time.[84] It provokes the question to what extent World War I introduced a more equitable balance between pure and applied mathematics.

> As far as mathematics was concerned, it would seem natural to think that the experiences of World War I would have produced renewed interest in applied mathematics. In the opening speech he delivered at the Strasbourg ICM, Picard warned his colleagues against this perceived threat: "Some say . . . that in years to come applications of mathematics will be the most studied and pure theory somewhat neglected. . . The times we are now living in have indeed become harder for mind workers, and the more optimistic of us sometimes ask whether our civilization will not be eclipsed. We therefore must not tire ourselves of repeating that in the final analysis the true source of all progress in the applied sciences lies in theoretical speculations."[85] The myth of pure, [disinterested] science—so potent, as we have seen, in the prewar period—needed to be reinvested with new meanings. In his closing speech, Picard however also argued that the world had completely changed between 1914 and 1920 and—odd in light of the above—that the mathematician now had to get out of his "ivory tower" (p. XXXII). Picard's injunctions may have seemed self-contradictory: how was one to resist utilitarianism, nurture theoretical speculations, all the while simultaneously striving to be more involved in society and industrial development? The most striking mathematical developments of the 1920s in Paris can be seen as so many attempts at resolving the conundrum.[86]

[82] See [Heywood 1920], p. 196.

[83] See the chapter on Weiss in [Crawford & Olff-Nathan 2005], pp. 197–204.

[84] About the techniques of soundranging, in particular the so-called Cotton–Weiss method, cf. [Aubin & Goldstein 2014], pp. 34–37; 144–149; 195; 315.

[85] See Proceedings ICM 1921, p. XXVII.

[86] [Aubin & Goldstein 2014], p. 159.

Heywood's article on the Strasbourg ICM passes in silence over the frequent appeals to patriotic feelings in the ceremonial speeches. He simply notes the concert where "the delegates had the pleasure of hearing *'s Elsasslied* sung by the mixed choir of the Concordia-Argentina Choral Society."[87] More receptive for the patriotic side of the Strasbourg event was the Portuguese astronomer Francisco Miranda da Costa Lobo (1865–1945) from Coimbra. He compiled and published a little brochure where French documents from the Congress are combined with his own personal recollections and the Portuguese texts of his addresses on behalf of the Portuguese delegation at the opening and the closing ceremony.[88] In a passage from his introductory text about the German University KWU at Strasbourg[89], da Costa Lobo illustrates the "devastating German mindset"[90] by the brutal attack on Hans von und zu Aufsess, whose summoning of a servant had been misinterpreted as the launching of a francophile protest during the opening banquet of KWU in 1872—a misfortune which the victim considered a bad omen for the German University.[91] As an astronomer he also decries the partly dysfunctional state in which the Germans left behind the observatory, because of "such a *collossal* German" cupola which could no longer be moved.[92]

We have gone into some detail about the Strasbourg events because they stood at the cradle of the first IMU, and thus mapped out what that IMU could, and could not achieve. As for the demise of the first IMU we will be a bit more sketchy, essentially taking the story as far as the Bologna ICM in 1928. The life and death of the first IMU hinged on the global political constellation and its evolution during the 1920s. We shall see in Chapters 5 and 6 below, however, that Mathematics International was in fact thriving outside of, and essentially independently of the IRC and the IMU.

4.4 The Waning Influence of the IMU

As quoted above (Section 4.3.2) the newly founded IMU at its very first assembly in Strasbourg decided to re-initiate the quadrennial rhythm of ICMs which had emerged before the war, and received offers to hold the next ICMs in the US and in Belgium; the US were given precedence for 1924.

[87] See [Heywood 1920], p. 196.

[88] See [da Costa Lobo 1921]. A copy of the brochure is preserved in folder M 119 323 at Bibliothèque Nationale et Universitaire, Strasbourg. About da Costa Lobo cf. [Leonardo et al. 2011].

[89] See [da Costa Lobo 1921], pp. 6–7.

[90] See [da Costa Lobo 1921], p. 5: *espírito devastador alemão*.

[91] The 71-year-old baron actually died from his injuries shortly afterwards, in Switzerland. Versions of the emblematic, oft-told story differ about who were his attackers; in da Costa Lobo's variant the baron was beaten up by two German professors of law. Cf. [Moerlen & Bechelen 1957], p. 7.

[92] See [da Costa Lobo 1921], p. 7; *a cúpula da casa onde está instalado é tão* colossalmente *alemã*.

> The 1924 International Congress of Mathematicians was planned for the United States—Harvard, Princeton, and New York were all possible locations. At the Strasbourg Congress (1920), when L.E. Dickson (University of Chicago), chair of the American Section of the International Mathematical Union (IMU), made the invitation, it was readily accepted. ...
>
> Dickson's invitation was not spur of the moment. Shortly before the Strasbourg Congress, he had given some thought to the matter and canvassed the opinion of all members of the American Section, who agreed unanimously that the invitation should be made. The difficulty, he expected, would be raising enough money because, in the aftermath of the war, Europeans would not likely be able to afford the cost of travel. With this in mind, before leaving for Strasbourg, he wrote to R[obert] S[impson] Woodward [1849–1924] of the Carnegie Institute in Washington, asking if the Institute might see its way to help financially. Although Woodward promised no money, he was optimistic that money could be raised closer to the event, and that various American universities would help offset costs by inviting prominent European mathematicians to give lectures in conjunction with the Congress.
>
> Nevertheless, circumstances surrounding Dickson's invitation for the 1924 Congress are murky. Was the American invitation given with the explicit condition that the meeting be open to all mathematicians from all countries including the Central Powers? This is hardly possible because the IMU would not have accepted such an invitation, which contravened the exclusion policy of the IRC. ... What seems likely is that the invitation was given optimistically, and the details glossed over, in the hope that by 1924 national passions would have given way to the usual cooperative instincts of mathematicians. This is confirmed in a letter from English mathematician G.H. Hardy to Gösta Mittag-Leffler in 1921 describing a visit Dickson had paid to Hardy on his way home from the Strasbourg Congress ... [93]

As time went on, the issue of excluding mathematicians from the Central Powers put increasing pressure on the AMS. However,

> a throw-away suggestion arose during a late-evening informal conversation at the December 1921 meeting of the American Association for the Advancement of Science (AAAS) in Toronto, which [John Charles] Fields [1863–1932] had organized, that the 1924 Congress might be held in Toronto. Fields was immediately keen and quickly went into action. But discussion dragged on among American mathematicians in the months that followed.[94]

In the end, Dickson gave up and the 1924 ICM was organized in Toronto, Canada, in accordance with the rules of the IRC and the IMU, i.e., without participants from Germany or the Central Powers. At any rate, Fields was not prepared to try and raise public Canadian money for German participants, and he was close enough to leading French colleagues to know how they felt on this matter.[95] The meeting was officially called "International Mathematical Congress."

> The Congress met in Toronto by invitation of the University of Toronto and the Royal Canadian Institute, its sessions being held in the buildings of the University. In its organization and the conduct of its proceedings it conformed to the regulations of the International Research Council, and the International Mathematical Union.[96]

---

[93] See [Riehm & Hoffman 2011], p. 129–130; references to the letters are given there.

[94] See [Riehm & Hoffman 2011], p. 131–132.

[95] Cf. [Riehm & Hoffman 2011], chapter 8: "The Politics of Avoidance", pp. 129–145.

[96] See Proceedings ICM 1924, front page.

Fig. 4.4 Leonard E. Dickson during his general lecture at the Strasbourg ICM in 1920, drawn by Eugène-Michel Maeckler.

4.4.1 John Charles Fields

John Charles Fields (1863–1932) from Hamilton, Ontario, Canada, played a major role for Mathematics International which in several ways is analogous to that played by Giovanni Battista Guccia and Gösta Mittag-Leffler, as we saw in Part I. Not as wealthy as these European counterparts—Fields's father had started his own leather store in Hamilton, Ontario—he would continue throughout his life to manage his resources well and live frugally to be able to afford his very frequent journeys to Europe.

He received only his basic education in Canada, and earned his PhD in mathematics from Johns Hopkins University in 1887. After initial teaching positions there and in Pennsylvania, he would spend the years 1892–1900 in Europe improving his mathematical culture; first in Paris, then six months in Felix Klein's Göttingen, and for the final five years in Berlin.

Back in Canada in 1903, he published a report on the Prussian University system.[97] Insisting on academic freedom, the role of the "seminary" and the encouragement of students to do research, Fields's account strikes similar notes as did Fréchet in his Strasbourg lecture discussed in Section 4.3.1 above. For Fields, what he had observed to a certain extent at Johns Hopkins, and then in Göttingen and Berlin would guide his resolute engagement for university reforms in his home country. This is also evident in his opening address to the Toronto ICM:

> We in Canada derive our earlier scientific traditions from Great Britain. More recently we have begun to feel the influence of continental Europe. The founding of the Johns Hopkins University in 1876 marked a new era in the history of universities and science in America. It meant the recognition of the place of research in the university. It meant acceptance of the fact that one of the functions of a university is to train men for research. It meant that the professor was to be encouraged to engage in research.[98]

Fields accomplished the herculean task—which actually started ruining his health—of coordinating the whole organization of the 1924 Congress, going back and forth between Canada and Europe. He returned from the last one of his numerous crossings of the Atlantic only a few days before the Congress began.[99] He was able to raise enough donations,[100] in particular also for travel expenses of European participants.

In the years after the Congress he personally edited the Proceedings, another gigantic workload executed in spite of his faltering health. The two hefty volumes finally appeared in 1928. Extra funds solicited for the printing left a surplus which was evaluated to be 2,500 CAD in February 1931.

By that time, the future of the IMU was already doomed (see Section 4.4.3 below) and rather than getting tied up in hopeless diplomacy about a possible rapprochement between Germany and the IMU, Fields now concentrated instead on his personal project of a Gold Medal for mathematics—or even two Gold Medals, one for pure and one for applied mathematics.

This is actually one of the analogies shared by Guccia, Mittag-Leffler and Fields, that all three of them wanted to sponsor an international medal for mathematics. Guccia realized his project, but the Guccia Medal was only awarded once, in 1908 at Rome—see Section 1.1.5.5 above. Mittag-Leffler had obtained the approval of the King of Sweden for a "Weierstrass Prize" to be awarded in 1916.[101] However, not only was the projected ICM in Stockholm canceled because of the war (see Section 3.4 above), but war-related financial losses made it even difficult for Mittag-Leffler to keep his journal *Acta Mathematica* running in those years.

---

[97] For this paragraph cf. [Riehm & Hoffman 2011], particularly chapter 3. The report is discussed there on pp. 35–40.

[98] See Proceedings ICM 1924, Vol. 1, p. 54.

[99] See [Riehm & Hoffman 2011], pp. 141–145. Cf. de la Vallée Poussin's remarks in his address at the opening, Proceedings ICM 1924, Vol. 1, p. 57.

[100] See Proceedings ICM 1924, Vol. 1, p. 71.

[101] See [Stubhaug 2010], p. 565.

Fig. 4.5 John Charles Fields.

Having secured by the beginning of 1931 the approval of the AMS for a recurring Gold Medal to be awarded at ICMs, Fields spent the Summer of that year meeting mathematicians in Europe. He obtained the backing of the German, French and Italian mathematical societies as well as of the *Circolo matematico di Palermo*.[102]

> Fields was clearly pleased with what he had done. ... [O]n 12 January 1932, he called together a meeting of the Organizing Committee [of the 1924 Toronto Congress!] to report on his reception by the mathematical societies of Europe. At this meeting, Fields proposed that the leftover money (now calculated to be CAD$ 3,209) should be held in trust by government, that medals once struck should be handed over to an accredited International Committee to be set up for the purpose of selecting winners, and finally that "the medal should be as international as possible so that to that end the name of no institution or country should be added to it." Further, "In coming to its decision, the hands of the International Committee should be left as free as possible. It would be understood, however, that in making the award while it was in recognition of work already done it was at the same time intended

[102] See [Riehm & Hoffman 2011], pp. 181–183; they wonder about the conspicuous absence of the LMS from this list.

to be an encouragement for further achievement on the part of the recipient and a stimulus to renewed effort on the part of others." This is the official document which was considered by the IMU council in Zürich in 1932.[103]

After Fields's death eight months later, his estate was added to the endowment of the medal.

4.4.2 "The Disagreeable Tempest which Raged at Toronto"

Let us go back once more to the Toronto Congress itself. Even though it took place only four years after the creation of the IMU in Strasbourg, it already marks the turning point in the short life of that union. When Fields had managed to divert the Congress to Toronto, he accepted the exclusion politics of the IRC and the IMU, if only since there was no viable alternative open to him at the time. The sizeable event, with 444 participants representing more than 30 nations[104] and a substantial harvest of lectures, was an uncontroversial success for Fields, for Toronto, for Ontario, for Canada, for the Americas, ... and by the same token also for the IMU.

And yet: "There was a great storm at Toronto over the question of admission or exclusion of Germans from international mathematical congresses." This sentence is from a long letter, dated 19 December 1924, which the American applied mathematician Edwin Bidwell Wilson (1879–1964) wrote to Émile Picard, one of the Honorary Presidents of the IMU who had not himself participated in the congress.[105] Wilson's epistle is a precious document precisely because its author had no particular sympathies for the Germans. In 1918 for instance he had published a little note in the *Discussion and Correspondence* section of *Science* entitled "Insidious Scientific Control" where one can read:

> In my opinion, whatever country takes care of the preparation and publication of the best reviews of progress in science, and of the best compendiums of scientific knowledge will inevitably be regarded by other countries as an essential for scientific development, and the language of that country will have to be taught to all young scientists. This, again, is subtle control, which may be used for good or bad, according as it is exercised for good or bad motives. That the government of Germany was alive to the possibility of this control seems patent; and that they expected their insidious control to be serviceable to them in swaying opinion in this country in their favor during this war is equally manifest from many points of view.[106]

---

[103] See [Riehm & Hoffman 2011], p. 183.

[104] See Proceedings ICM 1924, p. 48. The precise count of "nations" is a bit subtle, for instance with a League of Nations mandate like Samoa. However, the fact that a few nations, like Russia (RSFSR), Spain, India, and Georgia were officially represented in Toronto, even though they had not yet adhered to the IRC, would be mentioned as a precedent in the political debate about the organization of the 1928 ICM in Bologna; see Proceedings ICM 1928, Vol. 1, p. 7.

[105] For the complete text of the letter and more information about the author we refer to [Siegmund-Schultze 2011]. The title of the present section: "the disagreeable tempest which raged at Toronto", is a quote from the final paragraph of this letter.

[106] See *Science* 48, No. 1246, 15 November 1918, p. 492.

Wilson was thus sympathetic with at least one of the goals of the IRC and its scientific unions: breaking what was perceived as the German monopoly of scientific reviewing. Furthermore, his care for applied mathematics must have been similar to the attitude that Vito Volterra developed during the war.[107] In 1920 Wilson published a textbook on *Aeronautics*—a domain of research that the war had drawn him into—which chooses not to mention the name of Ludwig Prandtl in the brief discussions of boundary layers.[108] By the time of writing his letter to Picard, Wilson had switched from MIT to the Harvard School of Public Health, and from aerodynamics to biostatistics; in his talk at Toronto he presented a controversy between John Maynard Keynes (1883–1946) and Karl Pearson (1857–1936) on probabilities. He had a high visibility in the American scientific community, for instance as editor of the *Proceedings of the National Academy of Sciences*.

We now quote extracts from Wilson's long letter to Picard, starting with his own feelings and about an informal gathering of US mathematicians he was invited to in Toronto:

> If I may do so without impropriety or offence I should like to put before you the following considerations which occur to me. They are personal considerations. I have no official connection with any body which is a party directly or indirectly to this controversy and I don't want any connection with such bodies. I am a poor politician and I am not sure but that academic politics is a poor kind of politics.
>
> Let me say in the first place that I don't like the Germans. I never did like them. That is one reason that I went to France to study when almost all my friends told me I should go to Germany. Second, I do like the French which is another reason I went to France to study. I have some contacts with German science and have made acquaintances in past years both personally and by correspondence with a number of German scientists whom I regard somewhat highly for their scientific contributions, still I am not so eager to meet them at international mathematical congresses as to be led to favor any action which would result in the absence from such congresses of the French, among whom I have more friends and whom on the whole I should much prefer to meet.
>
> There are, however, a great many Americans who have practically no friends in France and have a great many friends in Germany. These persons even when they most strongly detest the conduct of the Germans during the war and when they most severely blame the German intellectuals for signing the famous document[109] that appeared in the early weeks of the war, nevertheless, desire the opportunity at international mathematical congresses to renew their acquaintances among German scientific men. There is a third group in America who are strongly pro-German who not only received their mathematical education in Germany but who so completely absorbed German *Kultur* that they have very little use for French and Italian culture, and would perhaps on the whole prefer an international congress with Germans present and French absent than to go without the presence of the Germans. Thus there are in the main three parties as I see it. A very small minority representing my own point of view who would prefer to do without the Germans if they could see the French; a much larger minority who would prefer to do without the French if only they could see the Germans; and a majority who will not be happy unless arrangements can be made whereby

[107] Cf. [Aubin & Goldstein 2014], p. 262. For Volterra see Section 3.3.1 and the introduction to Chapter 4 above.

[108] See [Wilson 1920]. Prandtl had presented his theory at the Heidelberg ICM in 1904. At the same congress, Wilson gave a talk on vector analysis in which he quoted another paper of Prandtl's twice.

[109] See Section 3.1 above.

> the congresses become thoroughly international in the sense that one may there meet both Germans and French. There was a conference on this matter, a purely informal conference, to which I was invited in Toronto. There was only one man in a group of one dozen of our leading mathematicians who was in favor of taking so strong a stand as to say that he didn't care whether the French stayed away from the congresses or not provided only the rules were so changed that the Germans could come. Everyone else who spoke, and there must have been 8 or 10 who did speak, said that the problem was one of getting both Germans and French to the congresses, not that the French and Germans might associate with each other but that the rest of us might be able to associate first with one then the other as we saw fit. It was the well nigh unanimous sense of the conference that any action which no matter how worded would actually result in the withdrawing of the French and Belgians from these congresses would be most unfortunate and that the real problem was to get both nations represented at the congresses not officially but through the presence of their leading scientific men. Inasmuch as this point of view was so nearly unanimous and inasmuch as I myself would prefer to have the congresses open to all nations I thought it best not to make the statement which I have above made to you, that so far as my own personal preferences went I should rather keep the French and do without the Germans provided I could not have both in attendance.[110]

Fig. 4.6 The first general group picture of an ICM: Toronto 1924.

Having so far expounded the situation purely in terms of personal, cultural or collegial affinities, Wilson nonetheless has to spell out the political implications:

> Now this is as I see it a very serious matter. Before the war the Germans were very numerous in their attendance on congresses. ... This means that for all those persons whose natural attachments... lie with German scientists any congress in which the Germans are not present is really no international congress at all. So long, therefore, as the rules of the International Mathematical Union or of the International Research Council prevent the attendance of Germans at international congresses we can't hope to have any whole-hearted participation in those congresses on the part of a good many American mathematicians ...
>
> ... We can do without our quadrennial mathematical congresses for a number of years if necessary. Or we can have them as we had one in Toronto without participation by the Germans. ... In due course of time it is inevitable no matter what one person or any group of persons may desire that the congresses shall be open to Germans, and it is further inevitable that in due time both Germans and French will participate in the same congresses although perhaps not with any very great intercourse between the representatives of these two nations, and further in due time though perhaps only after 30 or 40 or 50 years it is inevitable that French and Germans will participate in these congresses with more or less

[110] See [Siegmund-Schultze 2011], pp. 117–118. Siegmund-Schultze could identify the "one man in a group of one dozen of our leading mathematicians ..." as being the algebraic geometer Virgil Snyder (1869–1950).

> cordiality one with another just as between 1900 and 1914 . . . The real question, I suppose, that must be decided is whether through the International Research Council an attempt will be made to hasten the time when both nations will be at the congresses or whether the attempt will be made to delay that time and the decision though in some quarters regarded as highly important, will as a matter of fact not be vital for the long range future of scientific cooperation.[111]

In this way Wilson tries to avoid the impression of putting pressure on Picard or on the IMU. But neither *species aeternitatis* nor the human element help to disentangle the political problem, which however he is trying to evade. Thus, after two digressions of a more general historical nature, we finally read:

> Now as I see it the only hope of getting back to reasonably universal and cordial relations among scientific men lies in our exercising a great deal of good taste and charity and keeping out of political entanglements. In a certain sense the International Research Council is political. In this country our state department pays our dues and it would not do so if there were no political aspect to the organization of the International Research Council. I should expect that this political aspect would enable certain persons in power to continue the exclusion of Germans if they so determined and thereby to delay the resumption first of pleasant scientific cooperation between the various groups of scientists in this country with the French on one hand and the Germans on the other, and further delay the gradual re-establishment of amicable scientific relationship between the French and the Germans. I personally regard the organization of the International Research Council as possibly, though not surely, a bad thing for future international cooperation among scientific people. I personally believe that when relations between two parties are strained it is best to have nothing which will add to the group consciousness of either party and to have all arrangements so thoroughly informal and individual that each person of whatever nation comes not as a representative in any way of his nationality but as a scientist with his scientific interests. . . . [112]

So the political blame is finally heaped on the IRC, and Wilson's potential way out of the deadlock is to hope for politically neutral gatherings between scientists. However, in the top-down construction of Science International which had been instituted after World War I, the IMU was subordinate to the IRC. So the only way to fulfill Wilson's vision would be to more or less fall back on the way things were done before World War I: organizing each ICM by itself, independently of structures like the IRC and the IMU.

This is in essence what happened for the remaining ICMs till World War II: in 1928, 1932, and 1936. However, the road to such a renewed normality in Mathematics International turned out to be bumpy. Bypassing the IRC and the IMU required skilled drivers, all the more so as the landscape which opened up before them bore little resemblance to pre-war memories.

The first moves still happened in Toronto during the Congress, at the assembly of the IMU on 15 August 1924. Apparently as a result of the informal gathering that Wilson described to Picard, the US delegates filed the following motion, which was backed by Denmark, The Netherlands, Italy, Sweden, Norway, and the UK: "The American Section of the International Union request the International Research Council to consider whether the time is ripe for the removal of restrictions on

---

[111] See [Siegmund-Schultze 2011], pp. 118–119.

[112] See [Siegmund-Schultze 2011], p. 122.

membership now imposed by the rules of the Council." At the same IMU assembly, Salvatore Pincherle (1853–1936) from Bologna, Italy was elected new President of the International Mathematical Union at age 71, succeeding de la Vallée Poussin.[113]

Pincherle already served in another presidential position, on a national scale; he was the first president of the *Unione Matematica Italiana* founded in 1922. This national union had actually been conceived by Volterra top-down as the mathematics committee of the Italian Research Council placed under the auspices of the International Research Council, rather than emanating directly from the Italian mathematical community. As a consequence, instead of being elected, Pincherle was appointed president of the Italian *Unione* by the *Academia dei Lincei*, in turn presided over by Volterra between 1923 and 1926. This constellation, incidentally, would strain the relations between the new *Unione Matematica Italiana* and the *Circolo matematico di Palermo*. In the Spring of 1925, Pincherle was one of the two mathematicians who signed the fascist *Manifesto of the Italian intellectuals* which was drafted, following the first national *Congress of Fascist Institutions of Culture* in Bologna, by the Sicilian philosopher and former Minister of Public Education Giovanni Gentile (1875–1944).[114] Pincherle also joined the fascist party at the end of 1926, shortly after his first audience with Mussolini about the funding of the Bologna Congress,[115] at a time when Volterra had already fallen out with the regime.[116]

Since we have mentioned Gentile's manifesto, let us add in passing that the only other mathematician, besides Pincherle, who also signed it was the well-known statistician Corrado Gini (1884–1965). In 1926 he would become President of the newly created *Istituto Centrale di Statistica* (known today as *Istituto nazionale di statistica* or Istat). Gini would carry the peculiar alliance between sophisticated mathematical statistics and the Italian state—which we have pointed to for the period of the *Risorgimento* in Section 1.1.5.1—into the twentieth century, and all the way to Italian fascism.[117]

4.4.3 Bologna and the Marginalization of the IMU

As a mathematical researcher Salvatore Pincherle had contributed to the early history of functional analysis.[118] But he is principally remembered today for successfully mounting the 1928 ICM at Bologna. In the process he had to resolve the conundrum

---

[113] See Proceedings ICM 1928, Vol. I, p. 5. Cf. Proceedings ICM 1924, Vol. I, p. 65–66.

[114] See [Guerraggio & Nastasi 2005], pp. 67–73; 90–94. Cf. [Guerraggio & Paoloni 2013], p. 109.

[115] See [Capristo 2016], p. 294. Salvatore Pincherle was Jewish. He died in 1936, and would thus not be affected by the antisemitic fascist laws of 1938, the *leggi razziali*.

[116] On the growing influence of fascism on the Italian *Unione*, which would get more harrowing in the 1930s, see [Giacardi & Tazzioli 2018], [Giacardi & Tazzioli 2019], and [Giacardi & Tazzioli 2021].

[117] See [Prévost 2009].

[118] Cf. [Siegmund-Schultze 1982].

that Wilson had put before Picard. He did so, in the end, by pushing to the side the very IMU of which he was the president.[119]

At first it looked like European politics were moving in the right direction all by themselves. The bundle of treaties negotiated in 1925 at Locarno, principally by Germany's Foreign Minister Gustav Stresemann(1878–1929), Aristide Briand (1862–1932) from France, and Austen Chamberlain (1863–1937) from Great Britain guaranteed the German borders with France and Belgium and opened the way for Germany to adhere to the League of Nations. Germany was admitted as a new member to the League in September 1926, thereby bringing the treaties into effect. The same year Stresemann shared the Nobel Peace Prize with Aristide Briand.

Even before this, on 29 June 1926, the Assembly of the IRC decided to invite Germany, Austria, Bulgaria, and Hungary to join the Council and its subordinate scientific organizations. The politics of exclusion which had governed the very foundation of the IRC and the IMU thus appeared to have come to an end, and preparations of the 1928 ICM proceeded accordingly. The IMU followed its president and formally decided on Bologna as the location of the Congress,[120] and the other organization presided over by Pincherle:

> the Unione Matematica Italiana, on which fell the heavy task of preparing the Congress, decided in view of the new situation to take up the traditions of the International Congresses from before the war, and lift all exclusions depending on political considerations. The Congress was placed under the auspices of the University of Bologna. Its organization was put into the hands of a local committee consisting of professors from this university and distinguished citizens.[121]

Recall that Pincherle was professor at Bologna; he himself took the chair of the committee; his principal helping hand would be his Bologna colleague Ettore Bortolotti (1866–1947). The patronage of the Congress was extended to his majesty the King of Italy. The *Capo del Governo*, i.e., Mussolini, was declared Honorary President of the Congress.

However, as soon as the announcements of the Congress were sent, in their thousands, "to all parts of the world where a mathematical school existed", opposition arose, revealing that the attempt to return to prewar routines was premature. Actors on either side could no longer pretend to have overcome the political legacies of the war and its aftermath.

---

[119] One can still get a lively impression of the tensions Pincherle had to negotiate in the concise account he published in the Bologna Proceedings; see Proceedings ICM 1928, vol. I, pp. 5–10. More recent studies include [Capristo 2016], [Giacardi & Tazzioli 2021].

[120] Rather than Stockholm. The place of the 1928 Congress had been left open in 1924. Cf. [Lehto 1998], p. 44.

[121] See Proceedings ICM 1928, Vol. I, pp. 5–6; my translation.

Fig. 4.7 Salvatore Pincherle about 1900.

4.4.3.1 The German Reaction

The first objections came from a certain faction of the German mathematical community. In reaction to the disastrous outcome of World War I and the ensuing boycott of German science, a majority of German scientists developed a *de facto* political attitude which has attracted the attention of historians (of science), not least because of its peculiar mixture of partially contradictory principles. Viewing German science as one of the last remaining assets of the defeated nation was thus mixed with contempt for the new democratic regime; the fundamental conviction of the transnational and thereby international character of scientific knowledge went hand in hand with a refusal to engage in certain international cooperations; pleas for the apolitical nature of science could go along with the explicit consideration of national interests in scientific matters.[122]

In the early 1920s there was a German project of a blacklist for all foreign scientists backing the boycott of German science. The Göttingen number theorist Edmund Landau (1877–1938) was invited to contribute to this project and identify mathematicians who should be put on the list. He feared the difficulties of coming up with sufficiently reliable information about his foreign colleagues, considering that the whole project had to be "realized as seriously and carefully as it would necessarily be done by German scholars." Nevertheless, he did agree to provide specific information when asked about individual cases.[123]

A serious political cleavage of the German mathematical community became apparent during the second half of the 1920s on two different, yet related scenes: within the editorial board of the journal *Mathematische Annalen* and in the fight about whether or not to follow the invitation to Bologna. In both cases the Dutch and violently anti-French figure of Luitzen Egbertus Jan Brouwer led the camp of German nationalists. Siding with him faithfully in both fights was the Berlin mathematician Ludwig Bieberbach (1886–1982), who would afterwards turn into the infamous leader of a Nazi variant of *Deutsche Mathematik*, based on a racist theory of mathematical creativity.

The rows within the editorial board of *Mathematische Annalen* started in 1925, when a contribution by Painlevé, proposed by Einstein, to a special volume dedicated to the memory of Riemann was refused in view of Painlevé's utterances during the war. The fighting was cut short in October 1928—shortly after the Bologna Congress—when David Hilbert single-handedly fired Brouwer from the editorial board "in view of the incompatibility of our views on fundamental questions." In this *Annalen affair*, political differences merge with controversies about the foundations of mathematics, and the dramatic final coup was also related to Hilbert's health problems at the time.[124]

Insofar as the IRC had from its very foundation been the symbol of anti-German science politics, it is not surprising that it continued to function as a red rag for a majority of German scientists even when the Central Powers had been invited to join in 1926. As a matter of fact, much to the chagrin of German foreign policy makers in the Weimar Republic, the opposition of scientists across all fields would block Germany from joining the International Research Council, and the deadlock persisted even when, in 1931, the IRC transformed into ICSU, the *International Council of Scientific Unions*, whose chief difference with the IRC was that it accepted scientific unions as full members, on a par with nations.[125]

---

[122] This paragraph is freely adapted from [Schappacher & Kneser 1990], p. 54.

[123] See [Corry & Schappacher 2010], pp. 437–438.

[124] The richest account to date of the *Annalen affair* is available in [Rowe & Felsch 2019]. In addition to the correspondence of Otto Blumenthal (1876–1944), the editor in chief of *Mathematische Annalen*, some of Brouwer's correspondence, in particular with Karl Kerkhof (1877–1945), is also reproduced and commented there.

[125] Cf. [Schroeder-Gudehus 1966], pp. 255–265.

It was thus not so easy to smoothly return to a pre-war format of ICMs. The fact that the IMU—at the time still a subordinate structure of the IRC—was officially behind the invitations to the Bologna ICM in 1928 could raise suspicions about this invitation. National feelings in Germany and Austria could also be excited when Pincherle, despite countless exchanges with foreign colleagues, arranged to issue invitations only on behalf of the University of Bologna. One example of this was an excursion offered to Congress participants on 7 September 1928 to Riva di Garda and its electrical power station, in the Province of Trento, which until World War I had been part of Austrian South Tyrol.

Brouwer and Bieberbach won over all their Berlin colleagues, notably Erhard Schmidt (1876–1959) and Richard von Mises, to refuse the invitation, and the *Deutsche Mathematiker-Vereinigung* also abstained from sending a delegation to Bologna. However, David Hilbert and the Göttingen mathematicians, as well as quite a few other German mathematicians did take part in the ICM. In this way the postwar politics of exclusion of German mathematicians would even serve to refuel the decade long competition between Göttingen and Berlin as mathematical centers in Germany. In the end, the mathematicians from Germany participating at the 1928 ICM were 75 men and one woman. The latter, a German mathematician, gave a lecture in the Number Theory section. In the title of her lecture she used her favorite word when thinking about mathematics: *Auffassung*—see Section 6.1.2 below.[126]

A story told by Constance Reid has it that when the German delegation led by David Hilbert entered the hall of the Congress, there was silence first, followed by a standing ovation, and Hilbert delivered on the spot a short address which contained in particular the sentence: "Mathematics knows no races."[127] A handwritten text of such a little speech has actually survived in Hilbert's papers related to Bologna. All this has been scrutinized in [Siegmund-Schultze 2016]; there is no evidence that Hilbert actually gave this speech. Nevertheless, since one has the handwritten notes, one may ponder this text, and in particular wonder about the fact that in 1928 Hilbert insisted not just on the fact that mathematics could transcend national borders, but on the question of race. The most likely explanation for this is the growing anti-semitism that Hilbert could observe in the 1920s, in Göttingen and elsewhere.[128]

4.4.3.2 The French Reaction

German nationalists were not the only ones boycotting the would-be apolitical Congress at Bologna.

> The rules established by the IMU in the immediate postwar period only allowed "scientific groups belonging to member countries of the IRC" to be invited to the International Congresses. These rules—which were already breached at the Toronto Congress . . . —were now recalled, and in the most peremptory manner, when it became known that the invitations

[126] The title was: *Hyperkomplexe Grössen und Darstellungstheorie in arithmetischer Auffassung*, i.e., Division algebras and representation theory from the arithmetic point of view.

[127] See [Reid 1970], p. 188.

[128] Cf. [Siegmund-Schultze 2016], pp. 60–62.

> extended by the President of the Organizing Committee of the future Congress in Bologna included also Germany, a country which had not yet responded to the invitation to join the IRC. "This severe lapse", wrote the Secretary General of the IMU[129] in his letter of 29 May 1928 to the President of the Executive Committee, "renders all these invitations illegal. ... In view of the conditions under which these invitations have been issued it is no longer possible to consider the Bologna Congress as being held by the IMU. Hence, and after having conferred with the President of the IRC[130], it is impossible for me to officially invite our members to the Congress that the University of Bologna will have organized under the presidency of its Rector."[131]

In response to this, Pincherle—who after all was also the President of the IMU—wrote a letter to Picard[132] explaining that the politics of exclusion was no longer acceptable to many colleagues abroad; he mentioned explicitly The Netherlands, Denmark, Sweden, the US, and the UK. He added that this was also the point of view of the Italian government, whose head Mussolini "grants the congress his moral and material support."

In the end, leading French mathematicians like Picard and the IMU Secretary General Kœnigs did not show up at the Congress, but Pincherle had made sure beforehand that others did come, for instance Élie Cartan.[133] Altogether the French delegation counted 56 members.

In his letter to Picard, Pincherle had still expressed the hope that the IMU would hold a regular assembly at Bologna which could discuss the path to take for the future. But since Kœnigs did not cooperate, the gathering which Pincherle convened in Bologna would be considered as informal. Pincherle declared his resignation as President of the IMU on this occasion. In January 1929, William Henry Young—whom we first encountered in Section 1.3.3.3 above—took his place.

Young, his wife Grace Chisholm Young, and their daughter Cecily had participated in the Bologna ICM. He was the Vice-President of the Congress representing England; his wife acted as Vice-President of the broad Section I–B which spread from set theory to quasi-periodic functions. In the following years, Young would engage himself in ways that remind us of Fields's tireless trips (both men, by the way, were born in 1863), in order to keep the IMU alive.

> He decided to make a personal trip to the countries not yet belonging to his Union to persuade them to join it; but on his part it was much more than a visit by the President of an international organization. He saw it as his trip down the map of Europe to meet its leaders and convince them of the importance of his work. He spent part of the summer of 1929 in London arranging not only his visas but also audiences with Kings and Prime Ministers of the countries that he would be visiting. He made his trip in two parts, and at his own expense. The first lasted from September to December, and took in Poland, Austria,

[129] This was Gabriel Kœnigs. Cf. [Capristo 2016] for more comments on the exchange with Kœnigs.

[130] I.e., Émile Picard.

[131] See Proceedings ICM 1928, Vol. I, pp. 7–8; my translation.

[132] It is reproduced at length in a two-page footnote, Proceedings ICM 1928, Vol. I, pp. 8–9.

[133] See [Capristo 2016], pp. 300–304. The article also discusses hypotheses about Émile Borel's absence proposed in [Bru 2003].

> Hungary, Serbia, Roumania, Bulgaria, Turkey, Greece and Italy, while the second took place between April and June 1930, and included Germany, Denmark, Sweden, Finland, Latvia, Estonia and Czechoslovakia.
>
> It was a great personal success for him: he saw ministers and leading mathematicians in all the countries visited, and even secured a long audience with the King in Bulgaria. He also gave lectures in English, French, German or Italian on the aims and ideals of his Union, and took the opportunity to learn something of the languages of countries visited with which he was previously unfamiliar.
>
> A number of countries, especially in Eastern Europe, did in fact join the Union and the Council as a result of Will's efforts; but the main problem, which remained unsolved was, of course, the enrolment of the Germans.[134]

An apparently poorly prepared assembly during the 1932 Congress in Zürich would not manage to unwedge the IMU. An additional adverse effect may have been played by the fact that, according to the 'sunset clause', the Convention under which the IMU had been constituted would lapse on 31 December 1931, unless renewed.[135]

Thereafter IMU was *in limbo*, even if a few colleagues probably still hoped for a resurrection in 1936 which, however, would not take place.[136] At the same time world politics brutally transformed international relations. For example, Hitler would withdraw Germany from the League of Nations in October 1933, but in the sequel the Nazis would consciously try to showcase selected German scientists at international meetings. Mussolini's Abyssinian war started in October 1935, however, would lead to a boycott of Italy issued by the League of Nations and observed in particular by Norway, with the consequence that the Italian government would not authorize Italian delegates to participate at the Oslo ICM in 1936.

Pincherle's realistic and adroit politics had brought about the first ICM with more than 800 active participants, the biggest one to be assembled before 1950. Besides it could conveniently be claimed to be in certain ways 'apolitical', never mind that its coming about was a political feat *par excellence*. It was a resounding success and opened the gate for two other amazingly successful ICMs in 1932 (Zürich) and 1936 (Oslo). All this seemed to corroborate the old idea that mathematicians could very well organize ICMs without being framed by international institutions. This after all was what they had been doing since 1897.[137] However, Bologna demonstrated even more: that, backed by the peculiar coalition of a Locarno spirit and the

[134] See [Grattan-Guinness 1972], p. 175.

[135] We have mentioned this with respect to the IRC in Section 4.1, shortly before Section 4.1.1. The corresponding clause in the IMU statutes was Article 18: *La présente Convention est valable depuis le 1er janvier 1920 jusqu'au 31 décembre 1931. Après cette date, elle sera renouvelée pour une autre période de douze ans, avec l'assentiment des pays adhérents.*

[136] See [Lehto 1998], pp. 50–60. Cf. in particular Henri Fehr's attitude and the history of the ICMI, which we discuss in Chapter 9.

[137] The fact that the official title of the Bologna Proceedings ends with the line: "BOLOGNA 3–10 SETTEMBRE 1928 (VI)" has been occasionally read as a new count of the ICMs according to which Bologna would have been ICM number 6, so that the ICMs in Strasbourg and Toronto—controlled by the IRC and the IMU, and without a German presence—were not to be counted; see for instance [Curbera 2009], p. 89. In fact, the Roman numeral VI added to the year 1928 indicates the 'sixth year of the fascist era' in Italy: 28 October 1927 – 27 October 1928. This sort

national interests of a fascist regime[138], mathematicians from various nations could elude the rules of an International Council and its International Mathematical Union.

From its very conception in 1919—see Section 4.1.1 above—the first IMU had quite a narrow scope which never went substantially beyond ideas for schemes of reviewing the mathematical literature (which never grew to fruition)[139] and holding International Congresses.[140] This has provoked understandable criticism, for instance from Olli Lehto in his history of the IMU: "It is striking how few scientific activities the Union undertook. This lack of mathematical substance was a serious flaw. It played a role in the decline of the Union, which became increasingly obvious from 1928 on."[141] For instance, there was no IMU activity resembling the Encyclopedia project with its strong international aspect (the Encyclopedia itself—see Section 1.3.3 above—was based in Germany, and thus of course anathema for the IMU). Yet, at the Fourth General Assembly of the IRC, on 13 July 1928, the ICM Proceedings were generously acknowledged by Émile Picard as a kind of IMU equivalent to international publication projects of other scientific unions:

> Gentlemen, this is our fourth general assembly and the edifice we have erected in 1919 is now in the ninth year of its existence. I do not know what the future reserves for us, but it is certain that the creation of the IRC has been extremely useful. While before 1914 there were a lot of groupings whose tasks often overlapped, but which where unrelated to one another, the Council has succeeded in catalyzing the creation of a rather limited number of Unions, each dedicated to a corresponding branch of science. They work in the same spirit, and—albeit obeying certain general rules—enjoy great freedom of action. Their efforts, at least for some of them, have been very fruitful. It is with legitimate satisfaction that we are reading the important volumes published by the Geodesic and Geophysical Union, the Astronomical Union, the Chemical Union, and others. And even Unions dedicated to sciences which are much less prone to collective work have published highly interesting studies, such as the International Union of Mathematicians in its meetings at Strasbourg and Toronto.[142]

The Bologna Congress opened just three weeks after Picard's speech. The six big volumes of its Proceedings would no longer be claimed in the name of the IMU or the IRC. They mention the Council and the Union only as annoying political hurdles during the preparation of the Congress.

---

of calendrical political alignment had been mandatory in Italy since 1927. Personally, I am not aware of any public attempt to play with the numbering of ICMs before Hermann Weyl's exuberant speech at the Zürich ICM in 1932—see Proceedings ICM 1932, Vol. I, p. 71.

[138] The crucial importance of the fascist backing is highlighted by the fact that Pincherle thanked Mussolini for his help and informed him of the success of the Congress in a letter dated 11 September 1928, i.e., the very first day after the closure of the event. See [Giacardi & Tazzioli 2019], pp. 40–41; cf. the discussion in [Giacardi & Tazzioli 2021] of the echo of the Congress and its fascist connections in the *Bolletino* of the *Unione Matematica Italiana*.

[139] Cf. Section 6.3 below.

[140] The International Commission on Mathematical Instruction ICMI—whose creation in 1908 we have briefly mentioned in Section 1.4.1.2—played a slightly special role though. Cf. Chapter 9 below.

[141] See [Lehto 1998], p. 33.

[142] See [Schuster 1930], p. 2; my translation.

Based in part on his unsuccessful search for documents relating to the IMU in the IRC archives, Lehto concludes his reflection by writing: "In all, the old IMU had poor visibility within the International Research Council and was not well known among mathematicians."[143] In contrast to this, I have tried to show that both the IRC and the first IMU were in fact widely known among mathematicians in the 1920s, on both sides of the divide brought about by World War I. However, their reputation was profoundly and indelibly political, rather than scientific.

143 See [Lehto 1998], p. 33.

Chapter 5
Philanthropic Capital for Mathematics

The US-based activities that we briefly recall in this chapter are an important reminder, within the crystalline sphere of scientific endeavors, of the truly global transformation that World War I had wrought:

> In November 1918 Germany's planned economy surrendered in the face of a second even more powerful economic vision—a triumphant model of 'democratic capitalism.' At the heart of the democratic war effort stood the much-heralded economic potential of the United States. World War I marked the point at which America's wealth stamped itself dramatically on European history.[1]

After the war, American philanthropy was often able to provide what European states could no longer afford. This held true for countries on both sides of the war frontlines:

> I expected to find some outstanding differences between the victors and the vanquished in the late war, at least in so far as economic state, after-war national psychology, etc. might affect the higher education in these countries; in this respect I was quite wrong for some of the victors seem to be in quite as bad a state as any of the vanquished.[2]

The support of scientific projects was realized and acted out according to the principles of American philanthropy and guided by the US scientific perspective and expertise. Indeed, American philanthropy had begun its tremendous works of donation well before the first World War. It had already marked its durable imprint on the academic landscape of the US in the nineteenth century. In Chapter 1 we had several occasions, for instance, to mention the university at Baltimore, which had been endowed by Johns Hopkins. What interests us here, however, is how such large scale private donations to science went international after World War I.

As far as mathematics is concerned, rich new sources of international support began to flow in the mid-twenties, i.e., at about the same time as the relevance of the IMU for Mathematics International started to dwindle. Focussing on research training at the highest level they single-handedly set a new standard for the intercontinental framing of scientific excellence. Scientific ideals which had been built up and

[1] See [Tooze 2014], p. 200.

[2] Augustus Trowbridge as quoted in [Siegmund-Schultze 2001], p. 56.

N. Schappacher, *Framing Global Mathematics*,
https://doi.org/10.1007/978-3-030-95683-7_5

cultivated in the USA, largely with European examples in mind, now made a strong reappearance in the applications for funding submitted by Europeans. As humanity moved closer to World War II these philanthropic resources were malleable enough to be increasingly used to relocate refugee scientists in America.

5.1 The Rockefeller Philanthropies

In the context of the present book it is legitimate to focus on visible, international effects of US philanthropy for mathematics. It is also requisite, since no history of mathematics of the 1920s and thirties would be adequate if it failed to record the impact of financial support on international mathematical networking, especially that provided by the Rockefeller Foundation. The only systematic study to date of the Rockefeller Foundation's activities in the domain of mathematics is the book [Siegmund-Schultze 2001].

The vantage point of the history of mathematics does not project a fair image of American philanthropy in general,[3] nor does it duly capture the global scale of an organization like the Rockefeller Foundation—mathematics was lagging behind in globalization. Already *The Digital History* offered today on the Foundation's website gives a first impression of the true breadth of the activities at the time. Activities to improve health care, both in America and on other continents, stand out, but archaeology, literature and theatre also show up. The activities for mathematics, which did not have its own explicit subheading, fall into the category of *Natural Sciences*.[4]

Yet there was one period in the history of the Rockefeller Foundation when mathematics was treated in practice as if a dedicated line of expenditure for it existed. This was when the grants were handled by the Foundation's *International Education Board* (IEB), created in 1923. The time spell ended in 1931. Afterwards the financing of mathematics went over to the Rockefeller Foundation, which survived, unlike the IEB, but accepted only a relatively small number of mathematicians. Even so, among these later grantees one finds well-known mathematicians of the twentieth century, for instance the winner of one of the first two Fields Medals in 1936, Lars V. Ahlfors (1907–1996), the Polish logician Alfred Tarski, and the British algebraic geometer John Arthur Todd (1908–1994). On the other hand, applications in the 1930s by outstanding mathematicians such as Andrey Nikolaevich Kolmogorov (1903–1987) or re-applications by former IEB fellows like Stefan Banach (1892–1945) and Bartel Leendert Van der Waerden were dismissed, sometimes officially on the grounds that there was no specific program for this field of knowledge. On the other hand, as the

[3] For a general, political history of American philanthropy, see for example [Zunz 2012].

[4] See [URL 07]. Even though there is no special section dedicated to mathematics, browsing the site one does find a few related documents, for instance about John von Neumann, or on Vannevar Bush's *Differential Analyzer*.

1930s wore on, some mathematicians could profit from Rockefeller grants given to political immigrant scientists in the US.

Let us introduce the IEB in general by quoting from a portrait penned in 1941 by way of a summary of the book [Gray 1941]:

> The International Education Board was set up in 1923. The idea behind it was to help to make good some of the ravages of the War of 1914–18. The money, which amounted in all to nearly twenty-eight million dollars, was provided by John D. Rockefeller, jun., who ... imposed no conditions on the manner in which it should be spent, except that it should be used for 'the promotion and advancement of education throughout the world.' The inspiration with regard to the policy which should be followed came almost entirely from ... Dr. Wickliffe Rose [1862–1931][5].
>
> What is education? In Rose's mind it became for the most part, not the dissemination of certain accepted ideas and cultural patterns, for that he felt might well be left to the various national Governments, but the desire to forward the understanding of the natural world by the best possible means. The claims of educational training, particularly training for agriculture, were not overlooked, but they played a subsidiary part in the comprehensive scheme which he put forward for the support of the best research institutions and the most promising scientific workers, whose work was being held up for lack of funds. ... [6]

In other words, at least for the natural sciences, the meaning of *education* in the name of the IEB was narrowed down to the most advanced sense, i.e., education towards high-level research. The funding for institutes was conceived accordingly as providing solid structures for research training. In particular, despite its name, the International *Education* Board did not till the same soil as the ICMI did in the field of mathematics—cf. Chapter 9 below.

> No considerations of national prestige were allowed to stand in the way, and except for agriculture, no attempt was made to strike a balance between the competing claims of the different branches of science, for in Rose's view, 'all knowledge is inter-related, and if we help in any one field we help in all the others.' So it came about that the greatest scene of the Board's activities lay in Europe, including the British Isles; but a small number of individual projects in the United States received some of the largest grants, while smaller ones found their way to such places as South Africa, China, the Philippines and New Zealand.
>
> In all, 'fifty-seven universities, research centers, and other institutions were provided with new buildings, equipment, endowment and other material aids; and 603 individuals, chosen for their promise of future usefulness, were assisted in their higher education, given opportunity to study under world authorities in their chosen fields, introduced to new pastures of research under conditions which at the time seemed favourable to their development. Through grants for these various purposes, thirty-nine countries, representing Europe, Africa, Asia, Australasia, and the Americas, were aided.'[7]

Thus the IEB aid was essentially spent on two different kinds of projects: personal stipends and the funding of constructions for outstanding research centers. The question arises how IEB went about choosing the persons and institutions to be supported.

---

[5] On W. Rose, cf. [Siegmund-Schultze 2001], pp. 27–30.

[6] See [Weatherwall 1941], p. 398.

[7] See [Weatherwall 1941], p. 398.

A visit of Dr. Rose to Europe in 1923 initiated a scheme under which the whole world, but particularly war-worn Europe, was scoured for young scientific workers showing exceptional promise, whose studies were held up through lack of means. After careful scrutiny these were granted travelling fellowships for a year, which enabled them to profit by the best scientific experience available in the world in their own particular line. Within the five years, 1923–28, an exchange of workers and of scientific ideas took place on an unprecedented scale.

But this scheme of fellowships in science would have been held up by the cramped facilities existing in many of the leading research institutions. Realizing this, the International Education Board made available large sums to be spent upon buildings, equipment and endowment. One of the first institutions to benefit in this way was the Institute of Theoretical Physics at Copenhagen, under Niels Bohr [1885–1962]. . . . [8]

Starting with Rose's European journey, the IEB began—in the domain which interests us here—to design a map of mathematical Europe, or rather, of the Europe of mathematics and physics; of major persons and centers in the principal countries.

Both for the UK and for the overall constellation of European mathematics, Godfrey Harold Hardy was a key person to talk to.

The leading English mathematician Godfrey H. Hardy had been one of the first European scientists to be contacted by Rose during his trip to Europe in the fall of 1923. The two men met in Oxford, on December 23, 1923, and Rose got advice about promising mathematicians in Europe but no request proper from the English side. This changed when Rose got back to England, shortly before leaving Europe, and met Hardy in London, once again, on April 14, 1924.[9]

Hardy's role as correspondent for the IEB goes well beyond his own research fields, analysis and analytic number theory; it fits in with his outspokenness in favor of Mathematics International since World War I. Already during the war Hardy sternly refused to transport national preferences into scientific life. For instance, the Latin dedication of the joint book [Hardy & Riesz 1915]—the final manuscript had to be finished by Hardy himself; correspondence with his Hungarian coauthor was increasingly difficult—translates:

To the mathematicians (how many and wherever they may be): that they may soon again take up, as is to be hoped, the confraternity of their works which is currently disrupted, we, the authors, friends and foes at the same time, present and dedicate [this book].[10]

The war years put him at odds with most Cambridge colleagues. In 1919 he accepted the Savilian chair in Oxford. This is where Rose first met him to hear his views on mathematical Europe. It was also from Oxford that Hardy intervened in or commented on many of the correspondences about the exclusion policy of the IRC,

[8] See [Weatherwall 1941], p. 398. This article was written during World War II; on p. 401 one reads: "some of the work of the Board is already in ruins." The piece ends on a disillusioned note, timidly hoping for a brighter future after the war.

[9] See [Siegmund-Schultze 2001], p. 40; cf. pp. 247–249 for Hardy's note addressed to IEB at the second meeting, on behalf of the London Mathematical Society.

[10] Cf. [Corry & Schappacher 2010], p. 435. For Hardy's own account of World War I in Cambridge, see [Hardy 1942]. Cf. June Barrow-Green's chapter in [Aubin & Goldstein 2014], pp. 59–124.

the IMU, and the first post-war ICMs which we have mentioned in the preceding Chapter 4.

As of 1925 an office in Paris was established under the direction of the rich polyglot Augustus Trowbridge from Brooklyn, New York, who had obtained his PhD in Physics at the University of Berlin in 1897. His regularly kept diary ("log") is one of the central sources exploited by Siegmund-Schultze for the book [Siegmund-Schultze 2001]. Trowbridge was well-connected with the Paris scientific milieu. And for situating the merits and needs of European mathematics at large he could rely on reconnaissance missions undertaken by the leading American mathematician George David Birkhoff (1884–1944), the father of Garrett Birkhoff (1911–1996).

> ... George David Birkhoff travelled to Europe together with his family in the second semester of the academic year 1925/26 (probably starting in February 1926). He had planned a shorter stay in Europe within a sabbatical year, but stayed several months longer (until September 1926) on the basis of the support given by the IEB. Birkhoff chose France as his temporary home and country of departure for various trips to several European countries. In Paris he collaborated closely with American physicist Augustus Trowbridge, who was heading the European office of the IEB in the city. At the end of his journey, on 8 September 1926, Birkhoff submitted to Trowbridge a 12-page-long "Final General Memorandum for Dr. A. Trowbridge."[11]

In terms of physical constructions, these explorations resulted in two new buildings for mathematics, both granted by IEB on the same day in December 1926, and both inaugurated in 1928: the *Institut Henri Poincaré* (IHP) in Paris, and the *Mathematisches Institut* of Göttingen University.[12] Each one of them was apparently seen by the American donors as a contribution to a scientific campus. This vision fit reasonably well with the pre-existing buildings for physics, fluid mechanics, and chemistry in Göttingen near which the new Mathematical Institute was built. Also the *Institut Henri Poincaré* found itself close—in fact, very close—to other institute buildings (like chemistry and oceanography) that had recently been finished. This condensed Pierre and Marie Curie 'campus' is in the vicinity of Sorbonne University from the turn of the century, which is architecturally much more confined and squarely occupies full city blocks with internal courts. Lecture halls, a rich library, reading and seminar rooms, and also collections of mathematical models, were the visible assets for research training in both new institutes financed by the IEB.

Already in the Summer of 1924, Gösta Mittag-Leffler's application for IEB funds to insure the survival of the *Institut mathématique Mittag-Leffler* had been turned down. He and his wife had decided to set up a foundation around the extraordinary mathematical library in the generous villa at Djursholm, outside of Stockholm,

[11] See [Siegmund-Schultze 2001], p. 46; Birkhoff's memorandum is reproduced there on pp. 265–271.

[12] See Chapter V of [Siegmund-Schultze 2001]. I have also greatly profited from an inspiring lecture comparing both buildings from the point of view of the history of architecture, delivered by Bernd Hoffmann, Göttingen, at the eightieth birthday celebration of the IHP in 2008.

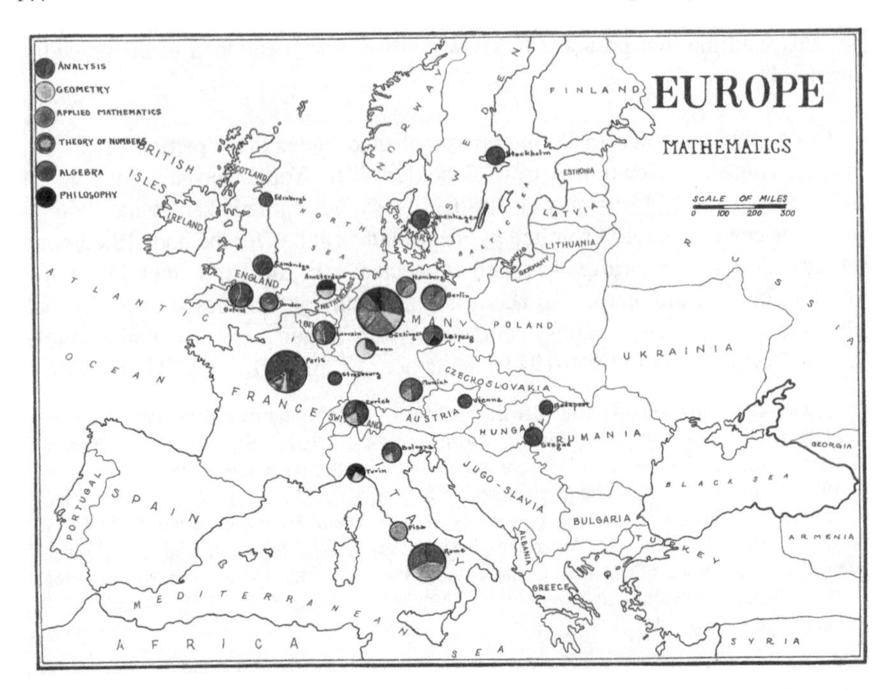

Fig. 5.1 Rockefeller map of mathematical centers in Europe, 1927. The colours indicate the various branches of mathematics: analysis, geometry, applied mathematics, theory of numbers, algebra, philosophy. Credit: [Arch. RAC].

Sweden. The *Institut mathématique Mittag-Leffler* had been formally created in 1919. The main reason for the IEB to abstain from helping this splendid site was apparently its isolated situation.[13]

The denial to fund Djursholm in contrast highlights the IEB's expectations to the effect that the grants for the IHP and Göttingen would create attractive and lively research centers. Paris was of course well chosen in this respect already because of the city's attractiveness for students, also from abroad; in fact, Rockefeller money also went into the construction of the central building of the Paris *Cité universitaire*, a complex of international student residences in the South of the capital. Mathematically, the newly founded IHP would play a particularly visible role in the development of mathematical statistics and probability theory.[14] As to Göttingen, the proximity between mathematics and physics seems to have played an important part in convincing the IEB to invest in this place. In Section 6.1 below about Emmy Noether's legacy, we will analyze the peculiar purely mathematical message which young researchers would pick up there and spread in the 1920s and early 1930s.

[13] See [Siegmund-Schultze 2001], pp. 178–180.

[14] See [Catellier & Mazliak 2012], and [Siegmund-Schultze 2001], pp. 169–175.

As to the individual grants extended to mathematicians, between 1924 and 1931, the IEB financed research sojourns of a total of 86 predominantly young mathematicians; three women and 83 men. Their fields of interests ranged widely, from logic, via all principal domains of pure mathematics, to applied fields like aerodynamics and statistics. Since Rockefeller grants for mathematicians became rather the exception after 1931, only the IEB period represents a fair measure of the internationalizing effect of Rockefeller money for mathematics. In terms of nationalities, 16 Germans, 14 Americans, among them 2 women, 11 Frenchmen, 6 men from Poland, 5 men and one woman from the USSR, 4 Austrians, 4 Czechs, 4 Hungarians, 4 Swiss; 3 men each from Holland, and the UK; 2 men each from Italy, Norway, Romania, and the Kingdom of Yugoslavia; and one mathematician each from Finland, Greece, and Japan received IEB grants.[15] So we are looking essentially at an affair between Europe and the US. Since they are so strongly represented, let us take a quick look at the IEB grant recipients from Poland and the USSR.

The Polish mathematicians who received IEB grants were the emblematic Stefan Banach, the topologist Witold Hurewicz (1904–1956), the analyst Szolem Mandelbrojt (1899–1983)—who had actually been based in France since 1920, would obtain French citizenship in 1927, and become Hadamard's successor at *Collège de France* in 1938—, the famous statistician Jerzy Neyman (1894–1981), the expert in fluid mechanics (and diplomat in his later years) Piotr Szymański (1900–1965)[16], and the analyst Antoni Zygmund (1900–1992). All of them went to Paris at least for part of their grant, except Hurewicz who spent the academic year 1927/28 in Amsterdam, hosted by L.E.J. Brouwer.

The Russian topologist Pavel Alexandrov (1896–1982) also spent a year (1925) in Holland, welcomed by Brouwer,[17] and in 1826–27 he was granted 8 months in Princeton,[18] invited by Solomon Lefschetz (1884–1972). Alexandrov's former teacher Luzin, in spite of the fact that he could no longer claim to be a young researcher, was finally granted a stay in Paris in 1928, one year after his former student Dmitrii Menshov (1892–1988). While Menshov was recommended by Arnaud Denjoy (1884–1974) and Paul Montel (1876–1975)—as well as by his teacher Luzin, Luzin was backed by Lebesgue; yet he had to try twice before he was admitted. In his French application, which is apparently difficult to translate, Luzin concentrates on set theory, adopting the point of view of *naming infinity*, which we have briefly touched upon in Section 2.1.1 above.[19] The same year Luzin also participated at the Bologna ICM where he sketched his take on the foundational debate.[20]

---

[15] See [Siegmund-Schultze 2001], pp. 288–301, for the total list of 130 mathematicians known to have either received IEB grants or to have been sponsored by Rockefeller grants later in the 1930s; see pp. 96–106 for remarks on the lucky and some of the less lucky applicants.

[16] See [Urbanowicz & Tijsseling 2016].

[17] For more background on this stay, in particular Pavel Urysohn's (1898–1924) work and tragic death and Emmy Noether's role, see [Rowe 2021], pp. 109–120.

[18] This is Princeton University. The *Institute for Advanced Study* did not exist yet—see Section 5.2 below.

[19] See [Siegmund-Schultze 2001], p. 250. Cf. [Graham & Kantor 2009], esp. pp. 205–211.

[20] See Proceedings ICM 1928, Vol. 1, pp. 295–299.

The only non-American woman among the IEB fellows, Nina Karlovna Bari (1901–1961)[21] had also been a student of Luzin's. She had profited from the opening of the universities for women in 1918 as a consequence of the Bolshevik Revolution and was in fact the very first woman to graduate from Moscow State University. Nina Bari would become full professor there in 1932. At the Bologna ICM in 1928 she presented in a sectional talk the peculiar result to the effect that every continuous function on a real interval is the sum of at most three functions of the form $f \circ \phi$, with both f and ϕ absolutely continuous. Thanks to the IEB, Nina Bari could spend nine months in Paris in 1929.

The remaining two IEB fellows from the USSR were: Abram Samoilovitch Besicovitch (1891–1970), who used his IEB fellowship as a stepping stone towards his future career in the UK, and the complex function theorist Vasilii Leonidovitch Gontcharov (1896–1955), who would later be known in the USSR for his elementary textbooks.[22]

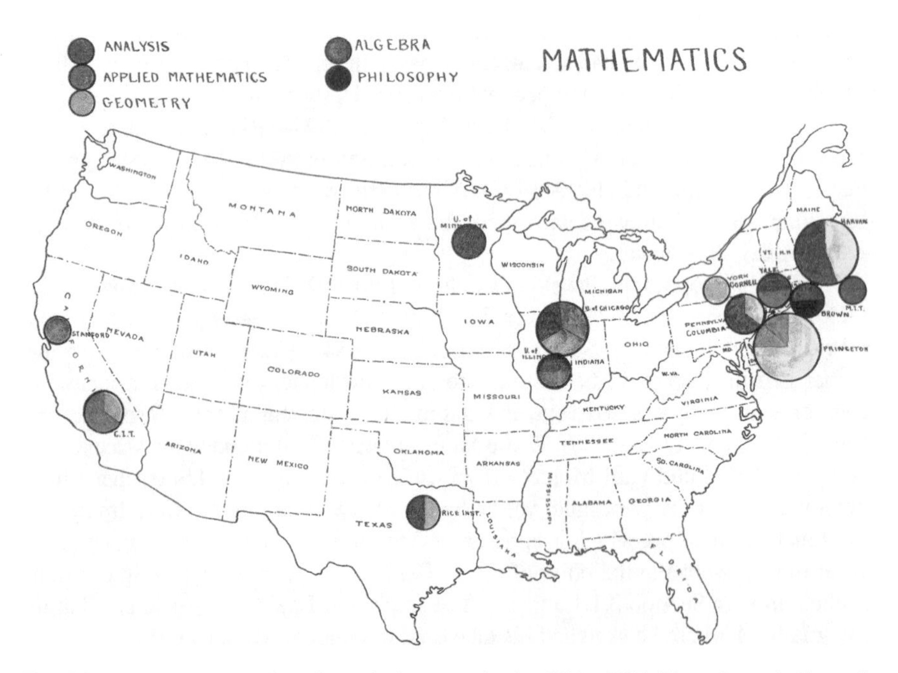

Fig. 5.2 Rockefeller map of mathematical centers in the USA, 1927. The colours indicate the various branches of mathematics: analysis, geometry, applied mathematics, theory of numbers, algebra, philosophy. Credit: [Arch. RAC].

[21] Both in the Italian Bologna Proceedings and in many of her publications in Western media, her name is transliterated as 'Bary', even though her Russian name ends with a single plain letter 'i.'

[22] Cf. [Siegmund-Schultze 2001], pp. 125–132, for a general discussion of IEB's perspective on Soviet Russia.

Before closing this subsection, we ought to remind ourselves that the Rockefeller Foundation was only the biggest actor in a broad field of philanthropic initiatives. This is obvious in the realm of mathematics when one looks at fellows of the *John Simon Guggenheim Foundation*, even though these grants, which started roughly at the same time as those of the Rockefeller Foundation, were limited to US citizens or residents. Their list for the second decade, 1935–1945, includes some of the most influential mathematicians of the twentieth century, such as Paul Erdős (1913–1996), Marshall Harvey Stone (1903–1989)—who would play a dominant role in recreating the IMU after World War II—, as well as the rewriters of algebraic geometry Oscar Zariski and André Weil.[23] Many of these later grants concern mathematicians who had emigrated to the US before. This will be put into perspective in Chapter 7 below. Actions of the Rockefeller Foundation in the 1930s will also be recalled coincidentally in Section 6.1.

For the time being, let us summarize for the record that the activities of the IEB and similar philanthropic foundations in the domain of mathematics led to the construction of two mathematical institutes, in Paris and in Göttingen, and provided generous help to a fair number of researchers. Since these actions were embedded in a very broad range of philanthropic activities, sweeping the spectrum of scientific domains and sometimes encompassing geographic regions which were not on the mathematical map yet, one may say that philanthropic convictions and a US-based analysis of the scientific world did more to promote a certain spirit of Science International between the two World Wars than any other international institution. Furthermore, they were major novel steps in pushing scientific practice towards the constant collegial interaction beyond local contacts that we take for granted today.

Indeed, foreign travel has been a recurring theme in our survey of the nineteenth century world of mathematics—see Chapter 1. For well established or wealthy actors such as Guccia, Klein or Mittag-Leffler it was a natural part of their networking strategies. Young researchers on the other hand like Betti, Brioschi, Casorati (Section 1.1.5.3) undertook their journey in order to discover the world of mathematics. Sofya Kovalevskaya had to leave her home country in order to study in the first place, and then again to embark on her academic career—see Section 1.1.7. But all those journeys were private initiatives. Typically, at least for the younger participants, participating in international congresses, with the resulting contacts and exchanges, had to be arranged privately. The idea of helping promising young mathematicians by systematically granting them the opportunity to spend time at a suitable institution only gained ground between the World Wars, as a relatively late ingredient of the ongoing professionalization of science. On the national levels it typically expressed itself through newly founded National Research Councils. The international dimension, at first covering mostly Europe and the US, was opened up by American philanthropic initiatives.

[23] See the complete list of Guggenheim fellows in mathematics up until 1945 in [Siegmund-Schultze 2001], pp. 302–303; see also pp. 138–139.

5.2 The Institute for Advanced Study, Princeton

Like the activities of the IEB, the founding of the Institute for Advanced Study (IAS) at Princeton in 1930 was also the result of American capitalist philanthropy. Here too a model institution for worldwide science came into being.

> [T]he Institute holds a special symbolism for mathematicians. The Institute for Advanced Study began in 1930 through the vision of Abraham Flexner [1866–1959]. Flexner was a figure of considerable influence during the first half of the twentieth century. He made his mark in 1910 with a scathing exposé of the deficiencies in American medical education. Flexner's revelations called for drastic action. Over a decade-long period he served as the architect of a Rockefeller philanthropic initiative that dramatically upgraded American medical schools.
>
> When Flexner retired from the Rockefeller Foundation, it was with the satisfaction that his career had been essential to the modernization of American medicine. Still, he had a distinctly different ambition that remained unfulfilled. As a long-time observer of higher education, Flexner was convinced that the United States should possess an exclusively graduate university with an ideal environment for research. There, a small faculty of geniuses would direct the studies of a few disciples while pursuing their own discoveries.
>
> With the power to direct millions of dollars to selected universities and hospitals, Flexner had accumulated a stunning collection of contacts among academic, business, medical, and political leaders. When department store magnate Louis Bamberger [1855–1944] and his sister Carrie Fuld [1864–1944] began seeking advice on devoting their fortune to the creation of a new medical school, it was inevitable that their consultations would lead them to Abraham Flexner. Out of these discussions Bamberger and Fuld decided to endow a graduate university with the 63-year-old Flexner as director.[24]

Flexner worked hard—trips to Europe included—to recruit outstanding scientists with international visibility on the faculties of the first 'Schools' he was planning for the Institute. Even though Flexner was personally as clueless about mathematics as Alexander von Humboldt a century before him, he set his mind on building a School of Mathematics. Solomon Lefschetz from Princeton University recommended to go after "the younger group of geometers. It is the most vital and promising of mathematical groups in the U.S., the one with the highest national and international standing. It includes [Oswald] Veblen [1880–1960] and [James W.] Alexander [1888–1971] of Princeton, [G.D.] Birkhoff and [Marston] Morse [1892–1977] of Harvard and also myself." Furthermore, Lefschetz remarked: "Hermann Weyl is the only mathematician anywhere definitely above these names. But as he occupies the most distinguished mathematical chair in the world (in Göttingen) I do not see him giving it up."[25] Whereas Birkhoff decided to stay at Harvard, Veblen joined the new institution in 1932, and the topologist Alexander followed in 1933. John von Neumann (1903–1957) was also hired in 1933. Morse would join in 1935. In terms of international luminaries, Flexner landed a brilliant success with Albert Einstein who arrived in 1933. It was Adolf Hitler's regime which finally decided the hesitating

[24] See [Batterson 2006], p. ix.

[25] See [Batterson 2006], p. 59, for these quotes from Lefschetz.

Weyl to leave Göttingen for Princeton in the Fall of 1933; the Nazi rule not only clashed with Weyl's democratic convictions after the 17 years he had spent in Zürich, it also potentially threatened his Jewish wife, and children.[26]

Thus within a few years, this new institute was one of the strongest mathematical centers in the world, and we shall see in later chapters how prominent a role it would play on the international scene of mathematics. But it is not so much the individual IAS which interests us, and how it continued to hold its eminent place for global mathematics. If we discuss its founding here, it is because this was in fact the birth of a new type of research structure, which would subsequently serve as a blueprint for similar centers founded all around the globe since the end of World War II. So what was this institute like, once its structure had crystallized? How can the IAS, in particular its School of Mathematics, be described once it was set up and functioning; for instance, when it had moved from its first, provisional quarters in Princeton University, to Fuld Hall in 1939? The original idea of a graduate school never materialized, and was apparently abandoned early on: "From the beginning the faculty identified and hosted scholars who had already received their doctoral degrees. These visitors, who became known as members, typically remained at the Institute for a year."[27]

Here is a concise description of this new kind of research site whose very first example was the IAS's School of Mathematics: It is a (relatively) independent academic structure (even though it may have collaborative ties with local academic institutions, for instance universities, nearby). Its goal is to support fundamental research in mathematics (and possibly also other domains) at the highest level of intellectual inquiry. A rather small permanent faculty selected for their outstanding research record guides the work, and each year up to about seven times as many visiting members are invited to join the Institute, from universities and research institutions throughout the world. Every researcher, permanent or visiting, is free to pursue their personal research agenda. Seminars are organized, talks are given, which reflect ongoing work. In addition, the institute offers numerous informal occasions for exchange.[28]

We will call an institute that fits this mold a *Locally-grounded Transnational Research Site for mathematics*, or LGTRS for short. This is our adaption, specifically for the domain of mathematics, of a notion that historians-sociologists of (predominantly experimental) science have coined to capture important elements of the professionalization of scientific research in the second half of the twentieth century:

> Before World War I, the threat of cognitive fragmentation came from the massive and rapid introduction of additional specialties. Today this menace is linked to the existence of several representation systems inside almost every major scientific discipline. Within

[26] For many more details packed in an entertaining narrative, cf. [Batterson 2006].

[27] See [Batterson 2006], p. x.

[28] Some of the formulations of this paragraph are slightly adapted clippings from the website [URL 08]. The approximate factor of 7 linking the number of permanent faculty and invited members corresponds to the current situation at the IAS. It is cited here as an indication only; I will not just use any precise value in order to rule out a potential LGTRS.

a given field, scientists choose between numerous alternative ways of representing their phenomena. ... Some groups emerge within disciplines whose principal loyalty lies with their chosen representation rather than the field. This dramatically affects the pattern of work since scientists sharing a representation system often forge more meaningful and stable intellectual relations with colleagues of the same representation than they do with their specialty or home-laboratory. Moreover, effective use of a new representation to solve a particularly thorny problem, or strategies to gain legitimacy for a novel representation system spur scientists to band together for a period as they shift for a short while their research to a new site. ... Today, scientists have ample opportunities to seize advantages lying outside their laboratories without the need to transfer to a new laboratory, agency or nation. Most research agencies provide funding and sometimes encouragement for short stays in an alternative laboratory. ...

In the 19th century scientific travel was a form of ambassadorship or assumed the form of brief laboratory visits intended as fact-finding missions. In the 20th century, however, scientists travel either to cooperate with colleagues or to carry out gritty research that is better done with resources located away from their customary base of operation. The logic of LGTRS is hence thoroughly functional.

This leads to the emergence of new ties. Scientists' bonds with their institutional base is supplemented by an additional network consisting of individuals and laboratories scattered around the globe. In some instances, involvement with transnational groups, projects, and institutions becomes overriding, thereby neutralizing affiliation with the home-setting. Here, the local/national coordinate system is countered by the appeal of LGTRS. ...

Yet, to portray the relations between LGTRS and nation-based research as antagonistic would be to misunderstand this recent and crucial phenomenon. LGTRS are not a professional, cognitive or educational alternative to national science. They constitute an incremental resource as scientists attempt to expand and multiply strategies and techniques for problem-solving. The LGTRS dovetail the local, regional and national endeavors. Scientists operate simultaneously and on the three planes in complete comfort and without the slightest sense of contradiction or alienation. The salient feature of this new aspect of research practice and organization is oscillatory movement of individuals away from and going back to their home-base. Centrifugal and centripetal trajectories succeed one another as required by the research projects.[29]

This analysis clearly takes into account the importance of experimental devices such as a Hadron Collider, a supermagnet or the like. It nonetheless also describes very well the crucial changes that have affected professional mathematical research as of the middle of the twentieth century. These novel features, which were first realized at the Institute for Advanced Study and may today seem banal (at least in the countries that are fully integrated into the world mathematical community), were intrinsically international.

The ambience in Princeton, which is still fairly cosmopolitan, was even more so in 1937. The Institute for Advanced Study did not yet have its own buildings; the University provided it with comfortable facilities in the old Fine Hall, to which Veblen had devoted so much care, but guests such as I were left to their own devices as far as housing went. Such stays are fruitful, but the experience has become such a common one that any remarks I could make would be superfluous. As planned, I gave a series of lectures on the topic of my future

[29] See [Crawford et al. 1992], pp. 28–30.

> paper in the Journal de Liouville[30], and it was no small boost to my ego to see Hermann Weyl among those who attended regularly. Through contact with Alexander, I tried to find out more about "combinatorial topology"... [31]

Flexner's original plan of a graduate school of mathematics for the IAS was fairly close to the kind of institutes that the Rockefeller Foundation helped building through its *International Education Board.* Already in this respect, and in spite of the narrow, elitist interpretation of the word *education* upheld by the IEB, neither the Rockefeller institutes nor the initial layout for the IAS were projects in the style of an LGTRS. The Göttingen Mathematical Institute financed by the IEB was an integral part of Göttingen University and actually contained a generous class room for graphical methods (*Zeichensaal*) open to students of all levels.

The *Institut Henri Poincaré*, on the other hand, in some ways resembled an LGTRS. In spite of initial difficulties in view of the economic situation, nine chairs were finally integrated into the plan for the IHP.[32] While graduate teaching did take place in its lecture halls, the most visible and novel roles of the IHP in the interwar years was to host regular seminars, and to welcome mathematicians from many different countries, albeit usually for one or several lectures rather than for a prolonged stay.[33]

We will pick up the global history of the LGTRS model after World War II in Section 8.3 below.

[30] This alludes to the paper *Généralisation des fonctions abéliennes*, labeled [1938a] in [Weil 1980].

[31] See [Weil 1992], p. 117.

[32] See [Siegmund-Schultze 2001], pp. 157–168.

[33] See [Siegmund-Schultze 2001], pp. 168, as well as the list of international lecturers at IHP on probability and statistics in [Siegmund-Schultze 2001], pp. 173.

Chapter 6
Mathematical Consolidation and Unification in the 1930s

The 1930s may well be the most difficult decade of the twentieth century to come to terms with. Given what humanity was about to be led into, i.e. World War II, one may be tempted to consider it a "morbid age":

> In his recent memoirs, the historian Eric Hobsbawm [1917–2012] remarked of the 1930s that 'we lived in a time of crisis.' Nothing very surprising about that. But I recall a conversation with him a few years ago, shortly before starting the research for this book, when he told me that he could remember a day in Cambridge in early 1939 when he and some friends discussed their sudden realization that very soon they might all of them be dead. This did strike me as surprising, and it runs against the drift of the memoirs, in which he argued that communists were less infected by pessimism than everyone else because of their confidence in the future. It is also very different from my own memories of life in Cambridge thirty years later in the late 1960s where, despite labouring under the shadow of the bomb and the threat of war in Europe during the second Czech crisis, students did not contemplate early extinction but preferred to listen to Leonard Cohen in rooms made mellow by too much smoke and cheap wine.[1]

For us who know how history went on, World War II looms over the thirties, but taking the decade from its very beginning the picture is hardly brighter. The great depression was sacking its ransom of losses and unemployed in most countries around the globe. A remarkably candid reflection on the state of the world is contained in the farewell address to the mathematicians gathered for the closing session of the ICM (11 September 1932), pronounced by the mayor (*Stadtpräsident*) of the city of Zürich, Ständerat Dr. Emil Klöti (1877–1963):

> Our city has done its best to present itself to our honourable guests in beautiful September sunshine so that we may hope it has won your respect, not only because it is so expensive, but also for the beautiful landscape. ... Looking at the city and its surroundings you may have thought that it must be home to a happy smallish people.
>
> But if you had had the occasion for closer contact with the inhabitants, it would have likely shattered this idea. True, Zürich is neither ugly nor poor. It even counts among the richest cities of the world, in which every inhabitant has a respectable average income. But you as mathematicians know what it means to take an average. ... [A]t least we may say that the standard of living of the lower classes is quite a bit higher here than in many other places.

[1] See [Overy 2009], p. xiii.

N. Schappacher, *Framing Global Mathematics*,
https://doi.org/10.1007/978-3-030-95683-7_6

> ... The crisis which had long spared our country is now spreading every day. The number of unemployed is growing ... and we look to the coming winter with apprehension. ... Thus also in our population you will find right now more unrest and discontent than placid happiness.
>
> Unfortunately the isolation of the individual states is not limited to the economic domain. It also encroaches on the people's mindsets. In these difficult times it is especially valuable if the various sciences cultivate and strengthen their international character ... and maintain collegial contacts ... regardless of race and nationality.
>
> I hope, and I am confident, that this ... Congress has lived up to the task and contributed to upholding the international spirit and to defending it against those movements whose aim it is to pervert the natural love for one's homeland into unabashed nationalism and chauvinism.[2]

This speech was given less than five months before Hitler became chancellor of Germany. And it was given less than six months before Frances Perkins (1880–1965) was sworn in as Secretary of Labor, the first woman ever to serve in a presidential US cabinet. She would remain in this position for 12 years, through June 1945, proposing and implementing essential elements of President Franklin D. Roosevelt's (1882–1945) *New Deal*.

Among the younger participants at the 1932 ICM in Zürich were Henri Cartan, Jean Delsarte (1903–1968) and Szolem Mandelbrojt—who each gave a sectional talk—, Claude Chevalley (1909–1984)—who did not give a talk himself, but both Emmy Noether in her plenary lecture and Helmut Hasse (1898–1979) from Marburg, Germany, alluded to his work[3]—as well as Jean Dieudonné (1906–1992) and André Weil—whose results were mentioned in plenary lectures, by Gaston Julia (1893–1978) and Francesco Severi, respectively. These six young men from France were the core of the group that would choose the collective pen name *Nicolas Bourbaki* for their joint project in 1935.

The way in which the Bourbaki project grew way beyond the analysis textbook initially intended, launching a comprehensive rewriting of mathematics as a whole, is an unmistakable symptom of 'Consolidation and Unification' which indicates that the overall image of the 1930s as a crisis ridden decade cannot be the whole story for the historian of mathematics. The objective of the current chapter is to try to put these symptoms of mathematical 'Consolidation and Unification' during the thirties into the context of what had happened in the previous decades, stressing both continuities and those novelties which set the 1930s apart. While doing this we will highlight the international dimension of the new decade, in spite of the IMU being essentially absent.

---

[2] See Proceedings ICM 1932, Vol. 1, pp. 76–77; my translation.

[3] See Proceedings ICM 1932, Vol. 1, p. 190, and Proceedings ICM 1932, Vol. 2, p. 19.

Encyclopedias. Before focussing on mathematics and mathematicians, let us recall that 'Consolidation and Unification' during the thirties was of course not reserved for mathematics, and that it could take on a wide spectrum of political colors. This is conveniently illustrated by encyclopedic projects various sorts of which flourished in those years.[4]

Among the examples following the classical model of alphabetically ordered learned dictionaries, there were monumental projects destined to express political regimes established after Word War I. The earliest and most impressive case of the genre is the (first edition of the) *Large Soviet Encyclopedia*; the sixty-six volumes of its first edition appeared between 1926 and 1933. The mathematician Veniamin Kagan (1869–1953), who held the chair of Geometry at Moscow University as of 1923, was part of the initial founding group of this project.[5]

Preparations for the Italian *Enciclopedia Italiana di scienze, lettere ed arti*, commonly known as *La Treccani*, also go back to the twenties, under the direction of the industrialist Giovanni Treccani (1883–1961) and Giovanni Gentile. The twenty-five volumes appeared between 1929 and 1937. The article on 'Fascism' in volume 14 "was written about 1931, signed by Mussolini but obviously composed by a number of hands including that of Gentile."[6]

An altogether different kind of project, and a token of the thirties' search for national orientation outside of Marxist-Leninist or fascist influence, was the *Encyclopédie française* initiated in 1932 by the French Minister of Education Anatole de Monzie (1876–1947) in collaboration with the historian Lucien Febvre, who had left Strasbourg in 1933 for a chair at Collège de France in Paris. Where the Soviet Encyclopedia and the Treccani negotiated their take on civilization with their respective regimes, Febvre's concept aimed at a modernist[7] tableau of the world and of humanity. The reader is not offered a sequence of quintessential entries in alphabetic order, but a collection of signed systematic expositions of problems and potential answers. Some of these essays are of a fairly technical nature. The whole product was realized as a loose leaf edition, ostentatiously open to improved later versions, so as to avoid the illusion of a finished, coherent account. The economic success of the undertaking left much to be desired.

The first volume of the *Encyclopédie française* is devoted to the 'Toolkit of the mind' (*Outillage mental*); its third (and last) part is dedicated to mathematics. A first edition of it appeared in 1937, edited by Paul Montel.[8] The chapters of this first edition

---

[4] The (German version of) Felix Klein's Encyclopedia of the mathematical sciences—see Section 1.3.3—also continued to produce fascicles in the 1920s and 1930s. However, it was no longer the vector of Mathematics International it had been before World War I. Therefore we will not go into the late phase of the mathematical *Enzyklopädie*. See, however, [Siegmund-Schultze 1993], pp. 98–101.

[5] Cf. [Mazliak 2018].

[6] See [Smith 1969], p. 412.

[7] Say, in the sense of Charles Baudelaire's 1863 essay *La modernité*: "By modernity I mean the transitory, the fugitive, the contingent which make up one half of art, the other being the eternal and the immutable." See [Baudelaire 1964], p. 13.

[8] Later additions by new authors, notably Jean Leray (1906–1998), were published in 1950.

were mostly penned by mathematicians of the older generation. One of the oldest among them, Hadamard, wrote in particular an introduction to mathematics which fills the first 110 columns of this part. After initial remarks of a more philosophical nature, the portrait of our science is loosely structured according to historical periods and their contributions to the shaping of various parts of mathematics. Poincaré's contributions are duly underlined; but Hilbert's formal axiomatics are prominently mentioned as well, and the concluding paragraphs allude to Fréchet's "abstract" spaces as well as to group theory, citing Galois of course, but also Sophus Lie (1842–1899). Hadamard also coordinated the more detailed essays on Analysis. These and the other texts on major subdisciplines of mathematics were mostly written by mathematicians of older generations: Ernest Vessiot (1865–1952), Élie Cartan, Émile Borel, Paul Montel, Maurice Fréchet, Jean Chazy (1882–1955), René Gosse (1883–1943), Arnaud Denjoy, and the Belgian algebraic geometer Lucien Godeaux (1887–1975) were all born between 1865 and 1887. The only younger authors were the Hungarian Béla Kerékjártó (1898–1946), as well as two youngsters: René de Possel (1905–1974) and Claude Chevalley who were also part of Bourbaki in the 1930s.

> The various sections have been written in gradually increasing difficulty so that the reader can go just as far as his knowledge and energy permit. It is well known that there are no "royal roads" into mathematics. Whoever ventures in this domain will inevitably bump his feet against stones along the way. But we have at least tried to indicate the roads which are the most direct and have the best layout, to discover this whole country.[9]

All the projects mentioned so far were national enterprises. A totally different, inherently international project of an *International Encyclopedia of Unified Science* (IEUS) was launched by the politically active economist and philosopher Otto Neurath (1882–1945), and further developed in the 1930s at international congresses inspired by the philosophy of the Vienna Circle. Only a torso[10] of this project would finally be published in twenty monographs which were meant to lay the ground for a unified science based on the principles of logical empiricism.

It would be interesting—although clearly outside the scope of the present book—to compare certain texts of this Encyclopedia with parts of Febvre's *Encyclopédie française*, such as for instance Maurice Halbwachs's essay "The numerical point of view" about how statistics provides evidence of social facts, which nonetheless remain extremely hard to interpret.[11]

Instead, let us just mention in passing that a fair number of mathematicians were part of a committee for the *International Encyclopedia of Unified Science* instituted at the Paris International Conference for Scientific Philosophy in September 1935: Élie Cartan, Federigo Enriques, Fréchet, Hadamard, Jan Łukasiewicz (1878–1956), Richard von Mises, and also Bertrand Russell (1872–1970). No-one from

[9] From Montel's preface to the Part on Mathematics of *Encyclopédie française*, p. I•50–7; my translation.

[10] Cf. [Dahms 1999].

[11] See [Halbwachs & Sauvy 2005]. For a presentation of the issue from the mathematical point of view, see [Brian & Jaisson 2007].

the Bourbaki group seems to have been involved with the IEUS. Claude Chevalley did give a lecture at the 1935 Congress in Paris, but the philosophical view on the language of science he presented there differed substantially from the perspective of the Vienna circle.

6.1 Emmy Noether's Legacy

Today Emmy Noether (1882–1935) has become a household name, even somewhat beyond the mathematical community, and there are now well-informed books in English that mirror up-to-date research about this exceptional mathematician and woman.[12] Readers may know that, in terms of mathematical achievements, she is most famous (a) for her two theorems in theoretical physics of 1918 which link symmetries, i.e., invariance under the action of a Lie group, to physical conservation laws, (b) for her work in abstract algebra, especially ring theory, in the 1920s, and (c) for playing an important part in applying the theory of division algebras to number theory in the late twenties and early 1930s. However, her impact on the development of mathematics between the World Wars was both broader and more profound than any list of theorems can convey.

Since Emmy Noether was a woman, her education and academic career in Germany was ridden with administrative hurdles, and often further complicated by anti-Jewish or political prejudices. Even influential men like Hilbert could not obtain for her the right to teach at Göttingen University before the end of World War I.[13] The position she finally obtained in Göttingen was but a caricature of her scientific standing; it was inferior to Sofya Kovalevskaya's Stockholm professorship about 40 years earlier (Section 1.1.7). The "only time Emmy Noether, the mother of modern algebra, was treated by a German authority just as her distinguished male colleagues, and not according to her inferior hierarchical position, occurred when the Nazi government put her on leave, forbidding her to teach at the mathematics institute of Göttingen University, by way of a telegram dated 25 April 1933."[14] See Fig. 6.1 for a page from the ministerial file about her dismissal in 1933.[15]

[12] See the two overlapping books [Rowe & Koreuber 2020] and [Rowe 2021]. These books will point the reader to the vast existing literature on many aspects of Emmy Noether's life and work that we will mention only in passing in our account.

[13] Emmy Noether, as a woman, was denied the right to obtain the *Habilitation*. The affair has been analyzed in [Tollmien 1990]; it is retold in English in [Rowe & Koreuber 2020], Section 2.1. Two volumes of Tollmien's new Emmy Noether biography adding new aspects to the story are expected to appear in 2021.

[14] Opening sentence of our marginal note [Schappacher & Tollmien 2016], which documents Hermann Weyl's unsuccessful attempt in 1932 to get Emmy Noether elected into the Göttingen Academy of Sciences.

[15] Overview of the questionnaire shown in Fig. 6.1. 1) Name, 2) Given name; 3) Date of birth: *23 March 1882*; 4) Nationality (current / at birth): *Bavaria*; 5) Date of habilitation: *4 June 1919*; 6) Career details; 7) Front line fighting in the World War; 8) Race of the four grandparents: *Jewish*; 9) Father or son killed in action in the World War: *No*.

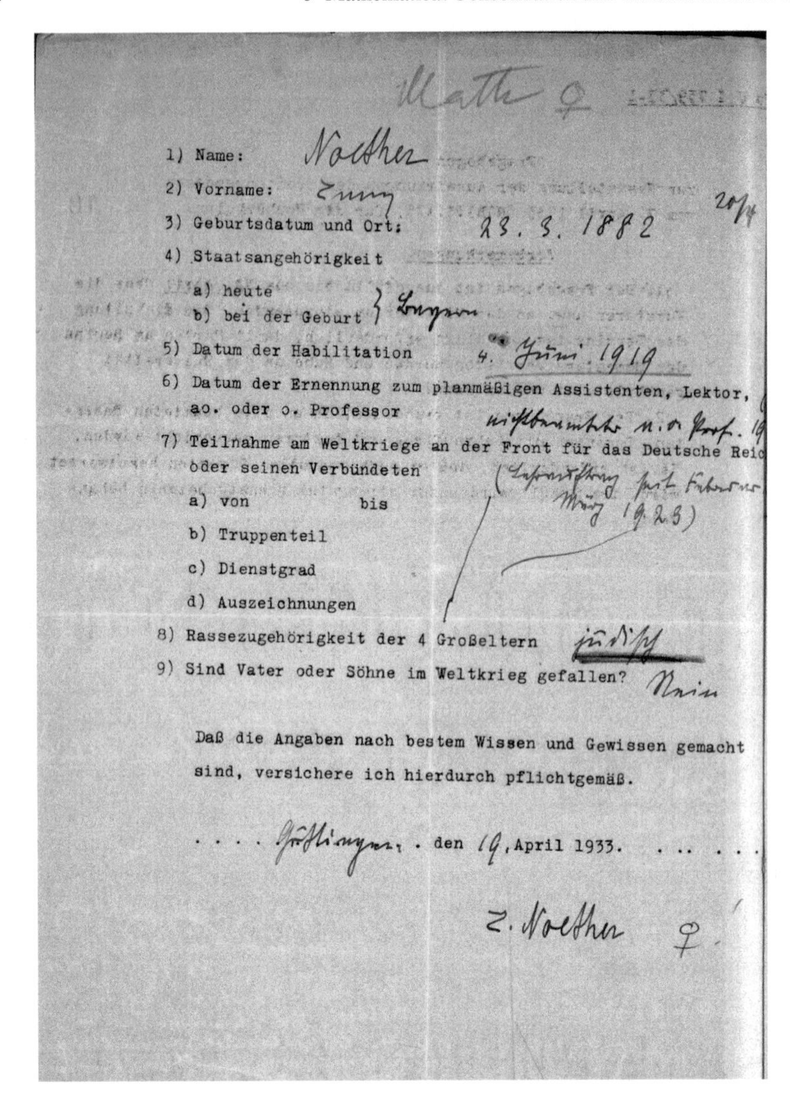

Math ♀

20/4

1) Name: Noether

2) Vorname: Emmy

3) Geburtsdatum und Ort: 23. 3. 1882

4) Staatsangehörigkeit

a) heute
b) bei der Geburt } Bayern

5) Datum der Habilitation 4. Juni 1919

6) Datum der Ernennung zum planmäßigen Assistenten, Lektor, ao. oder o. Professor nichtbeamteter a.o. Prof. 19

7) Teilnahme am Weltkriege an der Front für das Deutsche Reich oder seinen Verbündeten (Lehrauftrag seit Februar März 1923)

a) von bis

b) Truppenteil

c) Dienstgrad

d) Auszeichnungen

8) Rassezugehörigkeit der 4 Großeltern jüdisch

9) Sind Vater oder Söhne im Weltkrieg gefallen? Nein

Daß die Angaben nach bestem Wissen und Gewissen gemacht sind, versichere ich hierdurch pflichtgemäß.

. . . . Göttingen . den 19. April 1933. . . . . .

E. Noether ♀

Fig. 6.1 Emmy Noether's personal questionnaire of 19 April 1933, filled in by herself, with two female gender symbols added by the administration. Credit: [Arch. GStA].

In this section I argue that Emmy Noether's way of practicing mathematics was in fact a model for that 'Mathematical Consolidation and Unification' which characterized the 1930s. To see this one has to take a closer look at her personal approach to mathematics. Above we have called her the "mother of modern algebra." This is fair enough as a brief marker, but it could occlude the broadness of her influence; we need further questions to put us on track.

6.1.1 What is 'Modern' about 'Modern Algebra'?

A few months ago at a dinner, I asked a colleague from the humanities which label he would see fitting to characterize the 1930s. His immediate answer was: *modernism*. He explained his reaction by pointing to the omnipresence of certain art forms, in particular in architecture, even across very different political regimes at the time. Above I have also called Febvre's *Encyclopédie* 'modernist', albeit with a flashback all the way to Baudelaire. There is actually a certain ambiguity among historians of mathematics when it comes to extending the notion of modernism or modernity to the 1930s. For instance, Jeremy Gray's book on the subject [Gray 2008] stops before 1930, whereas for Leo Corry in [Corry 2004] both the Bourbaki enterprise and the slightly later takeoff of category theory constitute the very climax of modern algebra with its central concept of *structure*.

Recall from our brief discussion of Mehrtens's book [Mehrtens 1990] at the beginning of Chapter 2 that the most visible radical manifestation of modern mathematics was Hilbert's formal axiomatic approach to the *Foundations of Geometry* in his seminal book [Hilbert 1899]. So this notion of modern mathematics dates back to the turn of the century. However, Hilbert did not care much for the new algebra in his own publications; for instance, "Hilbert's *Zahlbericht* of 1897 made even less use of unifying notions from abstract algebra than one might have expected from a text written in the last decade of the nineteenth century."[16] Nonetheless, what Hilbert had done to geometry was heeded, for instance, by the Chicago school of Eliakim Hastings Moore (1862–1932) who transferred Hilbert's paradigm to the comparative study of a great number of different algebraic structures in the early years of the twentieth century.[17] Outstanding American mathematicians such as Leonard Dickson, Oswald Veblen, and Joseph Wedderburn (1882–1948, originally from Scotland) were in touch with this movement.

This first decade of the twentieth century is naturally included in *all* historical accounts of modernism, inside as well as beyond mathematics. But it is also sometimes considered a period of "unhappy modernity"[18], with mathematicians showing signs of "anxiety"[19]. This may incidentally remind us of the conflicting associations provoked by the thirties.

---

[16] Quoted from [Schappacher 2005], p. 704, where the statement is fleshed out further, and Minkowski's different attitude is cited.

[17] See Section 3.5 on "Postulational Analysis in the USA" in [Corry 2004], pp. 172–182.

[18] See the section *Une modernité malheureuse* in [Charle 2011], pp. 335–336. Incidentally, this conclusion of his Chapter 11 follows Charles's brief account of Daniel Halévy's somber, nietzeschean utopia *Historie des quatre ans 1997–2001* published in 1903, in which an epidemic kills 400,000 Parisians in three months, seemingly opening up a better future (for the survivors) based on science and a healthy diet; but this hope is then destroyed in turn.

[19] See in particular the analysis in [Gray 2004] of Oskar Perron's 1911 Tübingen lecture "On Truth and Error in Mathematics." Cf. [Gray 2008], Section 4.8.3, pp. 274–277, and the doubts expressed in [Schappacher 2012], pp. 234–235.

Other developments of modern algebra, before Emmy Noether left her mark on the budding discipline, include Ernst Steinitz's (1871–1928) systematic investigation of the (algebraic) field concept, published in 1910. The interest in studying this abstract notion of algebra was prompted in particular by Kurt Hensel's (1861–1941) introduction of p-adic numbers, which had produced a whole bunch of concrete examples of fields never encountered before. Shortly afterwards Adolf (Abraham) Fraenkel (1891–1965) started to investigate ring theory, coining as it were the algebraic notion of ring.[20] So modern algebra as a systematic study of algebraic structures existed before Emmy Noether joined in. It was driven by known *instances* of such and such structures—groups, rings, modules, fields—that had previously been researched in more concrete mathematical contexts, typically in the theory of equations, in geometry or in number theory.

In order to fathom Emmy Noether's role for modern mathematics, one has to pin down the kind of insights that the investigation of algebraic structures can yield. Modern Algebra obviously opens up a whole new kingdom of creatures, and thus the modern algebraist would appear like a new Carl Linnaeus, all busy inspecting and classifying even the most exotic genus in the new zoo, all the way from magmas to structures with n inner and m outer operations. Clearly some sort of stock taking had to be done and in certain cases this represented a formidable mathematical research program in itself—think of the Classification of Finite Simple Groups. But if this were all that is meant by modern algebra: a discipline that studies and classifies algebraic structures, calling Emmy Noether the 'mother of modern algebra' would be misleading.

Already Steinitz proceeded differently in his abstract theory of fields: mindful of traditional problems which could be seen to have involved certain types of fields—such as the resolution of polynomial equations—he checked in which way and how far the traditional theory could be generalized, and where it had to be modified in view of the new supply of structures, for instance for fields of finite characteristic. This new vantage point for looking at traditional results also allowed Steinitz to criticize Kronecker for having blurred the distinction between splitting a given polynomial into a product of linear factors, as opposed to splitting all polynomials, i.e., constructing the algebraic closure of a given field.[21]

Following Steinitz, Emmy Noether's contributions to modern algebra were oriented towards *either separating* aspects that had been previously undistinguished, *or else unifying* seemingly unrelated, or at best analogous theories, by accommodating them under the joint roof provided by a more general abstract theory.

---

[20] Cf. Chapter 4 of [Corry 2004].

[21] See [Steinitz 1910], pp. 169–170.

Her relentless ambition to unify shows for instance in her 1919 *Report on the arithmetic theory of algebraic functions of one variable, in its relation to the other theories and to the theory of number fields.*[22] She was asked to write this report for the German Mathematical Society in order to fill a deliberate gap which her father and Alexander Brill had left in their colossal report on the theory of algebraic functions written almost thirty years earlier.[23] Emmy Noether's exposition is not only more concise than the commendable scholarly prose of Brill and her father; she also pushes the juxtaposition of parallel theories to new levels of formal compatibility. All she was supposed to deliver was an account of various theories developed by others decades ago. However, while being even more careful than usual to always refer to the original papers, Emmy Noether also arranged things in a new, unifying manner.

It was well known at least since the 1880s that the arithmetic theory of number fields—i.e., finite extensions of the field $\mathbb{Q}$ of rational numbers—is in many ways analogous to that of function fields of one variable—i.e., finite extensions of the field $\mathbb{C}(z)$.[24] The *relation*[25] between the two cases is one of the foci of Emmy Noether's 1919 report. Instead of moving back and forth between the two cases, she tried to bind them together cogently under the umbrella of a single more general theory. However, this unifying theory did not really exist yet; its fully fledged exposition would be the object of her 1927 paper *Abstract Setup of the Theory of Ideals in Number Fields and Function Fields.*[26]

This later paper incidentally reminds us of the strong international dimension of modern algebra. Emmy Noether was not the only mathematician trying to develop a general theory of ideals in commutative rings; some of her insights had actually been anticipated by the Japanese mathematician Masazo Sono (1886–1969). She

---

[22] See [Noether 1983], pp. 271–292. – The reproduction of this report is an instance of sloppiness in the edition of Emmy Noether's *Collected Papers*: The title printed on p. 271 misses the last 8 words, and on the final p. 292 the last footnote of her report is not reproduced. Other mistakes which I stumbled across occur on p. 560, where the text reproduced is not that of her Bologna ICM talk, and on p. 636, where the printed title gets a word wrong.

[23] See [Brill & Noether 1892–93]; Kronecker's sudden death is cited to explain the omission of the arithmetic theories from this report on pp. I–II.

[24] Both a proposal for a unified treatment of the two cases by Leopold Kronecker and a treatment of the second case modelled on the first one by Richard Dedekind and Heinrich Weber were published in *Crelle's Journal* in 1882; for more background and details see for instance the discussion in [Schappacher 2010], pp. 3262–3267.

[25] The German word *Beziehung* is in the title Emmy Noether chose for her report.

[26] See [Noether 1983], pp. 493–528. The paper was written in 1925 and appeared in *Mathematische Annalen* in 1927.

eventually realized that she had had this precursor, and in her 1927 paper she refers at four places in detail to various articles by Sono published in English in the *Memoirs of the College of Science, Kyoto Imperial University* between 1917 and 1924.[27]

From the point of view of the genealogical thinking popular in parts of the mathematical community, Sono was the academic grandfather of two Fields medalists. Sono's student Yasuo Akizuki (1902–1984) counted among his own numerous thesis students Heisuke Hironaka (b. 1931; Fields Medal 1970) and Shigefumi Mori (b. 1951; Fields Medal 1990).

Anyway, reading her 1919 report we see Emmy Noether at work at a time when she could not simply allude to general notions of what we call today *noetherian rings* or *Dedekind rings*. Nonetheless, she was resolutely looking at things from the superior vantage point of such a unifying theory. Let us give an

Example. Consider the elementary statement about polynomials which says that if $f(x) = (x - a)^e \cdot g(x)$ with $g(a) \neq 0$, then the derivative $f'(x)$ is divisible by $(x - a)^{e-1}$ [and by no higher power of $(x - a)$, unless we are working over a field of characteristic p with p dividing e]. Somewhat analogous statements were known from the so-called theory of ramification, involving the 'discriminant' and 'different' of an algebraic number field, or a function field.

In Section 2 of her report, entitled *Parallelism and Differences in the Theory of Ideals of Algebraic Numbers and Functions*, Emmy Noether does not simply review, but tries to *explain* apparent differences of ramification theory between the two cases: "*Differences* in the further development [of the theory] are occasioned by special properties of the ground field. Thus the fact that in the case of algebraic numbers *elements of the ground field are at the same time exponents* implies that the ramification ideal may be divisible by a higher power than the $(e-1)^{\text{th}}$ of a prime ideal $\wp$ whose e-th power divides precisely p (in the case where e is divisible by p)."[27]

In other words, she explains the absence of the so-called wild ramification in function fields (of characteristic 0) by the fact that positive numbers are invertible in $\mathbb{C}^*$. But again, the general theory affording this explanation was not fully worked out yet at the time of the report. She would present it in a talk at the Prague meeting of the German Mathematical Society in 1929.[28] There she repeated that her approach goes "beyond formal analogies." Indeed, she developed an intrinsic reformulation, purely in terms of ideals, of the elementary statement about polynomials quoted above. And she gained significantly

[27] It would seem that she was not yet aware of Sono's work when she published her momentous first paper on the theory of ideals in *Mathematische Annalen* 83 (1921), see [Noether 1983], pp. 354–396. Cf. [Kümmerle 2021] who mentions Sono's work against the background of the establishment of Japanese mathematics, including mathematical journals.

greater generality; the "most essential result" of her work lay in those "completely invariant structure theorems which are also valid for algebraic functions in several variables."[29]

This example illustrates how "modern algebra" in the hands of Emmy Noether—instead of just introducing a plethora of novel structures—invariably strove towards a parsimonious hierarchy of mathematical theories. It is this peculiar explanatory power of generality which was acknowledged by Albert Einstein when he wrote to Hilbert in 1918 about Emmy Noether's work on differential invariants: "It impresses me that one can view these things from such a general standpoint."[31] Writing anachronistically one could say that Emmy Noether lay the ground for Alexander Grothendieck's unification of commutative algebra and algebraic geometry.

The basic toolkit for the modern, unifying expeditions into the thicket of traditional mathematics would quickly be covered in textbooks and monographs of which the outstanding example was Bartel L. Van der Waerden's *Modern Algebra, based on lecture courses by E.*[mil] *Artin* [1898–1962] *and E.*[mmy] *Noether* in two volumes.[32] The first edition appeared in 1930–1931. The special attention to reorganizing theories with respect to one another in this modern way of doing mathematics is crystallized at the beginning of the book in the *Leitfaden* that shows the various chapters and outlines their logical dependence in a flowchart.

At the beginning of the introduction, Van der Waerden offers without commitment various adjectives in quotation marks to characterize the new way of doing algebra: "abstract", "formal", "axiomatic." In the troubled times ahead these words could easily acquire a pejorative connotation. Hermann Weyl had to grapple with this in the petition letter which he contributed to the vain effort of stopping Emmy Noether's dismissal in 1933:

> She represents principally "abstract algebra." In this context the word 'abstract' does not at all indicate that this branch of mathematics is especially remote from life (*lebensfern*). The predominant tendency is rather to master the problems by visionary thought, by arranging a conceptual framework as appropriate as possible to the subject matter, instead of blind calculation.[33]

---

[27] See [Noether 1983], p. 278; my translation.

[28] A corresponding manuscript was published only after her death and is the last article reproduced in her *Collected Papers*; see [Noether 1983], pp. 690–710.

[29] See [Noether 1983], p. 692; my translation. The word 'invariant' here means that these theorems of modern algebra avoid any case distinctions according to the intended setting, like for algebraic numbers or algebraic functions. Cf. Nathan Jacobson's comments on this paper in [Noether 1983], pp. 15–16.

[31] Quoted from [Rowe 2021], p. 74.

[32] See [Van der Waerden 1930–31].

[33] From Emmy Noether's ministerial file kept at *Geheimes Staatsarchiv Berlin-Dahlem*, p. 39. Weyl's letter is dated 12 July 1933. Cf. [Rowe 2021], p. 145, who points to Zermelo's plea in 1929 to replace the expression "abstract algebra" by "general algebra."

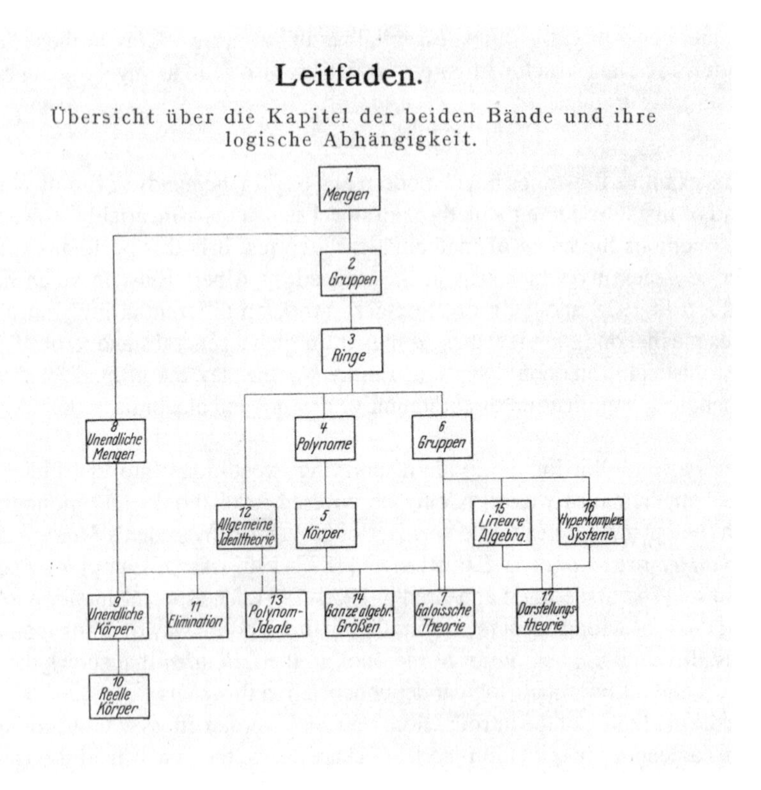

Fig. 6.2 *Leitfaden* from Van der Waerden's book *Moderne Algebra*, first edition.

As of the fourth edition of Van der Waerden's extremely successful textbook, in 1955, the epithet "modern" was dropped from the title. Van der Waerden motivates this in his 1955 preface by quoting from a book review by the algebraist Heinrich Brandt (1886–1954) who rejected the term 'modern' as being associated with fashions that come and go. Similarly, ten years later, Serge Lang (1927–2005) opened the preface to his textbook on *Algebra* by the following quote of Severi from 1949—certainly one of the more unlikely choices of a mathematician to quote about modern algebra:

> I prefer to call it 'abstract algebra' rather than 'modern algebra' because it will undoubtedly live for a long time and will thus end up becoming the ancient algebra.[34]

Both instances reflect the fact that after World War II the notion of modernity had largely lost the luster that it still had in the 1930s.[35]

---

[34] See [Lang 1965], p. V; my translation.

[35] See also [Corry 2004], p. 61, who cites the analogous case of Dickson's new edition of his book entitled *(Modern) Algebraic Theories.*

6.1.2 Emmy Noether's *Auffassung* and its Influence

We have seen that Emmy Noether's approach was about reorganizing mathematics, about new ways to look at theories, theorems, and problems. This is the intrinsic reason for Emmy Noether's broad influence on mathematical practice between the World Wars, beyond her own theorems.

It seems that Emmy Noether herself was not particularly gifted in explicating her philosophy of mathematical practice.[36] There is a word that she often used in this context: *Auffassung*. It is a rather poor choice for a word expressing a methodological tenet, because it has many plain meanings: opinion, point of view, conception, etc. This makes it quite a challenge for the translator, for example when she speaks about her *Arbeits- und Auffassungmethoden*:

> My methods are methods about how to work and how to view things; this is why they have penetrated everywhere, anonymously.[37]

We have already mentioned in Section 4.4.3.1 that the word *Auffassung* made it into the title of her ICM talk at Bologna in 1928. And in a letter to Hasse about her preparation for the 1932 plenary ICM talk in Zürich, she tells him that, for once, she has read Gauss, adding the peculiar remark that she "learned a lot from Gauss about *Auffassung*":

> I learned a lot from Gauss about how to view things; above all, that it is good to place the verification of the fact that the classes determined by factor systems are ray classes, at the end . . . [38]

Never mind the mathematical details; the quote shows that *Auffassung* expresses itself in proof arrangements. In her Zürich lecture she explained how she managed to smuggle non-commutative algebras into a study of (commutative) fields:

> *With the help of the theory of algebras one looks for invariant, simple formulations of well-known facts about quadratic forms or cyclic field extensions, i.e., formulations that only depend on structural properties of algebras. Once such formulations have been proved . . . a transfer of these facts to general galois field extensions is achieved. . . .* the passage to the non-commutative world is afforded by *viewing* field and group *simultaneously*, via their "crossed product."[39]

We learn from this quote that the rearrangements of proofs in the interest of the most adequate, 'invariant' insight, or *Auffassung*, can require the introduction of abstract intermediary objects used to assemble and treat data provided by the

---

[36] See for instance Paul Dubreil's description of an unsuccessful presentation of hers before a mixed audience of philosophers and mathematicians in Hamburg in 1930; Dubreil's account is translated in [Rowe 2021], p. 152.

[37] From a letter to Hasse of 12 November 1931; see [Lemmermeyer & Roquette 2006], p. 131; cf. their comments in endnote 8, pp. 132–133, and their proposed translation p. 8. The translation given here is mine.

[38] We quote from the translation in [Goldstein, Schappacher, Schwermer 2007], p. 557, where the original text is also given. The quote marks the end of Lemmermeyer's contribution to the book.

[39] Proceedings ICM 1932, vol I, p. 189; the italics are Emmy Noether's; my translation.

original problem. This is Emmy Noether's philosophy of mathematical practice in a nutshell. It is also the blueprint for seminal mathematical innovations during the interwar period. For instance, Emmy Noether encouraged Van der Waerden to start rewriting Algebraic Geometry from the point of view of her algebraic approach. This would become the first among several different fundamental rewritings of Algebraic Geometry in the twentieth century.[40]

In 1935, shortly after Emmy Noether's death at Bryn Mawr, Pennsylvania, and around the time of the first Bourbaki congress, Claude Chevalley wrote an essay about *Variations of mathematical style*. Even though the collective project is not explicitly mentioned, this text is in fact the first manifesto of Bourbaki, written and published even before the group had adopted their collective pen name.[41] It faithfully follows Emmy Noether's legacy, even though it is Hilbert's name, not hers, that Chevalley uses to label the new *style* of mathematics.

Fig. 6.3 Group picture of the Bourbaki Congress at Besse la Chandesse in July 1935. Standing from left to right: Henri Cartan, René de Possel, Jean Dieudonné, André Weil; Luc Olivier (biologist). Sitting: A. Mirlès, Claude Chevalley, Szolem Mandelbrojt.

Two recent mathematical practices (*styles*) are confronted in Chevalley's essay: Weierstrass's "style of the epsilons" from the second half of the nineteenth century—which achieves impeccable rigor at the price of reducing all statements to relations involving real numbers—is opposed to the freedom of the "Hilbertian mathematician"[42] whose objective it is to prove each theorem in an intrinsic way, i.e., without

[40] See [Schappacher 2007].

[41] See [Chevalley 1935]; the essay appeared in the Fall of 1935.

[42] We borrow this expression: *mathématicien hilbertien*, from [Chevalley & Dandieu 1932], p. 111. This earlier paper justifies the existence of an active mathematician with respect to the *personalistic* philosophy of Arnaud Dandieu (1897–1933) and Alexandre Marc (1904–2000) and its supreme value of creativity. That Chevalley was strongly committed to this thought in the 1930s shows in the more than 30 philosophical and political essays he published between 1932 and 1938.

introducing auxiliary objects or making arbitrary choices (like choosing a coordinate system to translate a geometric problem into a statement about real numbers) that are not germane to the statement at stake:

> ... constructive definitions from real numbers, even if they made rigorous proofs possible in the first place, often had the effect of seriously hiding the very nature of what they strove to define, or to unduly blur mathematical domains that are actually distinct. This leads to futile complications that can be observed in many classical proofs. They arise when methods are used that have nothing in common with the desired result, or one could say: methods which do not have the same transformation group as the result.[43]

The agenda of the Hilbertian mathematician as described by Chevalley involves precisely the sort of reorganization of mathematics which was Emmy Noether's central concern.

> Today almost all mathematical facts are categorized in more or less autonomous theories that have accumulated. The principal examples are the following: in algebra one finds the general theory of fields ... ; group theory; the theory of those systems having less restrictive algebraic relations than those of fields (like rings or algebras)—these are essentially due to Emmy Noether whose recent passing is deplored by the mathematicians—; in analysis, the theory of measure and integration, of topology, of general Riemann surfaces, Hilbert spaces, etc.; in geometry, projective and conformal geometry; the theory of Riemann spaces; combinatorial topology ... One should not imagine that these theories exist in isolation from each other. On the contrary, they are coalescing into more complex theories. For example, if one operates on certain elements between which both strictly algebraic relations (like multiplication) and relations of continuity exist, one obtains topological algebra, which provides richer properties than the juxtaposition of algebra and topology. There are numerous such complex coalescences of theories. On the other hand, very different theories may be transposed into one another provided they are based on the same axiomatic structure: for instance, probability calculus and measure theory.[44]

This new ordering of mathematics—and the energetic way in which it was practiced by Emmy Noether in her research, and presented here by Chevalley—is a far cry from any idea of crisis. It is not the kind of radical overhaul of mathematics that the foundational crisis of the 1910s and 1920s had suggested to mathematicians like Brouwer and Weyl—see Section 2.2. Ongoing research work on the one hand and an increasing number of systematic presentations and textbooks on the other including the enormous Bourbaki project—see Section 6.3 below—would completely remodel the world of mathematics from the inside.

6.1.3 Emmy Noether's International Network

The nineteenth century had professionalized science, and mathematics. In Part I of this book we have traced international aspects of this evolution. For instance, in the different editions or translations of Felix Klein's Encyclopedia, we were looking for interesting comparisons of national mindsets. The 1930s were different. Emmy

[43] See [Chevalley 1935], p. 379–380; my translation.

[44] See [Chevalley 1935], p. 383; my translation.

Noether's *Auffassung* prepared the way for fundamental mathematical renovation in the context of international teams of mathematicians, be that in Algebra, Topology, Arithmetic or Geometry. Professionalization continued, but differently, with stipends, jobs and institutes created by national or private sponsors. The Rockefeller Foundation, which we have presented in Section 5.1, was but one of several donors. The growing mobility options of young researchers started to show its effects on the world of mathematics. The founding fathers of Bourbaki for example profited from various stipends or invitations acquainting them with global mathematical centers of the time. These modern instruments of Science International would merge with the novel structural emphasis inside mathematics to build a new way of living mathematics which involved more direct personal interaction than ever before. The resulting consolidation and unification was the work of many, and it was an international enterprise.

> ... the research of Fräulein Noether and her students, van der Waerden, [Heinrich] Grell [1903–1974], [Wolfgang] Krull [1899–1971], who disseminate the theory of ideals all over algebra; research on the theory of sets and dimension ([Felix] Hausdorff [1868–1942], [Karl] Menger [1902–1985], Urysohn and the whole Polish school), on group theory ([Otto] Schreier [1901–1929], [Élie] Cartan). In all these domains only the use of axiomatics allows a perfect adjustment of the proof to its object. ... [45]

Chevalley's list of (groups of) mathematicians is euro-centric but extends beyond algebra. Emmy Noether fits into this picture, but her mathematical networking between 1925 and 1935 extended further, both geographically and in terms of mathematical domains. The group around Emmy Noether has been called an "International School in Modern Algebra."[46] She did build a school of modern algebra, and it was international. But just as the title "the mother of modern algebra" can mask her broader influence on mathematics, also her international contacts reached beyond the group of her (former) students.

Pavel Alexandrov, one of the designers of contemporary topology whom we have briefly mentioned in Section 5.1 above, was a frequent visitor to Göttingen in the 1920s, accompanied by his partner Urysohn until the latter's tragic death. A close friendship developed between Emmy Noether and Alexandrov. The latter arranged for Emmy Noether to stay in Moscow during the winter of 1928, following the Bologna ICM. Her lectures on Algebra there, given in German, provided direct inspiration for Alexandrov's own teaching and left their mark on the Moscow mathematical community.[47]

Alexandrov also formed a troika of friendship with Heinz Hopf (1894–1971) and Otto Neugebauer (1899–1990). As topology was in the air, the three of them adopted a simplex as a symbol of their friendship. This pops up time and again in their correspondence.

---

[45] See [Chevalley & Dandieu 1932], p. 105; my translation.

[46] For instance in the title of the fifth chapter of [Rowe 2021].

[47] See [Rowe 2021], pp. 131–133.

Another important connection of Emmy Noether's would reach out to Japan. These contacts were mediated by Teiji Takagi whose work on Hilbert's 12th problem and class field theory became well-known in Europe in the 1920s.[48] Takagi was one of the vice presidents of the Zürich ICM in 1932, and a member of the committee for selecting the first Fields Medals to be awarded in 1936.

Two of Takagi's students, Zyoiti Suetuna (1888–1970) and Kenjiro Shoda (1902–1977), spent time in Göttingen in the 1920s. Suetuna—who would succeed his teacher Takagi on his chair at Tokyo University in 1936—was more interested in analytic number theory. Shoda on the other hand would become one of Emmy Noether's students, adapting as well as possible to her way of teaching, talking, and doing mathematics.

> They remained in contact from a distance after 1929, when Shoda returned to Japan. Soon afterward, he began to write his textbook *Abstract Algebra*, which was first published in 1932 and then reprinted many times later. Its impact on mathematics in Japan has sometimes been compared with that of van der Waerden's *Moderne Algebra* on research in Europe and the United States. Kenjiro Shoda went on to a highly successful career, beginning in 1933 with his appointment as professor in the Faculty of Science at the newly founded Osaka University.[49]

In the Summer of 1933, Takagi and Shoda wrote letters of support for Emmy Noether to the ministry as part of the concerted effort to avoid Emmy Noether's dismissal. Takagi in his letter of 12 July 1933 insists on the great loss that her dismissal would mean to German science, and he adds, obviously alluding to the racist politics threatening Emmy Noether, that the loss would be all the more deplorable "as it would be due to a circumstance for which she is not to blame."[50]

Apart from the 14 petitions written by distinguished colleagues on behalf of Emmy Noether, which were sent that summer to the German ministry via the administration of Göttingen University, there was also a letter of her students which reflects the Göttingen political constellation at the time in tragic irony: The letter chooses to explicitly endorse the "National Revolution" of the Nazis in order to make the authorities well-disposed to the request. The students thus insist that "it is not by chance that all her students are Aryan; this is grounded in her *Auffassung* of the essence of mathematics which is especially germane to the Aryan way of thinking."[51] As it happened, among those who signed the letter were two young mathematicians from China: Chiungtze C. TSEN (1898–1940) and Wei-Liang CHOW (1911–1995).[52] Under Emmy Noether's guidance, Tsen had just proved that a function field in one variable over

---

[48] Cf. [Schappacher 1998].

[49] See [Rowe 2021], p. 128, and more generally the vivid, partly anecdotal description of Shoda's apprenticeship with Emmy Noether on pp. 125–128.

[50] From Emmy Noether's ministerial file kept at *Geheimes Staatsarchiv Berlin-Dahlem*, p. 35: . . . *zumal wegen eines Umstandes, woran sie keine Schuld trägt.*

[51] From Emmy Noether's ministerial file kept at *Geheimes Staatsarchiv Berlin-Dahlem*, pp. 13–14. Cf. [Rowe 2021], pp. 200–201. The letter is undated but was surely submitted in June 1933.

[52] I tend to follow the version of a Chinese name which is most commonly used in English publications; on the first appearance, I HIGHLIGHT the family name.

an algebraically closed field admits no non-commutative finite-dimensional central division algebras. Formally he got his doctorate only after Emmy Noether's emigration to Bryn Mawr.[53] Chow had only recently come to Göttingen and found the Mathematical Institute he was keen to study at in political agony—see Section 6.4 below.

6.2 Encounters, Workshops and Congresses in the 1930s

In the previous section we have focussed on Emmy Noether in order to present key elements of the mathematical renewal that asserted itself during the 1930s. The new era not only changed what mathematicians would know, or would have to learn, but how they worked, and indeed how they spent their days. Get-togethers outside of one's professional institution, for the immediate exchange of mathematical ideas and results, became more and more frequent; both individual face-to-face meetings and bigger conferences. We have already stressed at the beginning of Section 6.1.3 the increased mobility of young researchers which facilitated international exchange on an unprecedented scale.

> A meeting with the Polish mathematician Juliusz Schauder [1899–1943] reveals to [Jean] Leray he topological techniques required by his thesis. The consequence is a joint paper also published in 1934, worked out in two weeks, the year before, in Luxembourg's garden of Paris. Topological degree theory in infinite-dimensional Banach spaces is born, as well as the global theory of nonlinear elliptic partial differential equations. The Leray-Schauder degree and the Leray–Schauder continuation method remain the model for the whole development of nonlinear functional analysis and of fixed point theory. The founding paper, whose style is astonishingly modern, is one of the most quoted and used mathematical works in the twentieth century.[54]

Schauder had been able to come to Paris on a Rockefeller stipend.[55] His first meeting with Leray was mediated by Hans Lewy (1904–1988)—another mathematician who had fled Göttingen for political reasons—and took place over lunch in a modest restaurant in rue Soufflot, close to the *Panthéon* and *Institut Henri Poincaré*. As for Leray, he received a grant from the French *Caisse Nationale de la Recherche Scientifique*, the predecessor of today's CNRS (where only the initial letter 'C' has changed its meaning; it now stands for *Centre*). This allowed him to attended courses in Germany, at Berlin and Leipzig. During their joint work in Paris and in their subsequent correspondence Schauder and Leray would mostly converse in German. In 1935 Leray was invited to give the distinguished *Cours Peccot* lectures at *Collège de France* on the topological degree.[56]

---

[53] See [Ding et al. 1999], which also reviews Tsen's later work in China, before his untimely death.

[54] See [Mahwin 1998], p. 200.

[55] Cf. [Siegmund-Schultze 2001], *passim*.

[56] Cf. [Kantor 2000], as well as the booklet compiled for the event *Jean Leray et les équations aux dérivées partielles* at Université de Nantes 9 November 2006: [URL 09].

6.2.1 Specialized Conferences

As for specialized research meetings in mathematics—of more than just two coauthors, and outside of regular seminars—probably the first example that should be mentioned in the 1930s is the workshop on the arithmetic of skew fields held at Marburg in February 1931. It was Emmy Noether who prodded Hasse into organizing it; he had recently succeeded his former thesis advisor Kurt Hensel in the small university town. She jokingly called the gathering a *Schiefkongress*, i.e., 'skew congress.' Jacques Herbrand (1908–1931) from France was one of the young participants.[57]

That the 1930s were indeed the first decade of intense international conferences on specific mathematical subjects is nicely documented by the following passage from a brief overview of mathematical congresses.

> In 1933 at Geneva, a series of international conferences and congresses dedicated to the study of specialized research domains was created under the name *Conférences internationales de sciences mathématiques*. In October 1933 a conference on Quantum Theory was held. The subject in June 1934 was Mathematical Logic; the Theory of Electrons in Metals followed in October 1934; Topology in October 1935; Partial Differential Equations in June 1935; Theoretical Probabilities in October 1937; Applied Probabilities in July1939. This last conference was organized with the help of the International Committee on Intellectual Cooperation[58] which in 1938 had brought the "Conversations on the Foundations of Mathematics" to Zürich in 1938. In May 1934 a meeting on Tensorial Differential Geometry gathered in Moscow. This conference decided to form groups of scholars for a planned research work. The members of each group would stay in touch via scientific correspondence exchanging interesting problems. In August 1935 at Brussels, the first International Congress on Mathematical Recreation was held. In September 1935 at Moscow, the First International Congress on Topology was held, at the same time as the International Congress on the Philosophy of Science in Paris. Clearly, the need to bring together specialists from all over the world around specialized problems was strongly felt. Let us hope that this will not bring about a fragmentation of mathematics, but that, on the contrary, following Hilbert's view, the study of a particular subject can but strengthen to whole.[59]

Following Hollings and Siegmund-Schultze, we may slightly enrich the list, and put things into perspective:

> As well as specialised conferences, the 1930s saw several international conferences on a level lower than the ICM, for instance the 8th and 9th Scandinavian congresses (1934 in Stockholm, 1938 in Helsinki), and several conferences organised by the Germans and Italians, even after the outbreak of the war but necessarily with the participation only of their wartime allies. One specialised international conference on applied mathematics not mentioned by Wavre above is the Volta Congress in Italy (1935) on high speeds in aviation.[60]

The outstanding event among the meetings listed above by the Swiss mathematician Rolin Wavre (1896–1949) was without doubt the 1935 Moscow Congress on Topology. The reasons for this are its truly international format and the timely

[57] Cf. [Rowe 2021], pp. 157–159.

[58] Or ICIC for short, cf. Section 4.2 above.

[59] See [Wavre 1948], pp. 301–302; my translation.

[60] See [Hollings & Siegmund-Schultze 2020], p. 145–146, and the further references given there.

presentation of seminal ideas, which would redirect research on Topology and related domains for years to come. It was called the "First International Congress on Topology", but it was also the last such conference that took place in Stalin's Soviet Union.

Topology entered a new era in the thirties, emerging as a mathematical discipline of its own. As Moritz Epple has pointed out, this transpires when one compares the research orientations of topologists like Max Dehn, Kurt Reidemeister, and James W. Alexander. Referring to the well-known lecture about the professional aspects of science (*Wissenschaft als Beruf*) published in 1919 by the sociologist Max Weber, Epple links Alexander's professional situation in Princeton (full professor at Princeton University since 1928, then at the newly founded IAS) to the peculiar modern, impartial overview of various approaches to topology that Alexander surveyed in his plenary lecture at the Zürich ICM in 1932.[61] In this well-prepared terrain, the Moscow Congress planted new seeds. We quote from Hassler Whitney's (1907–1989) account of the event:

> The International Conference in Topology in Moscow, September 4–10, 1935, was notable in several ways. To start, it was the first truly international conference in a specialized part of mathematics, on a broad scale. Next, there were three major breakthroughs toward future methods in topology of great import for the future of the subject. And, more striking yet, in each of these the first presenter turned out not to be alone: At least one other had been working up the same material.
>
> At that time, volume I of P. Alexandroff & H. Hopf, *Topologie*, was about to appear.... Its introduction gives a broad view of algebraic topology as then known; and the book itself, a careful treatment of its ramifications in its 636 pages. (It was my bible for some time.) Yet the conference was so explosive in character that the authors soon realized that their volume was already badly out of date; and with the impossibility of doing a very great revision, the last two volumes were abandoned.
>
> ...
>
> What was the main import of the conference? As I see it, it was threefold:
>
> 1. It marked the true birth of cohomology theory, along with the products among cocycles and cycles.
>
> 2. The pair of seemingly diverse fields, homology and homotopy, took root and flourished together from then on.
>
> 3. An item of application, vector fields on manifolds, was replaced by an expansive theory, of vector bundles.
>
> Yet seven years later, a single paper of Hopf would cause a renewed bursting open of the subject in a still more general fashion.[62]

Concerning Whitney's first item, the omnipresence of cohomological techniques in all branches of geometry, topology, and algebra today is an important new large-scale development in pure mathematics of the last 75 years. One might even be tempted to call it a revolution of mathematical practice. It appears that the history

[61] See §104, *Topologie als Beruf* in [Epple 1999], pp. 348–352. I would translate the title of this section as "Topology as a profession", although English translations of Max Weber's lecture tend to have "Science as a vocation."

[62] From Whitney's contribution to [Duren et al. 1988], which is entitled *Moscow 1935: Topology Moving Toward America*, pp. 97 and 109–110.

of this profound transformation has not been properly studied yet.[63] The Moscow Congress was but an early milestone of this development. The cohomological practices have at least two different historic roots: the calculus of homologies inherited from Poincaré and the early combinatorial topology on the one hand, and the continuation of Emmy Noether's use of crossed products which would usher in group cohomology on the other.[64] Even Alan Turing (1912–1954) briefly tapped into this mathematical oil well in the thirties.[65]

Given the remarkable importance of the Moscow Congress, people have been interested in reconstituting the lists of participants and lectures presented. Part of the problem is to correctly identify each of the 24 mathematicians—22 men and two women—on the famous group picture.[66]

Fig. 6.4 Group picture of the First International Congress on Topology, Moscow 4–10 September 1935. Credit: [Arch. ETH].

[63] See, however, the upcoming volume on the history of the notion of *Duality* edited by Ralf Krömer.

[64] See for instance the remarks following equation (15) in [Teichmüller 1982], pp. 550–551, on the one hand, and the Appendix A to [Lefschetz 1942], written by Samuel Eilenberg and Saunders Mac Lane, on the other.

[65] See [Turing 1938]. I thank Johannes Huebschmann, Lille, for drawing this paper and its relation to Oswald Teichmüller's (1913–1943) cocycle to my attention.

[66] For this problem see [Apushkinskaya et al. 2019], pp. 7–9.

Another issue is to name all those that were present at the conference but do not happen to be in the picture.[67] For instance, both André Weil and Oscar Zariski were in Moscow during the congress and gave lectures at the University after the meeting, on Algebraic Geometry rewritten in a more modern algebraic language.[68] Interestingly, Weil was convinced that he owed his being invited to the congress to the fact that he had spent time with the Noether circle and Alexandrov a few years earlier.[69]

Having extolled the merits of this landmark Congress, it is only fair to report that not all participants were equally impressed, and some may have missed the important reorientation of the field that began to transpire in Moscow. Juliusz Schauder for instance reported to Jean Leray—frankly admitting that he himself was not a topologist—to have come away with

> the impression that there is currently a standstill in topology and that people are unable to attack fundamentally new problems. Almost all the talks revolved around new ways to define manifolds, properties of cycles, and so forth, which only roll out broadly (or in the best case, complete) in an uninteresting way what is already known. In general, these investigations are not tailored towards applications and the mathematicians that were present there are interested only in topology as such. Two talks, however, were kind of sensational in that they presented solutions to problems which may have some importance ... [70]

To summarize, a new type of mathematical networking took place in the 1930s, characterized by specialized meetings. Such conferences continue to mark the life of research mathematicians today. Before closing this section, let us review international mathematical congresses of the same decade that were not devoted to a specific mathematical topic or discipline.

6.2.2 ICMs of the Thirties

Only two ICMs could be realized after the Bologna ICM of 1928 and before 1950; they took place in Zürich in 1932, and in Oslo in 1936.

In addition to these there was also a curious International Congress organized by the French Mathematical Society (SMF) in connection with the Paris World Fair in July 1937: the *Réunion Internationale des Mathématiciens*. This faint reminder of

---

[67] See [Apushkinskaya et al. 2019], p. 5.

[68] See [Parikh 1991], p. 79, and [Audin 2011], p. 473. Cf. Weil's personal account of his stay in Moscow in [Weil 1992], p. 106–110.

[69] See [Weil 1980], p. 531.

[70] From Schauder's letter to Leray dated 2 December 1935 in [Arch. MIR]; my translation from German. I thank Christophe Eckes, Nancy, for having shared this source with me. The two talks that Schauder singles out—because others had told him of their importance—were given by Georg Nöbeling (1907–2008) and Witold Hurewicz.

the 1900 ICM, which had also taken place during a Paris World Fair, seems to have been decided rather on short notice, in January 1937, by the Council of the SMF. However, the mathematical presence in the 1937 World Fair was not an accident.

> In 1933, the French *Commission for Intellectual Cooperation*[71] and the *Confederation of Intellectual Workers* (CTI) . . . decided to entrust the General Commissariat of the international exhibition planned in Paris in 1937 to the deputy Adrien Berthod (1878–1944) who belonged to the Radical group. Berthod was a close associate of Borel, himself a deputy since 1924 and vice-president of CTI. Another vice-president of the CTI was . . . André Léveillé (1880-1962). . . . Berthod asked Léveillé to take care of the coordination of the intellectual aspects of the international exhibition. This included a section of Sciences, and in particular a sub-section called 'Scientific discoveries in their applications' chaired by the physicist Jean Perrin (1870–1942) with Borel as vice-president. A project was proposed, notably under the influence of the exhibition 'A century of progress' presented in Chicago in 1933, which focused on the presentation of technological advances and on the participation of visitors to discover and understand scientific matters by touching objects and participating in realtime experiments. . . . The [Paris] project was . . . [a] huge success. . . . [It led to the creation of] the *Palais de la Découverte* ('Palace for discovery'). It should be noted that the initiative was facilitated by the fact that the government in power in 1937 was still the *Front Populaire* government, in which Jean Perrin was the undersecretary of state for research.[72]

The conference organized by the SMF was also generously financed by the French Minister of Foreign Affairs Yvon Delbos (1885–1956). According to the Proceedings of the event, "more than 200" members of the SMF, French or foreign, "adhered" to this meeting.[73] Twelve lectures were given and later published in [SMF 1939]. They were due to the Frenchmen Denjoy and André Marchaud (1887–1973)[74], the Romanian mathematician Petre Sergescu (1893–1954)—one of the initiators, in 1929, of the Congresses of Romanian mathematicians; in 1937 he spoke on the mathematics in medieval Paris—, the Swiss mathematician and philosopher Ferdinand Gonseth (1890–1975), the Belgian algebraic geometer Lucien Godeaux, the Dutchman Johannes Gualtherus van der Corput (1890–1975), the Italians Tullio Levi-Civita and Vito Volterra (on the Lotka–Volterra equations in biomathematics); Marcel Riesz (1886—1969) of Hungarian origin, Stanisław Zaremba (1863–1942) from Poland, Richard von Mises—at the time based in Istanbul, Turkey, following his emigration from Nazi Germany—, and the British analyst Laurence Chisholm Young (1905–2000), son of Grace Chisholm Young and her husband William Henry Young.[75] In spite of well-known names in this list of speakers, the Proceedings of this meeting do not radiate the same spirit of the thirties we have encountered before.

So we pass to the two ICMs of 1932 (Zürich) and 1936 (Oslo). They offer sharply contrasting images, which are mainly due to the dismal political tide in the 1930s. Both ICMs took place in Europe. But whereas the only potential barrier to attend

[71] The French national commission affiliated with the ICIC—cf. Section 4.2 above.

[72] See [Mazliak 2020], pp. 52–53; see also the literature cited there.

[73] See [SMF 1939], p. VII.

[74] Cf. [Ferrari 2018].

[75] We learn from [Young 1981], p. 258, that Volterra could not attend the meeting because of bad health; his talk was given by a proxy. Young also tells us that attendance of the talks, which took place at *Institut Henri Poincaré*, varied considerably.

the Zürich Congress seems to have been financial considerations, the Oslo ICM four years later also suffered from various severe political restrictions: Soviet and Italian delegates were not cleared by their governments, and the German delegation was handpicked by the Nazi administration.[76]

On 31 October 1931 Hermann Weyl wrote to Ludwig Bieberbach about the possibility of canceling the 1932 meeting of the German Mathematical Society DMV in view of the upcoming ICM:

> I have recently met with the Zürich colleagues who are naturally worried about mounting the ICM in the current [economic] situation. It is likely that the Swiss Mathematical Society will skip its regular meeting next fall because of the ICM. Would it not be reasonable that the DMV also forgoes its annual meeting—assuming of course that Germans will not in fact be excluded from participating in the ICM because of difficulties to get foreign currency and cross the border? Under the current circumstances I can hardly imagine that there would be people who are so keen to go to congresses that they would like to participate in both events. Or would you mind cancelling a national in favor of an international Congress?[77]

The last sentence is of course an allusion to the fact that Bieberbach, along with Brouwer, had been one of the mathematicians leading the boycott movement in Germany against the Bologna ICM in 1928 (see Section 4.4.3.1 above). As it turned out, Bieberbach would actually attend the Zürich ICM and give a plenary talk there. That session was chaired by Takagi; Bieberbach was sandwiched between Élie Cartan and Emmy Noether. In his lecture he stressed both the connections of his subject—domains of functional operators—to virtually all fields of mathematics and its international dimension, or as he put it: "almost everyone of us could give this talk without having to fear that his own nation would be left out."[78] This was actually the only ICM that Bieberbach ever attended; a year later he would develop a racist theory of mathematical creativity and use it to back the dismissal of Jewish colleagues such as Edmund Landau.

This is one more instance where the Zürich ICM appears like a pleasant spell of sunny weather for Mathematics International, before the storm. But we already know (see Section 4.4.3.2 above) that even this friendly spirit did not help the IMU, whose image continued to be tarnished by the politics that had stood at its cradle in 1920. The IMU had played no role in preparing this Congress; but it continued to be listed among the International Scientific Unions subordinated to the IRC. When the

[76] For all questions related to the Oslo ICM we refer the reader to the very detailed and complete study [Hollings & Siegmund-Schultze 2020]. Siegmund-Schultze has also compiled a timeline of international mathematical conferences between 1933 and 1945 to monitor the participation of German mathematicians in them allowed by the Nazi regime; see [Parshall & Rice 2002], pp. 356–357.

[77] For the carbon copy of the letter, see [Arch. ETH], Hs 91:478 (Weyl papers); my translation. The end reads: . . . *Oder sind Sie dagegen, daß man zugunsten einer internationalen eine nationale Tagung ausfallen läßt? Herzlich grüßend . . .*

[78] See Proceedings ICM 1932, Vol. I, p. 162; my translation.

latter transformed itself into ICSU in July 1931 at Brussels, the IMU became one of the scientific member unions of this new body, thus enjoying greater autonomy than under the IRC.[79]

However, the IMU had no time to profit from this new status. The 'sunset clause' in its own statutes naturally put the discussion and vote about new statutes on the agenda of the IMU General Assembly in Zürich on 11 September 1932. This provided the occasion to influential members—the report by the secretary Georges Valiron mentions explicitly Oswald Veblen and Norbert Wiener from the USA, Harald Bohr (1887–1951) from Denmark, Jan Arnoldus Schouten (1883–1971) from the Netherlands, and George Neville Watson (1886–1965) from the UK—to declare the whole IMU to be superfluous. Defenders of the IMU—such as Bohuslav Hostinský (1884–1951) from Czechoslovakia, Stanisław Zaremba from Poland, and Rudolf Fueter (1880–1950) from Switzerland—could not weigh in sufficiently. The General Assembly finally decided the official liquidation of the IMU by 23 votes, against 16 votes in favor of continuing the IMU, and 5 abstentions. The financial assets of the union were to be temporarily deposited at the *Banque de France*.[80] Only the following motion was adopted in Zürich:

> *1. An international commission is formed in order to re-study the question of the international collaboration in the sphere of mathematics and to make propositions with regard to its reorganisation at the next congress.*
>
> *2. The actual president of the congress is charged to appoint the members of this commission.*
>
> The motion is unanimously adopted. On its basis, the president Prof. Fueter elected jointly with Cartan, Severi, Veblen, and Weyl the following commission: F. Severi (president), P. Alexandroff, H. Bohr, L[ipót] Fejér [1880–1959], G. Julia, [Louis Joel] Mordell [1888–1972], [Esteve] Terradas [1883–1950], Ch. de la Vallée-Poussin, O. Veblen , H. Weyl, Zaremba.[81]

In due course, the subsequent General Assembly of ICSU in 1934 acknowledged the loss of a member union, as well as a faint hope for 1936.[82] In 1936, however, the president of the commission created in Zürich was not even allowed to travel to Oslo. Thus it was really only through the activities on the teaching of mathematics entertained by ICMI that the IMU managed to send signs of life after 1932.[83]

After Bologna, the Zürich Congress was the second ICM in a row that was organized without the IMU. Given that the organization of ICMs was—and still is—the principal mission of the IMU, this fact must have been a major factor in the vanishing of the first IMU. This is also suggested by a comparison of what happened

---

[79] See [Lyons 1932], p. 29, and *Annexe X*, pp. 50–54. The transformation was motivated by the 'sunset clause' no. 23 in the statutes of the IRC which imposed renewal after 1931 (see Section 4.1). As noted before (Section 4.4.3.1) the main difference between the IRC and ICSU was that the latter accepted scientific unions as full members; cf. [Greenaway 1996], pp. 33–38.

[80] See the summary of Valiron's report of the General Assembly published by Henri Fehr in his journal *L'Enseignement mathématique* 31 (1932), 276–278. After this report, on page 278, Henri Fehr adds his personal regrets about this decision.

[81] See Proceedings ICM 1932, Vol. I, p. 61; my translation of the non-italicized part of the quote.

[82] See [Lyons 1935], p. 119: *L'Union Internationale des Mathématiques s'est dissoute, mais on pourrait revoir la situation à Oslo en 1936.*

[83] See Chapter 9 below. Cf. Chapter 10 of [Hollings & Siegmund-Schultze 2020].

to the IMU with the history of the International Astronomical Union (IAU). The IAU had been founded under the IRC rules in 1919 (a year before the IMU; see Section 4.1.1). Protests inside the IAU against the IRC politics of exclusion followed roughly the same timeline as within the IMU (recall in particular Section 4.4.2, the 'disagreeable tempest' at the Toronto ICM in 1924). The main difference between the two unions is that astronomers had various international cooperation projects running before World War I, which called for continuity, and some of them implied colleagues from countries excluded under IRC politics. Therefore already in 1922 the astronomers pushed for the modification of a clause in the statutes of the IRC to the effect that "The President of a Union shall have power to invite scientific men who are not delegates to attend a meeting of the Union, provided they are subjects of one of the countries adhering *or entitled to adhere* to the International Research Council." When Willem de Sitter (1872–1934)—who as a Dutchman counted as 'neutral' from the point of view of World War I—was President of the IAU in the 1920s, he would make generous use of this new clause, which allowed him to invite individual German astronomers as of 1926, when IRC membership was open to Germany—never mind that Germany would never join the IRC.[84] Also today the IAU differs from the IMU and other unions in that it counts among its members many individuals—which are no longer all men—, next to nations represented by scientific boards.

In contrast to the liquidation of the IMU in Zürich, the 1932 ICM itself was much more encouraging in other respects. The presence of young researchers foreign to the host country was notably stronger in Zürich than in Oslo. Here, by 'young', to fix ideas, we mean at most 30 years old, which happens to restrict attention to persons who did not have to serve in World War I. Indeed, we have already mentioned in the introduction to the current chapter the presence in Zürich of almost all the founding fathers of Bourbaki. Also from France there was Paul Dubreil (1904–1994) and his wife Marie-Louise Dubreil-Jacotin (1905–1972), who is only listed in the ICM Proceedings as a wife accompanying her husband. By the time of the Congress both were already directly influenced by Emmy Noether, but Marie-Louise Jacotin was still working in hydrodynamics. In 1952 she would become the first woman president of the French Mathematical Society SMF. From among Emmy Noether's entourage, Erna Bannow (1911–2006; as of 1940: Erna Witt), Max Deuring (1907–1984), Heinrich Grell, and Van der Waerden (at the time already professor in Leipzig and co-active in Heisenberg's research seminar) participated at the Zürich ICM. Two German women mathematicians participated, who would both become professors at German universities after World War II: the statistician Maria-Pia Geppert (1907–1997) and the geometer Ruth Moufang (1905–1977); the latter had spent the Winter 1931–1932 in Rome on a scholarship. The young Austrian mathematicians Gottfried Köthe (1905–1989), Karl Strubecker (1904–1991), Olga Taussky (1906–1995; as of 1938: Olga Taussky-Todd) and Egon Ullrich (1902–1957) were also present. To

[84] For details see [Blaauw 1994], Chap. 4, pp. 70–101; the quote is on p. 71. Cf. [Schuster 1923], p. 63. It is instructive to read Blaauw's account parallel to the case of the IMU as told in [Lehto 1998], Chap. 2, pp. 23–60.

end this selection of examples with two mathematicians who would play important roles in the future IMU twenty years later, the Zürich ICM was the first ICM that the young Marshall Harvey Stone attended. The young Number Theorist Shôkichi Iyanaga (1906–2006) was also present, although not a registered participant of the Congress, to meet up with his teacher Takagi.[85] Iyanaga studied in Europe between 1931 and1933, first in Hamburg and then in Paris. Even though this phenomenon of young participants does not extend to all nations represented at the Zürich ICM, it strongly highlights the kind of promise for the future that the 1932 Congress held.

However, to fulfil this promise many young mathematicians in the thirties needed considerable resilience. Indeed, if one looks for a feature of the Oslo ICM which somehow marks that Congress, in the same way that the presence of young researchers marked the Zürich ICM, and which also reflects its historical moment, it is probably the presence of many emigré mathematicians (mostly from Germany) at the ICM in 1936.

> The breakdown of the members of the congress by country (Proceedings ICM 1936, Vol. 1, p. 39) is additionally marred by the vaguenesses stemming from the uprooting and the mass emigration of mathematicians at that time, and from the principal problem of whether to count the country of origin of a mathematician or that of their present residence.[86]

Therefore, if used as a "lens through which the reader of this book can view the state of the art of mathematics in the mid-1930s"[87], the Oslo congress projects a somewhat lopsided image. In spite of the relentless efforts of the Norwegian hosts, the politics of the day were more evident than the fundamental rewriting of mathematics which was under way at the time.[88] This emerges for instance when the excellent list of mathematical presentations in Oslo[89] is confronted with the contemporary book production:

> Another benchmark for judging the extent to which the Oslo congress reflected the top level of mathematical research might be the representation there of authors of the then-newly founded Springer series *Ergebnisse der Mathematik*. The series had been started in 1932 by the editors of *Zentralblatt für Mathematik und ihre Grenzgebiete* around Otto Neugebauer in Göttingen and was continued under the latter even after his emigration to Copenhagen. By 1936, this series of rather short books on cutting-edge areas of mathematical research could claim several very successful publications, although their real influence would only be visible in the decades to come. By the time of the Oslo congress, 19 of these *Ergebnisse* ("result-reports") had appeared in print. One of the most influential of these was Kolmogorov's 1933 report on the foundations of probability [Kolmogorov 1933]. But the reports of [Reidemeister 1932], [Bohr 1932], [Veblen 1933], [Radó 1933], [Bonnesen & Fenchel 1934], [Behnke & Thullen 1934], [Heyting 1934], [Zariski 1935], [Deuring 1935], [Krull 1935], and [Koksma 1936] were probably no less influential in their respective areas of mathematical research. Many of the reports were referred to and used in Oslo ......

[85] See [Iyanaga 1994], p. 187.

[86] See [Hollings & Siegmund-Schultze 2020], p. 17; see p. 188 (and *passim*) for a more detailed discussion of the emigrants participating in the Oslo ICM.

[87] From the backcover text of [Hollings & Siegmund-Schultze 2020].

[88] See [Hollings & Siegmund-Schultze 2020], pp. 150–158, for an analysis of the overall mathematical profile of the 1936 ICM.

[89] See [Hollings & Siegmund-Schultze 2020], Chapter 9, for comments on all the plenary talks given at the 1936 ICM.

> Looking at how many of the 21 authors of the 19 *Ergebnisse* reports attended the Oslo congress, one finds just four: Harald Bohr, Heinrich Behnke (1898–1979), Werner Fenchel (1905–1988), and Oswald Veblen. Only one of them, namely Veblen, gave a plenary talk, which was in fact related to the content of his *Ergebnisse* report... Of the other 18 authors of reports that had appeared by 1936, four (Reidemeister, K[urt] Hohenemser [1906–2001], Deuring, van der Waerden) lived under political suspicion or racial discrimination in Nazi Germany with no realistic chance of, or support for, going to Oslo; one author was a refugee from Nazi Germany (P[eter] Thullen [1907–1996]), whose career as a mathematician was destroyed; two were Russians (Kolmogorov and [Alexandr] Khinchin [1894–1959]), and for this reason were unable to come. Thus, a mixture of political and scientific conditions gave rise to the rather low participation of *Ergebnisse* authors in Oslo.[90]

The Oslo ICM did point the way to the future in one respect that today dominates our very notion of an International Congress of Mathematicians: two Fields Medals were awarded for the first time on this occasion. The winners were the Finnish mathematician Lars Ahlfors and the American Jesse Douglas (1897–1965).[91]

6.3 Books, Journals; *Zentralblatt* and *Mathematical Reviews*

The long passage we just quoted, concerning the *Ergebnisse* series, already introduced a material aspect of the 'Consolidation and Unification' of mathematics in the 1930s. It also indicated that this phenomenon concerned the broad spectrum of mathematics—not just modern algebra, which we chose as our starting point in Section 6.1. Indeed, while *rewritings* of various sorts are a recurring feature of the historical development of mathematics[92], the 1930s were special in this respect, not only for the scale but also for the novelty of the reshaping of mathematical theories. The rewriting of mathematics which was in the air during the 1930s was as deep as it was extensive.

6.3.1 Books

N. Bourbaki's project *Éléments de mathématique*—the title alludes to Euclid and uses the word 'mathematics' (*mathématiques*) in the singular, supposedly to stress the unity of the mathematical sciences—is an obvious symptom of this spirit, and incidentally reminds us of encyclopedic projects of the time.[93] It properly started working in 1935. Over the years it would produce visible and influential results, both in terms of published pages and of the international impact on the image of

[90] See [Hollings & Siegmund-Schultze 2020], p. 147. The *Ergebnisse* volumes were conceived as up-to-date surveys of ongoing research, more prompt and flexible than the volumes of Felix Klein's *Enzyklopädie*; see [Remmert & Schneider 2010], Section 6.6.4, pp. 177–178.

[91] See [Hollings & Siegmund-Schultze 2020], pp. 225–230.

[92] Cf. [Schappacher 2010], Section 1.

[93] See the last part of the introduction to the present Chapter.

mathematics that it projected. Its effect on mathematical practice was considerably more radical than that of the volumes of the Mathematical Encyclopedia initiated by Felix Klein half a century earlier (see Section 1.3.3).

Here we are interested not so much in the weighty series of *fascicules* on our bookshelves, but in the historical environment of their initial conception. Indeed, there was a flow of similarly radical rewritings of mathematical knowledge undertaken during that decade, and continued later. Kolmogorov's seminal axiomatic rewriting of the "Fundamental Concepts of Probability Calculus", which was already mentioned in the long quote above, appeared in 1933 and was immediately recognized as the beginning of a new era of probability theory.[94] Back in 1900, taking what he had done to geometry as a model, Hilbert had challenged mathematicians in his sixth problem for the new century, to axiomatize physics, specifically the theories of probability and of mechanics. Ideas of how to do that had of course been around, but it seems as though 'the time was ripe' only in the thirties. The French probabilist Paul Lévy for instance, looking back in his memoirs, regretted not having had the courage to take Kolmogorov's step in the 1920s.[95]

Another instructive example is provided by the new field of Functional Analysis, a domain of mathematics that had been in the making for decades, building in particular on work of Volterrra and Fréchet.[96] Polish mathematicians such as Stefan Banach, Kazimierz Kuratowski, Hugo Steinhaus (1887–1972), and others then contributed to it in a decisive way, particularly after World War I.[97] However, functional analysis appeared on the world mathematical scene as a fully fledged mathematical discipline only in 1932, the year of the Zürich ICM, heralded by the virtually simultaneous publication of three fundamental books:

(i) Banach's treatise [Banach 1932], written in French, was the first volume in the new Polish book series *Monografje Matematyczne* whose editorial committee consisted of Banach himself, Bronisław Knaster (1893–1980), Kuratowski, Stefan Mazurkiewicz (1888–1945), Wacław Sierpiński, and Steinhaus. Looking back when he was 75 years old, Laurence Chisholm Young remembered this Polish book series:

---

[94] See [Kolmogorov 1933]; my own, literal translation of the German title. A Russian translation of this book by G.M. Bavli was published in 1936. An English translation by Nathan Morrison appeared in 1950 under the title *Foundations of the Theory of Probability*; it is based both on the German and the Russian version.

[95] See [Lévy 1970], pp. 67–68.

[96] For the early development, see for instance [Siegmund-Schultze 1982].

[97] Cf. [Kuratowski 1980]. For a first orientation about what is sometimes called the Polish school of mathematics, which encompasses also other domains such as set theory and logic along with functional analysis, see Chapter 13 by Zofia Pawlikowska-Brożek of [Goldstein, Gray, Ritter 1996], especially pp. 296–301; see also [Beeler & Norwood 2014].

> We see the impact of new methods and ideas that were replacing those in use since the early 1900's in Functional Analysis, Real Analysis, Topology, and Set Theory. I cannot begin to say how much, in those insecure years, those of the depression and close to "strength through joy"[98], the great Polish Journals and monographs meant to the mathematical world. I want to mention specially the 1937 "Theory of the Integral" by Stanisław Saks (1897–1942).[99]

(ii) The young Marshall Stone's English textbook [Stone 1932] published by the American Mathematical Society runs in many ways parallel to Banach's account; past a few introductory remarks and reminders, both textbooks quickly head for the modern axiomatic introduction of vector spaces, to then endow them with additional structures.
(iii) Finally, John von Neumann's book [Neumann 1932] focussed on the use of functional analysis for quantum mechanics. It appeared in German as Volume 38 of the *Grundlehren* series of Springer. This very successful textbook series, which is still a trade name of the Springer publishing house, had been launched already in 1921 following an initiative of Richard Courant's (1888–1972) in Göttingen. (The two volumes of Van der Waerden's *Moderne Algebra* had appeared as Vol. 33 and 34 of the *Grundlehren* series.)[100]

Apart from these systematic expositions of the new discipline, the impact of functional analysis was further strengthened at the time by new developments. A well-known example of these is the Leray–Schauder approach that connected the field closer to the budding topology.[101] Finally, we should not leave this topic without reminding the reader that both Saks and Schauder were killed by the Gestapo during the Nazi occupation of Poland.

6.3.2 Journals and Politics

The footprint of the mathematical life of the time is also visible in newly founded journals. As for functional analysis, the first issue of the Polish journal *Studia mathematica* appeared in 1929. The other well-known Polish journal *Fundamenta mathematica*, centered around mathematical logic, had been launched as early as 1920. Looking back on the Polish mathematical tradition in his opening speech at the 1983

[98] This is a sardonic allusion to the Nazi organization *Kraft durch Freude*.

[99] The quote, which follows a brief discussion of the Hahn–Banach Theorem, is from [Young 1981], p. 321. The 1937 edition of Saks's book, which is Vol. 7 of the series *Monografie Matematyczne*, was a substantial revision of the 1933 French edition, which had appeared as Vol. 2 of *Monografie Matematyczne*. The English of the 1937 edition was due to L.C. Young. An initial Polish version of the book had been published by Saks back in 1930.

[100] See [Remmert & Schneider 2010], Section 6.6, for a detailed comparison of various mathematical book series of German publishers between the World Wars. Cf. [Bergmann et al. 2012], pp. 467–476, in particular the documents reproduced there.

[101] We have briefly mentioned this collaboration at the beginning of Section 6.2 above.

ICM at Warsaw, Czesław Olech (1931–2015) would call the *Fundamenta* "the first specialized international journal of mathematics in the world" to be founded.[102]

In the USSR, a new era of mathematical consolidation and coming together was triggered, or forced, when the Presidium of the Academy of Sciences of the USSR moved from Leningrad to Moscow in 1934. We have highlighted above (Section 6.2.1) the international dimension of Soviet mathematics at that time by the example of the 1935 Topology Congress in Moscow. This process is reflected in the constellation of mathematical journals, which is summed up in Sergei Sergeevich Demidov's (b. 1942) historical note for the 70th birthday of the mathematical *Uspekhi Matematicheskikh Nauk*, i.e. literally, *Advances of the Mathematical Sciences*, founded in 1936:

> Of course, the active development of mathematical life was accompanied by an enlivening of publishing activities, including improvements of mathematical periodicals. In this respect the following situation had come into being by the beginning of the 1930s. The oldest Russian mathematical journal, *Matematicheskii Sbornik*, which had gone through a period of painful restructuring in 1930–1931 connected with the arrest and death of its editor-in-chief, Egorov, the president of the Moscow Mathematical Society, continued to be published. After the eruption in [the] country of the ideological campaign called the struggle against 'Egorovism', a new editorial board of the journal was organized headed by [Otto Yulyevich] Shmidt [1891–1956], who appealed to Soviet mathematicians to support their journal and break with the practice of publishing their best papers in Western journals. "Soviet mathematics should and must have a journal of international significance." In 1936 a new series of the journal began as an organ of the Mathematical Group of the Academy of Sciences of the USSR and mathematical research institutions and societies of the People's Commissariat for Public Education of the Russian Soviet Federal Socialist Republic. ... By the mid-1930s the situation with regard to efficient publication of new mathematical results had begun to go well. ...
>
> However, with the rapid development of mathematical research it became evident that a new type of publication was needed: a journal containing surveys on diverse areas of contemporary mathematics, information about events in Soviet and foreign mathematical life, and reviews and information about the newest works on mathematics and its applications. It was therefore decided at the Second All-Union Mathematical Congress to arrange for such a publication: *Uspekhi Matematicheskikh Nauk*.[103]

In economically difficult times, founding new journals could also be a way to enrich the local library of the editor's institution via exchange agreements with other journals from all over the world. A case in point for this business model was the creation of the *Abhandlungen aus dem mathematischen Seminar der Universität Hamburg*.[104] The first issue appeared in 1921; the University of Hamburg had only been created in 1919. Seminal papers, for instance by Emil Artin, Wilhelm Blaschke (1885–1962), and Erich Hecke, were published here.

> This handsome volume, illustrated by two portraits of Artin in his Hamburg days, cannot but recall, to the minds of those old enough to have had this experience, the times when one used to glance breathlessly through the table of contents of each new number of the *Hamburger*

[102] See Proceedings ICM 1983, Vol. 1, p. LII.

[103] See [Demidov 2006], pp. 793–794.

[104] See [Remmert & Schneider 2010], pp. 159–163.

> *Abhandlungen*, with the hope, seldom disappointed, of finding Artin's name there. Here they all are—the papers on the L-series, the law of reciprocity, the real fields, the hypercomplex arithmetic, the excursions into topology. The excellent photographic reproduction brings back even the distinctive typography of the journal; one misses only the texture of the paper, which somehow had become part of the magic.[105]

This quote should, incidentally, remind today's reader of the supreme importance of printed paper in pre-internet times. The consecutive issues of the principal journals arrived at the local library, where one regularly went to consult them. And requesting offprints—which one then had to file—was a standard way to interact with colleagues in whose research one was interested.

The business model of journal exchange was of course not new. It had been successfully practiced before, many times and all around the world. The internationally successful *Tôhoku Mathematical Journal* for instance was launched in 1911, a few years after the university of the same name was founded in Sendai, Japan. It could build on the strong research record of the Tôhoku Mathematics Department and attracted a number of international authors, in particular also in the thirties.[106]

We have given these examples—the list could be expanded in various directions —principally in order to illustrate bright aspects of the mathematical thirties. However, the decade also spurred national preferences in ways that would affect Mathematics International. In the remainder of this section we illustrate this other side of the 1930s with three quite different examples.

In 1938, the Tensor-Society (*Tenzoru Gakkai*) in Tokyo, whose members included mathematicians and physicists, founded the new journal *Tensor* (*Tenzoru*). The preamble of the first issue starts like this:

> So far, the White race has widely propagandized the idea that Japanese are rich in the ability to imitate, but poor in originality. Not only our common people had the tendency to believe this, but statesmen and non-scientists have turned it into public belief that this is a flaw of our national character. A few observations suffice to overcome this misconception: our scientific world has shown remarkable progress, our economy and industry is expanding globally, and our war airplanes are bombing the Chinese heartland. Now these are the applied sides of science, which easily meet the eye of the public. Regarding the theoretical sides, which are hard to understand for the public, and among them the fundamental research in the fields of mathematics and theoretical physics that can only be understood with utmost difficulty, it is also here that we Japanese have realized tremendous progress. In regard to creativity and also to excellence, I thus feel that we have proved with facts that we are by no means inferior to the white race, and this truly cannot be but a delight for our country.[107]

The genuine protest against white paternalistic attitudes and the plea for the pride of one's own achievements are put here into the context of Japanese politics and the ongoing war for dominance in Asia at the time. This reminds us of the Racial Equality Proposal that Japan had unsuccessfully tried to write into the Versailles

[105] Beginning of André Weil's review of Artin's Collected Papers; see [Weil 1979], Vol. III, p. 173.

[106] See [Kümmerle 2018].

[107] See [Kawaguchi 1938], p. 1. I thank Harald Kümmerle for drawing my attention to this journal and for the translation. Cf. [Kümmerle 202.].

peace treaty in 1919. British (particularly Australian) and US (particularly Californian) resistance had prevented this from happening.[108]

A more surprising link between a political constellation and a mathematical publication project is provided by the European scene of algebraic geometry in the thirties. Just like probability theory at the hands of Kolmogorov, algebraic geometry (cf. Section 1.1.5.4) was ready to be rewritten in the 1930s, along the lines of Emmy Noether's *Auffassung*, thereby generalizing the theory to situations over base fields other than the complex or the real numbers. Van der Waerden set out to do just this under Emmy Noether's influence as of 1926. But following his first personal encounter with Severi at the Zürich ICM, which must have been as inspiring as it was intimidating, Van der Waerden would subsequently tone down the modern approach in his articles on the subject. He published a pedagogically very valuable textbook [Van der Waerden 1939] as Vol. 51 of the *Grundlehren* series. In this book he essentially hides the gist of the new approach via abstract algebra. This compromise would please Italian critics. Around that time, Helmut Hasse dreamed of creating a German-Italian axis of algebraic geometry, parallel to the fascist Berlin-Rome axis, that was to manifest itself in a series of monographs. This led nowhere, and a first modern rewriting of algebraic geometry would be achieved during the 1940s in the US, by Oscar Zariski and André Weil.[109]

6.3.3 Review Journals and Politics

Yet another political disruption of the 1930s in the world of mathematical publishing concerned all of mathematics, not just a special discipline, and had global repercussions. It was at the origin of the fact, known to all active mathematicians, that today there are two global mathematical review journals: the *Mathematical Reviews* (with their corresponding website MathSciNet), and the *Zentralblatt* (with its website zbMATH.Open).

Before zooming in on what happened in the thirties, let us recall from Section 1.1 that Western mathematics underwent a process of professionalization in the nineteenth century which expressed itself in an increasing number of mathematical journals. As the publication of mathematical papers was globally adopted as the standard research practice, the need arose to classify and to survey the ever growing stream of printed works. Consequently, the task of classifying mathematical research was repeatedly mentioned as of the very first International Congresses—see Section 1.4.1.2. On a more political note, the problem of mathematical review journals came

---

[108] See [Shimazu 1998], Chap. 7, for a concise analysis of the political issues around this 1919 Japanese proposal, which was apparently neither meant nor discussed as a universal anti-racist statement at the time.

[109] This story of mathematical politics, or politicized mathematics is told in [Schappacher 2007]; see also [Schappacher 2006].

up strongly in the new IMU after World War I, with a view to breaking what was perceived as a German monopoly in this domain (Sections 4.1.1 and 4.3.2). Recall also the analogous mission of the ICIC (Section 4.2).

Several mathematical review journals had been launched in the nineteenth century. The German *Jahrbuch über die Fortschritte der Mathematik* was founded in 1869; its first volume appeared in 1871 and contained reviews of papers published in 1868. The French *Bulletin des sciences mathématiques et astronomiques* edited by Gaston Darboux was founded in 1870, and the Dutch *Revue semestrielle des publications mathématiques* started to appear in 1893. Around the turn of the century these three journals more or less complemented each other. Attention to classification of the reviewed articles varied both between the three journals and in the course of time. The principle of the *Jahrbuch* to compile volumes labeled by years that covered precisely all the papers published during that year was increasingly difficult to realize, and led to growing delays in the publication of its volumes, rendering it less and less useful for mathematicians eager to keep informed about ongoing research.

Around 1930, plans to improve the timeliness of mathematical refereeing mingled with the interests of the Springer publishing house, advised by Courant, on the one hand, and recent memories of the campaign led by Bieberbach and Brouwer against German participation at the Bologna ICM in 1928 (see Section 4.4.3.1 above) on the other. One result was the temporary fusion of the Dutch *Revue semestrielle* with the German *Jahrbuch*, which would last from 1932 to 1934. When this joint venture was still being negotiated, the new *Zentralblatt für Mathematik und ihre Grenzgebiete* was founded. It was directed by Otto Neugebauer, at the time Courant's assistant in Göttingen, and published by Springer; the first volume appeared in 1931.[110] From its inception, the *Zentralblatt* had to build up a largely new reservoir of referees because it was competing with the *Jahrbuch*. This group of referees was more international than those of the *Jahrbuch*, a difference that became increasingly visible with the effects of emigration of mathematicians from Germany for racial and political reasons. Neugebauer himself was forced out of Göttingen in 1934, and continued running the *Zentralblatt* from Copenhagen in the mid-thirties.

The situation got a lot more tense when World War II was already looming large. By 1938 influential German mathematicians such as Blaschke, who had arranged themselves very well with the Nazi regime, started to openly deplore what they perceived as a declining presence of the German language and German reviewers in the *Zentralblatt*. In October 1938, the publisher Springer was pressured into dismissing Tullio Levi-Civita from the board of editors of *Zentralblatt*, and to introduce a new rule according to which papers by Germans should no longer be refereed by 'non-Aryans' or émigré mathematicians.[111] The removal of Jewish coeditors, members of associations, etc. had been a key aspect of the Nazi science policy for some time. It had also been brought to Italy where the so-called racial laws (*Leggi razziali*) were about to be promulgated, which would forbid Levi-Civita to even enter his institute's library. Subsequently, Italy would be represented on the board of the *Zentralblatt* by

[110] See [Siegmund-Schultze 1993]; in particular the timeline recording events from 1928 to 1934, pp. 51–52.

[111] See [Siegmund-Schultze 1993], pp. 159–167.

Enrico Bompiani (1889–1975) and Francesco Severi. As soon as Neugebauer learnt of the ousting of Levi-Civita, he resigned as managing editor, along with Harald Bohr, Courant, Hardy, Jacob David Tamarkin (1888–1945), and Veblen.[112]

Meanwhile the idea of a new, US-based review journal for mathematics had already suggested itself in view of the increasing weight of American mathematics. Indeed, the Oslo Congress in 1936 had been the first ICM where the American participants outnumbered every other national group, and the US delegation had offered there to organize the 1940 ICM in Cambridge, Massachusetts—which would finally have to be delayed by 10 years.

The fascist *Gleichschaltung* of the *Zentralblatt* in 1938 naturally encouraged such plans for an alternative review journal, and they were further pushed on by incidents like Helmut Hasse's infamous reply in a letter of 15 March 1939 to Marshall Stone's remarks about the exclusion of Jewish reviewers for papers by German mathematicians:

> Looking at the situation from a practical point of view, one must admit that there is a state of war between the Germans and the Jews. Given this, it seems to me absolutely reasonable and highly sensible that an attempt was made to separate within the domain of the *Zentralblatt* the members of the two opposite sides in this war. I do not understand why the American mathematicians found it necessary there on to withdraw their collaboration in bulk. I do not know whether it was the intention, but it certainly has the appearance of taking decidedly and emphatically one of the two sides, and thus deviating from a truly impartial and hence genuinely international course.[113]

The *Mathematical Reviews* were actually launched in 1940, again under the direction of Otto Neugebauer, who from Denmark had emigrated further to the US, and was now working at Brown University.

6.4 Three Journeys to the West

The bright moon and the cool, clear dew,
Though in each corner not one speck of dust.
Sheltered fowls roosted in the woods;
A brook flowed gently from its source.
Darting fireflies dispersed the Bloom.
Wild geese spread word columns through the clouds.
Precisely it was the third-watch hour—
Time to seek the Way whole and true.

The Journey to the West[114]

---

[112] Cf. [Israel & Nastasi 1998]; on the *Zentralblatt* affair, see specifically p. 324–325, as well as p. 349, endnote 142.

[113] Quoted from [Siegmund-Schultze 1993], p. 164.

[114] See [Journey 2012], Vol. I, p. 119.

The year 1935 marks the founding of the Chinese Mathematical Society, which would publish the first volume of its journal in 1936.[115] In spite of this encouraging moment, the words 'consolidation and unification' in the title of the present chapter seem rather out of place when one looks at the tormented history of China in the 1930s, and indeed during most of the period of the Republic of China, 1912–1949. In the nineteenth century, China had been mistreated by Western powers with colonialist ambitions. The painful question of how to deal with these humiliations (*guochi*) continued to haunt intellectual and political discussions in China for decades.[116] When it was decided to learn from Western science in order not to be at the mercy of foreign military technology any longer, China was lagging behind the Japanese reorientation towards the West during Japan's Meiji and Taisho eras.

A number of institutions of international cultural exchange were created in China.[117] In the Academic realm let us mention here only Tsing Hua (Qinghua) University in Beijing. It was founded as a College in 1911. Its endowment was a transmutation of reparation payments imposed on China by the US under the Boxer Peace Protocol (1901). The sum due had been lowered by President Roosevelt's administration; the saved amount was partly funneled into a scholarship program for Chinese to study abroad. In 1928 Tsing Hua University became a national University. By the middle of the 1930s it was the major hub of scientific excellence and exchange with the West. During the second Sino-Japanese war (1937–1945) it was relocated twice, jointly with other universities.

In the present section we compare the early careers, and journeys, of three well-known mathematicians from China, all born between November 1910 and October 1911. The choice of these three mathematicians is rather elitist; they were of course neither the first nor the only research students of mathematics from China who went to Europe or the US. We have chosen them here to illustrate both the existing international career options, and how these could be jeopardized by the historical constellation of the thirties and forties. Also their lifelines intersect in interesting ways. In chronological order of birth, we shall discuss HUA Luogeng (1910–1985)[118], Wei-Liang CHOW, and Shiing-Shen CHERN (1911–2004).[119]

Hua Luogeng (also transcribed as Loo Keng) from Jintan—at the time but a small town in Jiangsu Province—grew up in very modest circumstances. For instance, the family could not afford to let him finish his College degree in Shanghai. He was mostly self-taught, by dint of relentless reading and working at home in Jintan. A typhoid fever resulted in a lifelong paralysis of his left leg. Hua did profit from books as well as general pedagogical guidance provided by two local teachers. One of them

[115] See [Dauben 2002], pp. 277–280.

[116] See [Cohen 2004], Chap. 6.

[117] Cf. again [Dauben 2002], in particular pp. 275–277.

[118] These are the dates usually cited. In [Arch. IAS], Nr. 56883 (Hua Luogeng's file), the *Application for Stipend for 1947/48* form indicates 11 October 1909 as his birthday.

[119] As before I tend to use the version of a Chinese name which is most common in English publications: At the first mention of a person, I HIGHLIGHT the family name.

was WANG Weike (1900–1952), a true intellectual who had taken part in the student protests that initiated the *May Fourth Movement* in 1919. In the mid-20s Wang spent a few years in Paris, where he studied astronomy and mathematical physics and worked under the direction of Marie Curie. Back in China, Wang Weike's later achievements include a translation of Dante's *Divine Comedy* into Chinese.[120]

The kind of mathematics that Hua was taught, respectively taught himself, was essentially Western, initially based on the curriculum which the so-called *Self-Strengthening Movement* in China had produced since the second half of the nineteenth century in a multilayered cultural transfer from England, France, Germany, the US, and also Japan.[121] Through a lucky chain of coincidences people took notice of Hua and his very first publications, on Sturm's theorem and on the impossibility of solving the general quintic equation by radicals. Thus, in spite of the fact that he had no graduate exam—the first degree he earned in his life would be an honorary doctorate from Nancy University in France in 1980—he was invited to join Tsinghua University, at first as a clerk; he arrived there in August 1931.[122] The inspiring atmosphere at the new place made him meet other mathematicians and shifted his mathematical interests from algebra to the Hardy–Littlewood circle method. It also forced him to improve his English. By 1934 he was promoted to lecturer at Tsinghua.

The Tsinghua Mathematics Department was Hua's door to the West. Jaques Hadamard and Norbert Wiener visited Tsinghua for several months in 1935/36.[123] It was apparently Wiener who suggested that Hua should spend time with Hardy in Cambridge, England. Hua did go to Cambridge in 1936–1937 on one of the Tsinghua scholarships mentioned above. Although he again lacked the money to crown his stay with a doctorate degree from Cambridge University, the contacts and inspirations he received there tremendously helped his standing in the world of mathematics. He returned to China a bit earlier than planned, worried by the Japanese invasion.

> While his earliest publications had shown his mathematical interests to be wide-ranging, his efforts had begun to focus on Waring's Problem as early as 1934 and during the Cambridge period he laid the foundations for his enduring contributions to additive number theory. It is quite astonishing to realize that these [less than] two years, then three important months during 1945-46 with [Ivan Matveevich] Vinogradov [1891–1983] in Russia and barely five years in the USA, from 1946-1950, were all the time that Hua spent at major mathematical centers in the west (three years at Princeton and two at Illinois); yet during this period he embarked on his researches in matrix geometry, in functions of several complex variables, equations over finite fields, automorphisms of symplectic groups, while yet he continued to make fundamental contributions to analytic number theory.[124]

On 1 October 1949, Mao Zedong proclaimed the People's Republic of China in Tiananmen Square. On 10 December 1949, Hua wrote from Urbana, Illinois, to Wei-Liang Chow, Department of Mathematics at Johns Hopkins University, Baltimore,

---

[120] See [Wang 1999], p. 24.

[121] I am not aware of a detailed analysis of young Hua's sources. The *Self-Strengthening Movement* and its continuation in various threads is discussed in [Dauben 2002], pp. 256–276.

[122] See [Wang 1999], p. 41.

[123] Cf. the detailed account of his stay given by Wiener in [Wiener 1956], Chapter 10.

[124] See [Halberstam 1986], p. 63.

announcing the end of his peregrination: "Dear Wei-Liang, It has been a long time since we have last been in contact. How are you? ... I have decided to go back to China soon. Please let me know what you think of my decision.[125]

The addressee of the letter, Wei-Liang Chow, or Zhou Wei-Liang was born on 1 October 1911 in Shanghai into a well-to-do "high mandarin family in China, which recognized early the need of westernization."[126] His great-grand-father, Zhou Fu, was a mandarin and governor of Jiangsu-Zhejian. The sons of Zhou Fu were successful industrialists. The grand-sons of Zhou Fu—including Chow's father, Zhou Da—were industrialists or scholars. At least seven of the great-grand-children of Zhou Fu, including Wei-Liang Chow, were well educated.

> Except for a very brief period, I never attended schools or colleges in China. Beginning at the age of five (1916) I was taught the standard Chinese classics by an old Chinese tutor and at the age of eleven I was taught to read and write English. However, I discovered very soon that the ability to read English provided me with the opportunity to acquire the knowledge about almost any subject I wanted to learn. Since the curricula in most Chinese universities at that time were modeled after those in the American universities and many of them often used books written by American professors, it was not difficult for me to find out the most commonly used text books in America on most subjects. Thus in this way I taught myself all sort of subjects from mathematics and physics to history and economics. This situation lasted from 1924 to 1926 when I succeeded in persuading my father to send me to study in the United States. At that time my main interest was political economy, and economics was still my major subject of study when I entered the University of Chicago in October 1929. However, during the next two years I began to have some serious doubts about taking economics as my major.[127]

Before moving to the University of Chicago, Chow had enrolled at Ashbury College in Wilmore, Kentucky, and then at the University of Kentucky at Lexington. But he graduated from the University of Chicago. After losing his interest in economics during the great depression, he moved into physics, but would not stay put.

> I happened to read the book called *[A Course of] Pure Mathematics* by the famous English mathematician Hardy. This book opened the door to mathematics for me, although I was at that time still studying applied mathematics, hoping eventually to study physics. In summer 1931 I discussed studying mathematics with a graduate Chinese mathematics student who got his Ph.D. at Chicago and then spent a year in Princeton. He was very enthusiastic about Princeton (he attended the lectures of John Von Neumann there) and he advised me to go to Princeton or even better to go to Göttingen in Germany which he thought was then the world center for mathematics. Therefore, with only a vague idea of studying mathematics, I went to Göttingen in October 1932. Although I had previously taken a course in German at the University of Chicago, it took me about three months to learn the German language sufficiently to enable me to understand the lectures.[128]

---

[125] See [Arch. JHU], Wei-Liang Chow papers, MS.0762, box 1. Hua's eminent role as leading mathematician of the People's Republic of China falls outside of the present chapter. It is treated in the sources we have quoted.

[126] In Chern's words; see [Yau 1992], p. 6.

[127] See [Chow 2002], pp. 481–482.

[128] See [Chow 2002], pp. 482–483.

This is how Chow came to Göttingen. Unlike Hua Luogeng, Chow did not need to solicit a scholarship to attain a world mathematical center in Europe. In June 1933, he signed the petition of Emmy Noether's students trying to avert her dismissal—see Section 6.1.3. As Chow himself would put it in 1990:

> [A]t the beginning of 1933, something happened in the German politics which would soon change drastically not only the university at Göttingen, but the entire Germany and in fact eventually the entire world, namely Hitler and his Nazi party came to power. . . . Thus the world mathematics center I hoped to come to study was essentially depleted.[129]

Rather than staying in Göttingen, Chow then divided his life for the next few years between Leipzig and Hamburg: Leipzig because he worked on a thesis in Algebraic Geometry with Van der Waerden; he obtained his doctorate from Leipzig University in June 1936. In the process, their exchanges resulted in the seminal, still famous joint paper [Chow & Van der Waerden 2002] which introduced what is known today as the Chow form of a projective algebraic variety. As for Hamburg, it was not Emil Artin's lecture courses, some of which he did follow, that primarily attracted Chow to this city, but a young lady, Margot Victor (19??–2001), whom he had first met there in the Summer of 1934. They eventually married in Hamburg, a month after his Leipzig doctorate. Thereafter the Chow couple moved to China; Chow worked at the University of Nanking until the beginning of the second Japanese-Chinese war.

Emil Artin's wife was Jewish. This is why Artin lost his Hamburg chair in the Summer of 1937. The Artins emigrated to the US, at first to the University of Notre Dame. The Victors were a Jewish family in Hamburg. An uncle of Margot's, Hans Victor, was a businessman who transferred his business to the US in the thirties and settled in Newark, New Jersey. Margot's parents eventually left Germany as well and found themselves almost penniless. To be able to support them, and to simply survive the second Japanese-Chinese war, Chow abandoned his mathematical work and started a business in Shanghai.

Back in 1934–1936, Hamburg was still a very bright spot on the world map of mathematics; we have mentioned the local mathematical journal *Abhandlungen* in Section 6.3 above. Shiing-Shen Chern would publish his thesis in this journal in 1936. Chow and Chern first met in Hamburg when they were both thesis students. After his first student years at Nankai University, Chern had moved to Tsinghua University in 1930 where he was a student of the differential geometer Dan SUN (1900–1979), who had obtained his PhD in Chicago in 1928, and whom Chern would later remember as "at that time the only mathematician in China publishing research papers."[130]

> In the spring of 1932 Blaschke visited Peiping and gave a series of lectures on "topological questions in differential geometry." It was really local differential geometry where he took, instead of a Lie group as in the case of classical differential geometries, the pseudo-group of all diffeomorphisms and studied the local invariants. I was able to follow his lectures and to read many papers under the same general title published in the *Hamburger Abhandlungen*

[129] See [Chow 2002], p. 483.

[130] See [Yau 1992], pp. 2.

Fig. 6.5 Chern (left) and the Chow couple in Hamburg, 1936. Credit: International Press Boston.

> and other journals. The subject is now known as web geometry. With this contact and my previous knowledge of Blaschke's books on differential geometry, I decided to go to Hamburg as a student when a fellowship was made available to me in 1934. . . .
>
> Hamburg had a strong Department, with professors Blaschke, Artin, and Hecke, and junior members including E[rich] Kähler [1906–2000], H[ans] Petersson [1902–1984], H[ans] Zassenhaus (1912–1991).[131]

After his doctorate in Hamburg—which he obtained in 1936, the same year that Chow got his PhD in Leipzig—Chern went to Paris for a year during which he studied with Élie Cartan and followed the *Séminaire Julia*, which was essentially run by the Bourbaki group at the time and whose overall subject that year was Élie Cartan's work. By the time he returned to China in the Fall of 1937, Tsinghua University had already been moved once because of the war, to the Southwest of China. In the next few years Chern and Hua were two of the mathematicians teaching at this 'Southwest Associated University'.

---

[131] See [Yau 1992], pp. 3–4.

One of the difficulties of those years was to supply researchers in China with recent journals or offprints from the West. Chern and many others who had contacts abroad organized as much as they could in this respect.

On 22 April 1942, Oswald Veblen of the Institute for Advanced Study in Princeton reported to Frank Aydelotte Aydelotte, F.(1880–1956), who had succeeded Flexner as the Institute's director in 1939, about his exchange with Chern, concluding that "Chern is the most promising Chinese mathematician who has thus far come to our attention." What he then suggested reflects the circumstance that, at that time, China was an ally of the US in the ongoing war against Japan:

> In the present circumstances our recommendation is that an attempt should be made to bring him to the Institute for a couple of years. The problem of bringing him here and of returning him to China at the end of his period might be referred to the Chinese Embassy in Washington, and the funds might be sought from one of the Foundations. We feel that Chern seems to be a man of such unusual quality, and the need of China for the development of such men so immediate, that there should be a good chance of carrying out a program of this sort. I enclose a short outline of Dr. Chern's career, which he supplied himself, and may add that he is now a professor at the National Tsing Hua University, which was moved from Peiping to Kunming. Where it is now I don't know.[132]

Chern did manage to come to the US, not without the help of US Army transport planes, and spent 1943–1945 at the IAS.

In the Spring of 1943, Hermann Weyl invited Hua to the IAS, in particular with a view of putting him into contact with Carl Ludwig Siegel (1896–1981). This plan took many letters, preparations, and emotions to finally materialize.[133] In September 1946 Hua was ready to take off, leaving his family behind in Shanghai for the time being. "Colleagues and fellow mathematicians in Shanghai came to a farewell dinner hosted by S.S. Chern and Zhou Weiliang. The Minister of Defence Chen Cheng and the Minister of Education Zhu Jiahua also offered their congratulations and gave him some farewell gifts."[134]

It was again Chern who convinced Chow to give up his life as a businessman and return to mathematics. It was also Chern who mentioned this perspective to Lefschetz, who in turn wrote to Weyl at IAS. In March 1947, the Chows arrived in San Francisco and made their way to Princeton. In 1949, Van der Waerden turned down a job offer from Johns Hopkins University and suggested his former student Chow in his stead. Chow would continue to teach and work at Johns Hopkins until his retirement.

As to Chern himself, he would combine a brilliant American career—at Chicago first, then in Berkeley—with missions in China, particularly the Chern Institute of Mathematics of Nankai University founded in Tianjin in 1985.

---

[132] See [Arch. IAS], Nr. 56484 (S.S. Chern's file). Most of this letter is also quoted by Yibao Xu in [Parshall & Rice 2002], p. 298.

[133] See [Arch. IAS], Nr. 56883 (Hua Luogeng's file).

[134] See [Wang 1999], p. 134.

Meanwhile, mathematics education in China on the whole proved its resilience during these trying times:

> The real test of Chinese mathematics . . . came during the 1930s when it established itself, professionally and institutionally, in ways that no longer depended upon foreign inspiration or support for its continuation and success. Remarkably, even at the beginning of the war with Japan—when many of the country's best faculty and students were evacuated to Southwest Associated University in Kunming—mathematics was still taught, research was still published. After World War II, mathematicians had much to contribute to the new People's Republic of China, and thanks to the foundations laid during the first half of the century, they were prepared to do so. Although many individuals and institutions played their parts, it was the colleges and universities, the Chinese Mathematical Society, and the journals it supported that gave modern mathematics in China a presence and an institutional stability that would ensure its persistence and enable it to withstand its greatest test to come later in the century—the Cultural Revolution.[135]

[135] Final sentences of Dauben's chapter in [Parshall & Rice 2002], p. 281.

Chapter 7
Forced Migration and World War II

In Section 1.4, I summarized the situation before 1914 like this: "In the last three decades before World War I, attention to national distinctions and feelings of national pride or imperial supremacy were extremely common, but by and large they *peacefully coexisted*—if I may put it like this—with increasing contact and collaboration among scientists from different empires or countries."

As World War II was approaching, the situation was much more antagonistic. Everybody could either feel directly—or would find out—that

C'est un peu en se barbarisant qu'on se nationalise.

That is: Focussing on the nation renders us somewhat barbaric. This extraordinary statement was pronounced by the renowned Romanian historian Nicolae Iorga (1871–1940) at the 1933 International Congress of Historians in Warsaw, when he expounded the idea that "the nation, particularly in Southeastern Europe, was a late phenomenon. ... Its appearance ... marked the end of the Middle Ages, which was characterized essentially by universal ideas."[1] His lucid comment linking nationalism and barbarism is all the more remarkable as Iorga himself not only enjoyed an international reputation as historian; he was also a right-wing, antisemitic nationalistic politician in his home country. He did criticize the Romanian fascist 'Iron guard', though. They assassinated him in November 1940.

The 1933 International Congress of Historians in Warsaw was actually surprisingly harmonious, in particular also between the Polish hosts and the German delegation.[2] One of its influential members was the Göttingen medievalist Karl Brandi (1868–1946), a personality who still managed to somehow combine a positive international spirit with strong nationalistic convictions all the way to a certain sympathy for the Nazi government. Back in Göttingen, though, Brandi was threatened by the ancient historian Ulrich Kahrstedt (1888–1962), an outright Nazi. In January 1934, Kahrstedt gave a public speech that implicitly called upon students to batter to death all members of the German delegation to the International Congress, and culminated in the declaration:

[1] See [Erdmann 2005], p. 173.

[2] Cf. [Erdmann 2005], Chapter 10, pp. 149–161.

N. Schappacher, *Framing Global Mathematics*,
https://doi.org/10.1007/978-3-030-95683-7_7

> We reject international science; we reject the international Republic of Letters; we reject research for the sake of research. In our country, medicine is taught, not in order to increase the number of known bacteria, but in order to keep the Germans healthy and strong. In our country, history is taught, not in order to say what has really happened, but in order to let the Germans learn from the past. In our country, the natural sciences are taught and learned, not in order to discover abstract laws, but in order to sharpen the toolkit of the Germans in their competition with other peoples.[3]

In the preceding chapters of Part II we have seen how selectively international nationalism was forced into Science International by World War I, and dictated the hapless episode of the IMU in the 1920s. By 1932, in beautiful neutral Zürich, it could perhaps still appear to participants of that sunny ICM as if those recent problems were now overcome. The way to a new buzzing international network of mathematicians seemed all the more open as many of the younger participants had already profited, for example, from Rockefeller grants. Transcontinental, open mathematics with an exciting new agenda was in the air.

However, as of May 1933, the Rockefeller Foundation had to reorient its activities towards emergency programs for refugees fleeing Europe.[4] By the end of the thirties, an international reshuffling of mathematicians of unheard dimensions was under way. The new mobility was migration induced by politics. In the world of mathematics, this meant for example that lofty research hubs had to also function as employment agencies. The ensuing war had even stronger effects on the mathematical profession.

7.1 Global Redistribution of Scientists in the 1930s and 1940s

In Section 6.4, we have presented a triptych of international mathematical careers that originated in China. For each of the three mathematicians, it was the IAS Princeton that paved the way to a university position in the US. The IAS was rooted in the same setting that had also fuelled the Rockefeller Foundation: joining philanthropy with the idea of scientific excellence and the need for research centers beyond universities—see Section 5.2. In the thirties and forties, the IAS and the Rockefeller Foundation, along with many other institutions, had to face an increasing number of scientists in emigration.

It was in this context that Hermann Weyl—himself an emigré who had left Göttingen for the IAS Princeton in the Fall of 1933 because of the Nazis—was called upon to ponder the fate of the French *Bourbaki* group in a letter dated 22 March 1941 to the Rockefeller-sponsored New School for Social Research in New York. At the time, part of France was under German occupation, and André Weil and Claude Chevalley were already in the US. It seems that André Weil had taken the initiative to secure a bicontinental future for the Bourbaki project. This furnishes an extreme but instructive case where issues of migration, the interest of a small

[3] See [Wegeler 1996], Section 3.2.2, pp. 147–162; Kahrstedt's whole speech is reproduced on pp. 357–368; my translation of a passage on pp. 367–368.

[4] See [Siegmund-Schultze 2001], Chapter VI.

but select group of mathematicians, and the fate of a major rewriting project of mathematics converge in Princeton, in Weyl's hands. In discussing his small list of French mathematicians, Weyl offers his personal reflections on the evolution of mathematics.

> Dear Doctor Johnson:
>
> André Weil told me that he had spoken to you about the *Bourbaki* enterprise. Under this collective pseudonym a set of young French mathematicians has started to publish a number of volumes concerning the basic disciplines of mathematics.
>
> The accent in classical mathematics lay on calculus, and for everything related to calculus the great French *Cours d'Analyse* by Camille Jordan, [Édouard] Goursat (1858–1936) and others, have in the past played a very vital part in mathematical training all over the world. But in the last twenty-five years the emphasis has shifted to other fields like topology and algebra and it has become necessary to lay the foundations deeper. … [T]he time seems to have come when integration and a certain degree of standardization should and could be attempted with a fair hope of success. Sometimes such integration has been brought about by an individual work of genius of such paramount importance that nobody working in the field could evade its influence. Systematic efforts undertaken by a group expressly for this purpose are less sure of success; their achievement will perhaps do no more than solidify one school adhering to a special brand of abstract ideas without finding acceptance among other schools, or the foundations laid might soon prove too narrow, etc. I see these dangers and am therefore less enthusiastic about the enterprise than the entrepreneurs themselves. But there is an urgent need, and as far as I can see *Bourbaki* is trying very earnestly and intelligently to find the best and simplest way to arrange the fundamental ideas and to fix the nomenclature. Plan and execution of each volume are discussed in full detail by the whole group, and before the manuscript is finished, it will have been rewritten by three or four authors. It seems certain that no single member of the group could have accomplished what they have done by pooling their mathematical intelligence.
>
> So far two small volumes, on abstract sets and topology, have appeared in French in the *Actualités Scientifiques et Industrielles*; preparation of the material for three more volumes is far advanced. But now the group has been broken up by the war; three of its leading members—André Weil, Chevalley, and Henri Cartan—are, or will soon be, in this country. A *conditio sine qua non* for the continuation of the work would be the bringing over of at least two more members, and this is the reason why I write to you about it.
>
> In October last year I sent Dr. Warren Weaver [1894–1978] a list of French mathematicians, mostly younger men, whom one could consider as candidates for the rescue action undertaken jointly by you and the Rockefeller Foundation.[5]

This list, which Weyl enclosed with the letter, runs as follows:

(1) Arnaud Denjoy, (2) Henri Cartan, (3) Jean Leray, (4) René de Possel, (5) Jean Delsarte, (6) Claude Chabauty (1910–1990), (7) Charles Ehresmann (1905–1979), (8) Charles Pisot (1910–1984), (9) Jean Dieudonné, and (10) Ervand Kogbetliantz (1888–1974).[6] As to the last person of this list, let us mention in passing that the

---

[5] See [Siegmund-Schultze 2001], pp. 284–285.

[6] See [Siegmund-Schultze 2001], p. 285, footnote. To put this action into perspective, one should also bear in mind the activities launched in 1940 by Henri Laugier (1888–1973) and Louis Rapkine (1904–1948)—see [Dosso 2006].

Armenian, Moscow-trained mathematician Kogbetliantz was already a refugee in Paris since the early 1920s.[7]

The time for triage has come, and Weyl has to explain his choice:

> It so happens that all the Bourbaki collaborators are on it. The two men whom Weil considers indispensable for continuation of the work are Jean Delsarte and Jean Dieudonné. Delsarte speaks no English and distrusts his linguistic abilities. It would be much better to place him in the French Catholic university in Montreal than anywhere in the United States. Things are different with Dieudonné who was a Proctor Fellow of Princeton University from 1927 to 1929.
>
> In my opinion an invitation to this country to any democratic-minded foreign scholar who is threatened by the (let us hope short-lived) Nazification of the European continent should depend first of all on his scholastic standing, and then on its adaptability. The fact of his being indispensable for work like that of the Bourbaki group, however meritorious, should hardly play a decisive role in the selection. However, all the young French mathematicians (2)–(9), except (2) Henri Cartan and (3) Jean Leray, are of nearly equal rank. (2) is placed, (3) out of reach. Hence, if there is a possibility of bringing over to America one more young French mathematician, I should find it justifiable to concentrate on Dieudonné, and to try to establish Delsarte in Canada.[8]

The war situation reflected in this letter, and also the peculiar case of the Bourbaki group that Weyl was treating here, certainly make this document very special. Nonetheless the letter illustrates crucial aspects that any attempt to historically account for the scientific migration in the thirties and forties has to balance. This book is not the place to give such an account. All we do in this section is highlight the extent of the phenomenon by scattered thoughts and examples.

On the one hand, migration in general, and forced emigration in particular, involves both professional (in particular, scientific) and personal matters. What is more, the private aspects tend to be more pronounced than in ordinary career affairs, often dramatically so. Emigration is all about leaving a former life behind and letting yourself—and your family—in for a new cultural environment. An adequate account of emigration must therefore not restrict itself to extracting an 'objective' general map or measurement of the displacements, but give personal life stories their due share. This being said, integrating personal elements into a comprehensive study of migration phenomena is not only a stylistic challenge; it faces the well-known difficulties of any biographical endeavor (for short: one never knows enough about a person).[9]

---

[7] Much more on what we do—and what we do not really—know about the eventful life of Kogbetliantz and his wife between the World Wars is summed up in a recent paper by Laurent Mazliak and Thomas Perfettini in [Mazliak & Tazzioli 2021], pp. 307–355. Their chapter also gives an overview of Russian refugee mathematicians in Paris in the 1920s and 1930s.

[8] See [Siegmund-Schultze 2001], p. 285. As Siegmund-Schultze also duly notes, Henri Cartan would finally not come to the US during the war, and Leray remained in a German POW camp.

[9] Cf. the standard reference about the history of mathematicians fleeing from Nazi Germany [Siegmund-Schultze 2009]. There the sequence of chapters follows the overall plan of the book, but the author adds 'D'-sections, presenting documentary sources, and 'S'-sections with individual case studies, to various chapters.

On the other hand, choosing the opposite approach, say, for the forced emigration of mathematicians instigated by the European fascist regimes before and during Word War II, may suggest the rather cynical conclusion that the fascist pressure actually worked hand in hand with a global, genuinely international 'consolidation and unification' of mathematics. Indeed, taking applied mathematics in the US as an example, Richard Courant, himself an emigré mathematician, would joke about this later. When showing visiting colleagues the main building of the Courant Institute of Mathematical Sciences in New York—Warren Weaver Hall, which was built in the early sixties—he remarked that he principally owed this wonderful institute to two influential men: John Rockefeller who gave the money, and Adolf Hitler who provided the talent.[10]

The second point of view tends to only count migrations which can be considered scientifically successful; it passes over victims in silence. Otto Blumenthal for instance, the editor in chief of *Mathematische Annalen*, was attacked by Nazi circles on both political and racial grounds as early as 1932; he was dismissed from his chair in 1933. In 1939, aged 63, he did emigrate—but only as far as Holland, where he was arrested in 1943. He died in the concentration camp at Theresienstadt in 1944.[11]

The scientific effect, or 'success', of emigration is a complex amalgam resulting from the encounter of the emigré with the host country. Taking Argentina as an example, the Spanish mathematician Julio Rey Pastor (1888–1962) had been present in Buenos Aires on a part-time basis since 1917, and permanently since 1927. He contributed immensely to the improvement of mathematics in Argentina, but his presence in Argentina may probably still be described best from the point of view of the Spanish metropolis interacting with the periphery.[12] The Catalan mathematician and engineer Esteve Terradas i Illa, however, is a case of emigration. He chose not to return to Barcelona after having participated in the Oslo ICM in 1936, because of the Spanish Civil War. He spent several years teaching in Buenos Aires and in La Plata, Argentina. But he eventually did return to Spain; his emigration was temporary. Terradas's case shares with Rey Pastor's a continuing exchange with Spain about returning to Europe.[13] An emigrant to Argentina who was there to stay was the Italian mathematician Beppo Levi (1875–1961). He lost his chair in Bologna in 1938 due to the racial laws (*leggi razziali*), shortly before his retirement. With his wife and daughter he went to Rosario, Argentina, where he would play the central role in building up the Mathematics Department.

> The founding of this institute at Rosario, upstream from Buenos Aires, took place at a time of cultural expansion of several provincial Argentinian cities, mainly Rosario, Córdoba, and Tucumán. A relative prosperity helped in the development of more substantial groups of professionals, mainly lawyers, medical doctors, and engineers, who promoted local cultural

[10] Harold M. Edwards told me this anecdote during my first visit to Mercer Street. Cf. [Siegmund-Schultze 2001], p. 210.

[11] See [Bergmann et al. 2012], pp. 88–89 and *passim*.

[12] See Eduardo Ortiz's account of mathematical relations with the "Iberian periphery" in the nineteenth century, in [Goldstein, Gray, Ritter 1996], Chapter 15, pp. 323–343.

[13] See [González Redondo 2002].

Fig. 7.1 Beppo Levi, about 1930. (Courtesy Laura Levi.)

> activity in these cities and invited leading intellectuals and artists from Buenos Aires to lecture or visit there. These professionals were financially better off, and their clients were richer yet. Societies, orchestras, art galleries, and publishing houses began to emerge in this period in Rosario.[14]

Beppo Levi was formally reinstated in his Bologna post in July 1945, but decided against returning to Italy because he was approaching the Italian age limit of 75 years for retiring, and he cared a lot for what he had built up in Rosario.[15]

As far as German mathematicians looking for a country of refuge are concerned, Siegmund-Schultze provides a truly global survey:

> Examples from various host countries show how widespread economic problems and political resentment, such as anti-Semitism, made acculturation difficult. Some countries, such as Austria and Poland, had to be ruled out as host countries from the outset, since they offered

[14] See [Schappacher & Schoof 1996], p. 67, based on information from Eduardo Ortiz. Cf. [Ortiz 1988].

[15] See [Levi 2000], pp. 75–77. Laura Levi also stresses her father's interest and contact with the physics community, and corrects accordingly the caption of the group picture reproduced on p. 67 of [Schappacher & Schoof 1996], which in fact shows the 1948 meeting of the *Asociación Física Argentina* (AFA).

> similar, if not quite as extreme, political conditions as Germany. Others, such as Italy and the Soviet Union, also ruled by dictatorial regimes, served nevertheless and somewhat surprisingly as temporary host countries. Hopes harbored by Turkey to profit from the German immigration for its own science system failed due to Hitler's expansion policies and the death of Kemal Atatürk in 1938; both circumstances forced the refugees to go on to safer places. Australia was a rather less attractive option for emigrants because of the rudimentary state of mathematics there at that time. Although some authorities involved in emigration tried to use Australia to ease the situation in other host countries, only two mathematicians finally ended up there before the end of the war.[16]

These general indications are then detailed according to countries, or continents in the particularly rich corresponding 'D' section of his book.[17]

Forced emigration was an important factor, if not the initial source for putting what would later become the state of Israel on the global map of mathematics. The Zionist movement had inspired the founding of the Hebrew University (HU) and its Mathematical institute—today called the Einstein Institute of Mathematics (EIM)—at Jerusalem in 1925. Edmund Landau gave a talk at the opening of the HU which actually reflected "the way the Zionist cause was inextricably linked to, and determined by, European political agendas" of the 1920s, in particular regarding Science International.[18]

> During the fourth decade of the twentieth century, with the voluntary emigration and enforced expulsion of scientists and scholars from Nazi Germany, new centers of mathematical research were created. The great nineteenth-century German scientific heritage, which had hitherto slowly pervaded Europe and abroad, now dispersed to new intellectual havens. Former students of the German academic system carried their heritage to new harbors to anchor their scientific expertise, and implement their intellectual traditions from Istanbul to New York and Buenos Aires. Displaced mathematicians were part of this migration. Although it took place at roughly the same time, the founding of the EIM at HU belongs to a different kind of phenomena. The EIM was less the outcome of the push of anti-Semitism and Nazism, and more a result of the pull exerted by the Jewish national movement.[19]

The quasi economic push-and-pull model is one of the lenses through which migration phenomena have been investigated.[20] However, multiple methods and questions should always be kept in mind, for instance, if a loss or gain of people also meant a loss or gain for science, and so forth.[21] Let us return to Shaul Katz's account:

> It was the Zionist vision that drove a few dozen scholars and scientists, most of them European, to prefer the new university in Jerusalem opened in 1925, over their mainly European *alma mater*. There is no other overwhelming explanation for Landau's coming for a short period to Jerusalem in 1928, followed by the arrival of [Adolf Abraham Halevi] Fraenkel in 1929. And it was a sort of mathematical idiosyncrasy of Landau, coupled with a certain variety of European national movement, Zionism, that embraced wholeheartedly

[16] See [Siegmund-Schultze 2009], p. 103.

[17] See [Siegmund-Schultze 2009], pp. 104–148.

[18] See [Corry & Schappacher 2010], p. 427; the claim quoted is elaborated in this article.

[19] See [Katz 2004], pp. 226–227.

[20] See [Lee 1966].

[21] See [Ash 2011].

> pure science and its promotion as one of its exalted cultural ideals (a kind, so they tended to believe, of national transformation of the biblical "From Out of Zion Goes Forth Torah") that begot the pure-mathematics trajectory of EIM. Concomitantly, Landau, Fraenkel, and [Mihály-Michael] Fekete [1886–1957] were proud intellectual inheritors of this variety of the Berlin tradition that not only conceived pure mathematics as a sublime neo-humanistic ideal, but also in parallel also disdained applied mathematics. Therefore EIM maintained the cultivation of pure mathematics only. Since the framework of European migration of the 1930s does not suggest itself as a proper comparative historical one for EIM case, a more general family of phenomena with more historical depth and geographical width invites attention. It is the comparative perspective of the process of implementation of Western science outside Europe.[22]

It was not my intention to confuse the reader with scattered examples—to which one may also add the three men discussed in Section 6.4—, approaches, and remarks. But the far-reaching global reshuffling of mathematicians and mathematical centers of the 1930s and 1940s was as dramatic as it is complicated to sort out. Mapping out the whole migrational reshaping of the terrain of Mathematics International in the thirties and forties would require yet another book.

> This exodus was, to be sure, a source of a tremendous upsurge in the internationalization of mathematics, especially in the sense of new and unexpected personal encounters and oral communication. Still, this type of internationalization was shaped in a peculiar way by emigration patterns. It was not necessarily healthy or natural when compared to the secular, long-term internationalization of mathematics that had been well under way in the decades before. Without entering into the foggy field of counterfactual history, it is important to focus on the losses for the various national cultures in mathematics in Europe that were brought about by the expulsions not just in Germany but also in other countries such as Poland, Hungary, and Austria. These losses were more than the sufferings of the refugees and the deaths of the victims.[23]

7.2 What World War II Meant for Mathematics

The Second World War was of "a far greater magnitude than the preceding world war, it was to engulf a larger area, bringing with it the horror of systematic genocide exemplified by the Holocaust. Over and above territorial considerations, the very future of civilization was at stake." Its theaters included Western Europe from April to June 1940, the German invasion of Russia as of June 1941, and Japan's overrunning of the whole of South-East Asia.

> The Japanese bombing of the American naval base at Pearl Harbor on 7 December 1941 enabled President Roosevelt to surmount the pacifism widely supported by the American public and lead the United States into the war. Until then, the American participation was limited to providing equipment to Great Britain and Russia under the Lend Lease Programme. Despite the Americans' superior weaponry and their contribution to the defeat of Germany in May 1945, the conflict with Japan appeared likely to endure. To curtail it, the United States resorted to atomic weapons in August 1945. The resistance movements in occupied

[22] See [Katz 2004], p. 227.

[23] From Siegmund-Schultze's chapter in [Parshall & Rice 2002], p. 339.

> France, Belgium, Norway, Greece, Yugoslavia, Poland and Russia and in South-East Asia were the protagonists of a conflict, which was a key feature of the war despite its lower profile.
>
> [Another theater] could be added: the battles in North Africa, which continued with the landings in Italy and the collapse of the Fascist dictatorship.
>
> The Second World War also differed from the preceding war by doing away with the dividing line between civilians and combatants. The bombing of Warsaw, Leningrad, Rotterdam, London and Coventry by the Germans, the Allied bombing of Berlin, Hamburg and Dresden, and finally the atomic bombs dropped on Hiroshima and Nagasaki, all targeted civilian populations. If those who died of hunger are included, the civilian death toll probably numbered approximately fifty million.[24]

In Chapter 3 above, in order to capture the impact of World War I on Mathematics International, we had to address not only the new role of mathematics and mathematicians in warfare, but also the violent nationalism fired by the war, because this spirit stood at the cradle of the IRC and the IMU in 1919–1920. Different but analogous observations apply to World War II. The role played by science was even more pronounced than during the First World War, and there was a greater variety of mathematical applications, many of which would flourish over the following decades. World War II also prepared a new global political landscape: The Cold War, which would reshape the professional structures for mathematical research and determine the first decades of the IMU after its renewed birth in 1950–1951.

Before going into this, let us start with a peculiar episode from the German occupation of France in 1940.

7.2.1 Searching for the Hiding Place of the IMU

During World War II, Harald Geppert (1902–1945)—the elder brother of Maria-Pia Geppert (Section 6.2.2)—was in charge of both German review journals, the *Jahrbuch* and the *Zentralblatt* (Section 6.3.3). The *Jahrbuch* would not survive the war; but in the first war years it was still trying to squeeze the delay between the publication of the papers and their reviews. The *Zentralblatt*, on the other hand, had just lost a number of its referees in the fight that had precipitated the foundation of *Mathematical Reviews*. Thus Geppert was trying to fill those gaps with mathematicians recruited in the large part of Europe that had come under German control by the end of 1940. Irrelevant and piecemeal as it may seem at first, this endeavor would be an important element of the attempt to re-order Europe under German domination, as far as mathematics was concerned. Individuals who were invited to write some of these much needed reviews for the *Zentralblatt* would not only get paid, but would get access to recent literature in their field, which was otherwise hard to obtain in times of war. The extent to which, say, a French mathematician had

[24] All quotes in this preamble to the present section are from [Gopal et al. 2008], p. 6.

accepted, or not, to write reviews could therefore become a key issue after the war when it came to judge if he behaved like a *résistant* against the Germans, or rather like a *collaborateur* during the occupation.[25]

Against the double background of his responsibility for the *Zentralblatt* and ongoing political discussion about the would-be German re-ordering (*Neuordnung*) of science in Europe, Geppert was sent on an official mission to Paris in December 1940. His explicit agenda, however, was to search for hidden signs of life of the IMU. Now that the Germans controlled Paris, they wanted to make sure to extinguish whatever might still smolder of that anti-German international construct. Thus on 3 December 1940, the minister confidentially ordered Geppert to travel to Paris in order to investigate what Geppert himself had alerted the ministry to in the first place:

> While preparing for a re-ordering of international scientific cooperation in the international unions, associations, etc., I was led to examine the International Mathematical Congresses and the former *Union internationale de mathématique*. It has come to my attention that there exists in Paris an *Institut Poincaré*, which also organizes international meetings in the domain of mathematics, which are different from the International Mathematical Congresses that take place regularly. It seems that this *Institut* perpetuates on its own account the *Union Mathématique*.
>
> I herewith order you to undertake before long an official journey to Paris in order to assess directly on the spot the importance that has to be attributed to the activity of the *Institut Poincaré*. I point out that the extent of your findings may be of fundamental importance for my future decisions.[26]

After his return from Paris to Berlin Geppert, in an attempt to respond to the object of his mission, submitted a survey of the ICMs that had taken place since World War I, based on the various ICM Proceedings. We quote starting with the Zürich ICM:

> The following International Congress took place in 1932 in Zürich. The IMU is mentioned neither in the invitation nor during the Congress. But at the final session an international commission is formed—its only German member was the Jew[27] Hermann Weyl—"in order to re-study the question of the international collaboration in the sphere of mathematics and to make propositions with regard to its reorganization at the next congress."[28] Obviously, this commission was to ensure a future substitute for the IMU, which was still in existence.
>
> The next ICM took place in 1936 at Oslo. It was again called without any intervention by the IMU. However, the Union suddenly appears in the minutes of the final session of the Congress, where Prof. Gaston Julia reports on the activity of the international commission mentioned before. After several meetings over the years the commission has determined that the creation of a truly international organization of mathematicians encounters unsurmountable difficulties and must therefore be delayed. Whether this means that the activity of the union has to be considered terminated, or whether it continues to be alive because of the lack of a truly international organization, is not clear from the minutes. From the

---

[25] See the detailed analysis in [Eckes 2018], which also connects the review issue to Geppert's and Hasse's vain attempts to free certain French POWs.

[26] Quoted from [Siegmund-Schultze 1993], p. 179; my translation.

[27] In fact, not even the Nazi administration claimed that Weyl was Jewish.

[28] See Section 6.2.2 above.

> German side, professor Blaschke, Hamburg, has participated in these meetings. The next International Congress of Mathematicians was planned to take place in 1940 at Princeton, USA, but has been adjourned because of the war.
>
> Two questions thus remain to be settled: that of the creation of an international organization of mathematicians, whose need is documented by the events described; and the organization of the next International Congress of Mathematicians, which will be called by the American Mathematical Society.[29]

About five years after Harald Geppert's suicide (Berlin, 4 May 1945) both questions were settled: in the US and in particular thanks to Marshall Stone.

7.2.2 Mathematics for the War

In Section 3.3 we briefly described the effects of the First World War on mathematicians and on mathematics. We have seen in particular that applied research topics imposed by the needs of the battlefields would modify the appearance, the context, and thus finally the substance of mathematics. And we have seen in the Italian example (Section 3.3.1) how the organization of military research during the First World War would create structures of scientific policy that outlived the war.

Thus prepared what to look for, we now turn to World War II. Scientific man and woman[30] power was mobilized for the new war effort on a considerably larger scale than during World War I.[31] Note that the enrolment in scientific work for the war could save the life of a young man who would otherwise be sent to the front; leaving a relatively safe place in a decoding unit in Berlin to volunteer for frontline fighting could amount to suicide, as in the case of the fanatic Nazi Oswald Teichmüller (1913–1943).[32]

The domains for which mathematicians were in high demand during World War II cover a substantially broader spectrum than in the previous war, and include a few recent, budding subdisciplines of mathematics. Here is a rough overview of the main areas:[33]

---

[29] Geppert to Ministry, 29 December 1940; my translation. For the German and French archival sources of copies of this report, see [Eckes 2018], pp. 299 and 305.

[30] The presence of women in science for World War II was not limited to the numerous women computers; see for instance Kathleen Williams's chapter "Improbable Warriors: Mathematicians Grace Hopper and Mina Rees in World War II" in [Booß-Bavnbeck & Høyrup 2003], pp. 108–125.

[31] This seems obvious, for instance if one looks at the whole spectrum of mathematical domains that were pushed during WW II. However, I have not been able to find reliable estimates from the various countries of, say, mathematicians enrolled in war research in the forties.

[32] Cf. the reflections about the adjacencies between Teichmüller's work on (quasi-)conformal maps and ongoing aerodynamic research in [Epple & Remmert 2000], pp. 291–293.

[33] See Siegmund-Schultze's schematic overview, with references, of mathematical war work in Germany, the US, USSR, UK, Italy, France, and Japan in [Booß-Bavnbeck & Høyrup 2003], pp. 63–74.

- Aerodynamics/hydrodynamics, especially problems near super-sonic speed and air foil design.
- Ballistics of torpedoes, anti-aircraft gunnery, and rockets.
- Cryptography. Among the countless cryptography units in all countries at war,[34] Turing's work at Bletchley Park has received the greatest attention in the literature on World War II.
- Development of early electronic computers.
- Operations research.
- Game theory.
- Cybernetics.[35]

Some of these domains really took off only after the war; game theory for instance. In the United States, the Manhattan Project working on the Atomic bomb, and continuing later with the H-bomb, naturally enlisted mathematicians. It required heavy numerical calculations. Mathematical problems arose in this context from

- Gas dynamics, and from
- Statistical approaches of various kinds, in particular the Monte Carlo method introduced (after the war) by Stanisław Ulam (1909–1984) and Nicholas Metropolis (1915–1999).

In the US, John von Neumann was the central figure, almost the incarnation of mathematical war research. The organizational setup of mathematical war research in the US and its consequences will be discussed in the next Section 7.2.3. The broad panorama of mathematical fields and the great number of mathematicians enlisted for war-related research, and its continuation after 1945, make it impossible to present an overall account. We visit a few examples instead.

For the period of World War II itself, an interesting contrast between Germany and the UK transpires from the report on *Applied Mathematical Research in Germany, with Particular Reference to Naval Applications* by the *British Intelligence Objectives Sub-committee* (BIOS), based on investigations made in June–August 1945 by John Todd (1911–2007), G.E.H. [Gerd Edzard Harry] Reuter (1921–1992), Friedrich G. Friedlander (1917–2001), Donald Harry Sadler (1908–1987), A. Baxter (?) and Fred Hoyle (1915–2001). We quote from the general observations at the beginning of the report:

> 2. There is no possibility of 'controlling' mathematical research, i.e. preventing work being carried out on 'war' subjects. It is abundantly clear from our observations in Germany and from information obtained from U.S.A. (and, to a much less extent, from our experience in U.K.) that almost any top-class mathematician practising in the most abstract fields can very quickly make substantial contributions in the mathematics of technology.
>
> ...

[34] See for instance [Weierud & Zabell 2019] for the German case.

[35] This term for the new science born out by his war research was coined by Norbert Wiener only in 1947. See [Galison 1994].

> 4. Nevertheless we feel that the mathematicians in U.K. made a bigger contribution to the war effort than those in Germany. On the one hand a considerable number of younger mathematicians in Germany were actually put in the fighting services, on the other, those in Government Departments and in Industry did not appear to work as conscientiously as the majority of those similarly placed in U.K. As evidence of this may be mentioned the fact that members of this party were continuously being asked to take with them, manuscripts prepared in 'Sparetime', for submission to editors of mathematical journals here or in U.S.A. Very few of the English mathematicians had energy left for such activities.[36]

And on the war work of the Number Theorist Helmut Hasse, who during the war was in charge of a research group at the High Command of the German Marine Forces (OKM), the committee notes:

> H[asse] seemed to have an exaggerated opinion of the value of his trajectory work, which, in our opinion, though elegant, is of little practical value. He stated he had forgotten all the details of his work but said they could be extracted from the OKM documents which he understood to be in our possession—he asked that we should send him copies of his own reports! It was considered unnecessary to encourage him to remember details of the work, as it appeared that in his position as administrative head of FEP III be was content to leave all technical matters to Prof. Karl Willy Wagner [1883–1953], and devote his energies to rather unpractical matters.[37]

Kolmogorov's work on the probability theory of firing techniques provides another, different example of a well-known mathematician's work occasioned by the war. It would fill a special volume of the *Proceedings of the Steklov Mathematical Institute* published in 1945.[38]

More historical research, in particular also comparative research, on the nature of mathematics for the War in various countries is still a desideratum for the history of mathematics in the twentieth century. Indeed,

> during the war(s) a lot of at least potentially applicable theoretical work was done in various countries—whether they were involved in the war effort or not—that escaped attention of men such as Norbert Wiener abroad and was likewise not noticed due to the communication blackout during much of the war(s) and even later in the Cold War. Mathematical work or mathematics-related engineering work that was potentially war-important, such as done in France by É[mile] Borel on game theory and émigrés W[olfgang/Vincent] Döblin [1915–1940] and F[elix] Pollaczek [1892–1981] on Markov chains and queuing problems, or in Germany by K[onrad] Zuse [1919–1995] on digital computers, was not, for various historical reasons, actually . . . transferred into the war effort and therefore partly or temporarily ignored in the countries that would write the history of the war and set the norms for the scientific enterprise after 1945, especially the United States.[39]

[36] See BIOS Report 79 (1945), pp. 2–3.

[37] See BIOS Report 79 (1945), pp. 48–49.

[38] See A.N. Shiryaev's account of it in [Booß-Bavnbeck & Høyrup 2003], pp. 103–107.

[39] From Siegmund-Schultze's chapter in [Booß-Bavnbeck & Høyrup 2003], p. 28.

7.2.3 How World War II Reshaped the World – the Case of Mathematics

World War II brought a tremendous impetus to mathematics. Indeed the military interest in all the areas of mathematics we have listed above led to the creation of applied research groups and of new specialized research institutes, in all countries at war. These institutes, and the whole organization of war research would reconfigure the professional setup of scientific disciplines. Let us start in the European countries under fascist rule:

> Aerodynamics, the scientific basis of aviation, represents one of the most significant successes in the mathematization of the technological sciences in the twentieth century. At the same time, ballistic problems concerning projectiles and missiles, in the air and under water, were tackled with the help of mathematical methods on an increasingly larger scale. During World War II, new coding and decoding projects required mathematical support. In some countries, this process had started already in or right after World War I.
>
> In all these areas, the traditional university system proved insufficient for the organization of specific mathematical research extensive enough to meet both armament and warfare interests. As in many other scientific and technological fields, research institutes outside the university system were founded with state, military, and industrial participation. Mathematicians either significantly shaped, or even entirely supported, these institutes.[40]

What happened after the war to those newly created structures, and to the whole war administration of mathematical research, would of course depend on the country and on individual circumstances. To mention a peculiar example known to many mathematicians, the "Mathematical Research Institute" at Oberwolfach, Germany, is today a conference center of international reputation. But it started out in November 1944—very late in the day as far as World War II was concerned—as a *Reichsinstitut* with the mission to coordinate mathematical war research in Germany.[41]

What was the long term effect of the war for mathematics? Looking at individuals, there were surely a number of mathematicians who had been enlisted in military research during World War II, but who would later look back on this period as a passing spell in their professional life, after which they took up (as soon as this was materially possible) their previous work more or less where they had left it. Looking at nations, the strongest and most influential long time repercussion of the war effort on the development of the mathematical profession seems to have taken place in the USA. There a certain divide opened up after the war, between those who returned to pure mathematics the way they had practiced it before—typically, in the axiomatizing spirit of 'Consolidation and Unification' inherited from the thirties—and those who followed up the type of applied problems they had worked on for the nation at war and ended up establishing more than one new mathematical speciality. To do this the latter could avail themselves of new employment patterns inherited from the

[40] See [Epple et al. 1995], p. 132. This paper then goes on to compare various research structures for aerodynamics and mathematics in the two fascist states Italy and Germany.

[41] See [Remmert 2020] for the history of the Oberwolfach institute in the first years after World War II.

administration of war research. However, pure mathematics could also profit from these rich new funding facilities. There was nonetheless a parting of the ways in the US mathematical community about how to position oneself with respect to the Cold War and the corresponding advent of *Big Science*.[42] Since the US would become the leading nation for Mathematics International after 1945, what happened there would also affect mathematical communities in other countries as well as international organizations. It was principally through the development in the US that World War II influenced the kind of mathematics showcased by the IMU and at the ICMs.

> The Second World War has brought about in the United States important changes in mathematical practice, in the scientific, intellectual, and social networks of mathematicians, bringing them into closer contact with physicists, engineers, economists, and specialists of the social sciences, as well as with military officers and politicians. The mathematicians were confronted with various concrete and pressing problems for which solutions, or rational, formal approaches were urgently wanted. At the end of the war an important part of the mathematicians returns to their traditional academic universe, taking up the research they had briefly interrupted. In the mathematical community at the universities and its international institutions a certain ideology of pure mathematics develops and seems to become dominant at the end of the 1950s. This 'purism' in which part of the community tries to shelter is in part a reaction against the American tradition of utilitarianism. It also has to be linked with the political context of the Cold War and the climax of McCarthyism. The mathematicians which represent this tendency consider having already paid their due to the global conflict; they now want to be able to dedicate themselves to the most abstract fundamental research, far from all preoccupations with politics or applications. However, there are also other mathematicians, other groups which have emerged during the war and whose interests as well as social and professional networks continue to hold their own, independently of the purist mathematicians.[43]

One could have imagined that World War II would create a sort of transparency between pure and applied mathematics, which would then likewise reshape the professional situation of mathematical research in society and politics. But this did not happen, neither during the war nor afterwards.

During the war, Warren Weaver directed the *Applied Mathematics Panel* that was created

> to coordinate the services of mathematicians and to serve as a clearinghouse for mathematical information pertinent to the war. ... Weaver's panel supervised an effort that employed close to three hundred people, including such mathematicians as John von Neumann, Richard Courant, Jerzy Neyman, Garrett Birkhoff, Harold Hotelling [1895–1973], and Oswald Veblen; wrote several hundred technical reports; and spent nearly three million dollars. The panel encouraged new developments in statistics, numerical analysis and computation, the theory of shock waves, and operational research, and served as a training ground for mathematically-minded workers in fields like economics, one of the more famous being the eventual Nobel Prize winner Milton Friedman [1912–2006]. The panel also promoted the institutionalization of applied mathematics through its support, e.g., of Brown University's Program in Applied Mechanics, Jerzy Neyman's Statistical Laboratory

[42] *Don't forget your mittens!* Laurie Anderson.

[43] See [Dahan 2004], p. 50; my translation.

at Berkeley, and Richard Courant's group in applied mathematics at NYU. When ground was broken for the Courant Institute at New York University in 1962, Warren Weaver was there to wield a shovel for the building that would bear his name.[44]

And yet,

judged in terms of its larger ambitions—the central coordination of wartime mathematics—the panel failed. Furthermore, the success it did achieve split the nation's mathematicians into angry factions. ... [The panel's] forgotten trials and tribulations illuminate both the uneven development of American mathematics at the outbreak of World War II as well as the imperial ambitions of those who, like Vannevar Bush [1890–1974], James [Bryant] Conant [1893–1978], and Warren Weaver, took the lead in the mobilization of wartime science.[45]

Adding to the places just mentioned Los Alamos, Aberdeen Proving Ground and CalTech, and also Princeton, we are looking at a list of the main centers of applied mathematics launched in the US during the war where mathematicians, physicists and engineers rubbed shoulders.

The most significant reconfiguration which emerges from these works, both on supersonic flow and on nuclear questions, concerns hydrodynamics, computers and numerical analysis. This reconfiguration shatters the established hierarchies between 'pure' and 'applied'. It blurs the borderline between what clearly belongs to mathematics and what does not belong to mathematics and would normally have been classified in the domain of engineering science or physics. Von Neumann emerges as someone who has realized this recomposition of interests for himself early on. From the beginning of the 1940s he convinces himself of the importance of hydrodynamics for all the physical sciences and for mathematics and of the fact that it requires a radically new development of methods and of computational capacity. When the project of an electronic computer gets under way, von Neumann, [Herman H.] Goldstine [1913–2004] and their collaborators explain that the economy of the machines absolutely calls for a profound remodelling of numerical analysis and for the elaboration of new algorithms. Also the program of digital meteorology chosen as a priority full scale application for the Princeton computer is an example of this reconfiguration of interests and practices.[46]

After the war, there was widespread

concern that the vitality and flourishing of wartime research would dissolve in the postwar period. The scientists would go back to the kind of work they did before the war with the consequence that the research cooperation within the military-university-industry complex, which had proved itself so productive during the war, would simply disappear. Not surprisingly there was a shared belief that the USA had to be strong scientifically in order to be strong militarily. ...

The National Science Foundation was not established until 1950 and in the meantime the military services initiated different channels for supporting scientific research. There were two primary places where the new mathematical techniques that emerged during the war became the subject of military funded basic research, Project RAND and the Office of Naval Research (ONR).[47]

[44] See [Owens 1989], pp. 287–288.

[45] See [Owens 1989], p. 289. See also [Parshall 2015], pp. 295–302.

[46] See [Dahan 2004], pp. 54–55; my translation.

[47] See [Kjeldsen 2003], pp. 133–134.

Fig. 7.2 Julian Bigelow, Herman Goldstine, J. Robert Oppenheimer, and John von Neumann in front of MANIAC, the Institute for Advanced Study computer, 1952. Credit: [Arch. IAS].

Several new mathematical disciplines grew from this peculiar constellation in the wake of World War II. We mention Operations Research—in particular Nonlinear Programming—and Game Theory.

> The ONR was established within the US Navy in 1946 to ensure the continuation of the vitality and thriving of scientific research done during the Second World War. During the first four years of its existence it was the main sponsor for government supported research in the USA. It continued the practise of the war organisation Office of Scientific Research and Development (OSRD) that had been the vehicle for the mobilisation of civilian scientists during the war. Like OSRD, ONR supported scientific projects through contracts with scientists working in the universities, projects of which many were proposed by the investigators.
>
> The logistics programme of ONR originated in 1948 as a result of the mathematician George B. Dantzig's [1914–2005] work with so-called programming planning methods in the US Air Force during and after WW II. An Air Force programme was a huge logistics schedule for Air Force activities. During the war Dantzig had worked on these programmes and taught Air Force staff how to calculate the programmes. The methods they used were slow and inefficient. It took more than 7 months to set up such a programme. After the war Dantzig went back to work for the US Air Force Headquarters where he functioned as mathematical advisor. Together with a group of Air Force people he worked on programming planning problems. In October 1947 the Princeton people became aware of this work because Dantzig visited John von Neumann, in von Neumann's capacity as a consultant for the Air Force, to

> discuss the possibility of solving such an Air Force programme. At this point Dantzig and his group at the Air Force had built a mathematical model for the programming problem, a model they first called programming in a linear structure and soon after became know as a linear programming problem. John von Neumann had just completed the first book on game theory with Oskar Morgenstern [1902–1977] and he suggested that Dantzig's programming problem was equivalent to a so-called finite zero-sum two-person game. This connection to game theory provided the linear programming problem with a mathematical foundation in the theory of systems of linear inequalities and the theory of convexity.[48]

It is remarkable how seamlessly the history of Operations Research slides from World War II to the Cold War. This is illustrated by the US airlift operation Vittles during the Berlin Blockade in 1948–1949.[49]

As another illustration of the same general process let us quote from Kjeldsen's summary of how "game theory became the main subject of mathematical research at the RAND Corporation":

> According to the historian [and economist] Philip Mirowski [b. 1951], the disregard shown by economists brought von Neumann to search for another 'home' for game theory. Given the time, the place, and the concept of optimal strategies for winning a game, which fitted perfectly into the war context, and given von Neumann's multiple connections, reputation, and influence within the military-science complex during the war, the military context was an obvious choice. Project RAND in Santa Monica, California became the most important home for game theory. This project originated in March 1946 by the initiative of Army Air Force Chief of Staff Henry H. 'Hap' Arnold and Donald Douglas, the president of Douglas Aircraft. In the beginning the project functioned as a subsidiary of Douglas Aircraft but in 1948 Project RAND became a free-standing nonprofit corporation, a so-called 'thinktank'.
>
> In the first decade after the war RAND was the center for mathematical research in game theory. The first mathematicians working there were recruited mainly from the Applied Mathematics Panel. ... This group at RAND was the first established group of game theorists and they all either came from the war work or had connections to mathematicians who had been involved with OSRD. The group at RAND held lengthy summer sessions in game theory and collaborated with another military financed project—the logistic project—in Princeton.[50]

Another, analogous example of continuity from war work to fundamental scientific reorientations of the 1950s and 1960s is Norbert Wiener's conception of cybernetics as analyzed in Peter Galison's penetrating study.

> What we have seen in Wiener's cybernetics is the establishment of a field of meanings grounded not through zeitgeist but explicitly in the experiences of war. For however far telephone relaying technology or A.N. Kolmogoroff's statistics had come before the war, it was the mass development and deployment of guided missiles, torpedoes, and antiaircraft fire that centralized the technology to scientists and engineers. To the thousands of servicemen who used and faced this new generation of weapons, the 'human' character of self-regulating machines seemed all too human. After all, trying to shoot down a Junkers JU

[48] See [Kjeldsen 2006], pp. 34–35. Cf. [Kjeldsen 2019], pp. 147–155.

[49] See Chapter 2 of the inspiring book [Erickson et al. 2013], pp. 51–80, which first focusses on the same scientific development as the last quote, complementing it at the end by a look at developments in the USSR.

[50] See [Kjeldsen 2003], pp. 135–136; see also pp. 146–14 for a discussion of how this institutional fixation may have influenced the development of the young theory.

> 88 heading for London or a V-1 buzz bomb doing the same thing was not all that different. A skipper trying to dodge a self-guided torpedo could be excused for referring to the device as 'trying' to kill him, as could the pilot ascribing airfoil self-adjustment to the work of 'gremlins.' And in the specific case of Wiener, [Julian] Bigelow [1913–2003], Weaver, and their colleagues, it is perhaps understandable that the pilot of an enemy plane could be said to 'behave like a servo-mechanism.' While prewar behaviorists might have cautioned against the ascription of internal states, war made it impossible; reading the hidden enemy meant reading his actions. In the mechanized battlefield, in those life-and-death confrontations with an enshrouded enemy, the identity of intention and self-correction was sustainable, reasonable, even 'obvious.'[51]

The mathematical landscape that resulted from the new actors and attitudes had repercussions on the way mathematicians would approach classical fields such as analysis, which is after all

> one of the oldest branch[es] of mathematics, especially linked to the study of nature, physics, and engineering science. Various conceptions of analysis and what its teaching should be strongly opposed those of pure and applied mathematicians. In the 1940s and 1950s, the emphasis put by the former on functional analysis was enormous. For Bourbaki, this was justified by the general state of confusion in mathematics at the time. In fact, except for Laurent Schwartz [1915–2002], none of its members was really an analyst. Bourbaki labored towards a conception in which algebra, analysis, and topology would form a single unified domain giving rise to vast syntheses at increasing levels of abstraction. Traditional branches of analysis were considered bleak and limited in their ambitions. When he tackled nonlinear oscillations, Solomon Lefschetz noticed that differential equation theory was deemed the most boring topic possible. L[ennart] Carleson [b. 1928] has described the reigning state of mind regarding classical analysis: 'There was a period, in the 1940s and 1950s when classical analysis was considered dead and the hope for the future of analysis was considered to be in the abstract branches, specializing in generalization.' Writing in 1978, he went on: 'As is now apparent, the rumor of the death of classical analysis was greatly exaggerated and during the 1960s and 1970s the field has been one of the most successful in all mathematics.'[52]

We leave this chapter with an example from the other side of the Cold War, of a long-term development of a war-related mathematical problem, whose solution would provide a central result of the theory of optimization.

> In 1970, at the World Congress in Nice, Prof. [Lev Semenovich] Pontryagin [1908–1988] gave a plenary talk on differential games, which was motivated by pursuit-evasion strategies of aircrafts for a very simplified model of behavior. During the after-talk discussions, A. Grothendieck put a rhetorical question to Pontryagin. He said that though the listeners witnessed a beautiful piece of mathematics, still he would like to know whether the speaker feels himself morally responsible for supporting military trends in the society. Pontryagin's answer was quite definite and blunt. He was convinced, he said, that, on an intellectual level, any intellectual problems could be discussed openly in a developed society, and if we would follow to the logical end Prof. Grothendieck's recommendation, we should be prohibited from speaking openly about some topics of abstract Algebra, since Cryptography, which has much deeper correlations with military problems than the differential game considerations he spoke about, is completely based on the theory of finite fields.

---

[51] See [Galison 1994], p. 263.

[52] See [Dahan 2001], p. 242; the author goes through further milestones of the story in her text, which we do not follow up here.

> Lev Semenovich Pontryagin was one of the leading figures in 20th century algebraic topology and topological algebra, but in mid-1950s he abandoned topology, never to return to it, and completely devoted himself to purely engineering problems of mathematics. He organized at the Steklov Mathematical Institute a seminar on applied problems of mathematics, often inviting theoretical engineers as speakers, since he considered a professional command over the engineering part of the problem under investigation to be mandatory for an adequate mathematical development. . . .
>
> Pontryagin was led to the formulation of the general time-optimal problem by an attempt to solve a concrete fifth-order system of ordinary differential equations with three control parameters related to optimal maneuvers of an aircraft, which was proposed to him by two Air Force colonels during their visit to the Steklov Institute in the early spring of 1955. Two of the control parameters entered the equations linearly and were bounded, hence from the beginning it was clear that they could not be found by classical methods, as solutions of the Euler equations. The problem was highly specific, and very soon Pontryagin realized that some general guidelines were needed in order to tackle the problem. I remember he even said half-jokingly, 'we must invent a new calculus of variations.' As a result, [a] general time-optimal problem was formulated. . . [53]

[53] See Gamkrelidze's Chapter in [Booß-Bavnbeck & Høyrup 2003], pp. 160–161. The chapter goes on to explore the meaning of Pontryagin's Maximum Principle all the way to its geometric bearing.

Part III
Seventy Years of Globalization: 1950–2020

> We are holding the Congress in the shadow of another crisis, perhaps even more menacing than that of 1940, but one which at least does allow the attendance of representatives from a large part of the mathematical world. It is true that many of our most valued colleagues have been kept away by political obstacles and that it has taken valiant efforts by the Organizing Committee to make it possible for others to come. Nevertheless, we who are gathered here do represent a very large part of the mathematical world. I will also venture the much more hazardous statement that we represent most of the currents of mathematical thought that are discernible in the world today. I hope that this remark will be dissected and, if possible, pulverized in the private conversations that are so valuable a part of any scientific meeting.

Oswald Veblen at the opening of the 1950 ICM at Cambridge, Mass.[54]

In this concluding part of the book we turn to the second IMU, the way it has increasingly asserted itself at the heart of Mathematics International over the past seventy years, and the public image of mathematics it continues to shape today.

[54] Proceedings ICM 1950, Vol. 1, p. 124. The quote in the section title of 8.1.4 is taken from this passage of Veblen's address.

Chapter 8
Seventy Years, Eighteen ICMs, and One IMU

Turning now to a period which coincides with my own life, I cannot deliver a historical account; the minimum distance to many of the events studied is missing. This is one of the reasons why the style of this third and final part of the book will become increasingly different from that of the preceding chapters.

War can be the heyday of specialized developers, constructors, and industrialists; but it chiefly tears things down, kills and wounds people, and changes the face of the world for those who survive. We have described in Chapter 4 how the cultural expression of World War I dealt a fatal blow to earlier conceptions of Science International, and gave birth instead to an International Research Council (IRC) whose primary inspiration seems to have been that of eclipsing the former enemies. The Scientific Unions created in 1919 and 1920, like the IMU, had to align their basic political orientation with this new mother structure IRC.

Compared to this, the consequences of World War II for Science International may appear to have been much milder, at least if looked at from the point of view of mathematicians starting their career after 1945 in the Western World. This is partly an illusion, especially as it fails to take into account the new division of the world, which affected all walks of life, and thus also Science International. We start this chapter by reviewing this new world, in order to be able to situate the new IMU in its original Cold War environment.

8.1 A New IMU and an ICM in Another World

Comparing what happened after 1945 with developments after the end of World War I is a good way to start. The obvious analogies never reduce to sameness. Just as in Chapter 4, the best way to proceed is top down, starting with the highest institutions on the international scene that were newly framed as nations emerged from World War II.

N. Schappacher, *Framing Global Mathematics*,
https://doi.org/10.1007/978-3-030-95683-7_8

8.1.1 United Nations, International Tribunals

Right after 1918, politics in general and Science International in particular were marked by the exclusion of the central powers. In November 1945, the *International Military Tribunal* convened in Nuremberg, Germany, to judge key actors of the Nazi regime. The trial was famously opened by the Chief U.S. Prosecutor Robert H. Jackson (1892–1954), who pointed out that the tribunal was "novel and experimental", created "to utilize international law to meet the greatest menace."[1] This International Military Tribunal, the formal constitution of which had been carefully drafted during the Summer of 1945 in London, was a seminal event in the history of law, and incidentally reminds us of the origin of the word 'international' in Jeremy Bentham's philosophy of law.[2] An analogous International Military Tribunal for the Far East was created a year later in Tokyo, Japan, pursuant to a proclamation by U.S. Army General Douglas MacArthur in January 1946. Both the Nuremberg Trial and the Tokyo War Crimes Trials predated the *Universal Declaration of Human Rights*, which would be adopted by the United Nations General Assembly on 10 December 1948, in Paris. The United Nations themselves had actually been constituted—following in the footsteps of the much challenged League of Nations—shortly before the Nuremberg Tribunal began.

A particular reason to mention the Nuremberg Trial of 1945–1946 here is that it was followed by other Tribunals that would have bearings on the way we judge scientific research. In 1946–1947, the first Nuremberg Doctors' trial looked into the practice of Nazi medical researchers. Even though it took place, not before an international, but a US military court, and even though it was often misunderstood as a trial of pseudoscientific practices, the Doctors' trial emerges with hindsight as a turning point in the ethics of science, well beyond medicine. During the inquiries, internal values of scientific quest and quality were explicitly found to be contrary to ethical values, typically the human rights of the test persons.

> Allied investigators of German military medicine were confronted by the choice of exploiting captured personnel and documents for weapons research, or prosecuting war crimes. The Allies had a high regard for the ability of German aviation medicine to solve problems of high-altitude flight. The Atom bomb required knowledge about the hazards of radiation, and German chemical weapons and nerve gas might be deployed against the Japanese and then against the Soviets. The British and Americans feared that German scientists would opt wholesale to work for the Russians. The Allies faced a conflict between exploiting German medical know-how and prosecuting its criminality.
>
> The American Medical Association (AMA) and British Medical Association (BMA) were concerned that releasing news of the German atrocities would undermine public confidence in medical research. Formulating new ethical standards became a priority to ensure the future viability of research-based clinical medicine. The Nuremberg Code on the conduct of human experiments promulgated at the close of the Trial was a response to such concerns. The consent of the research subject, and the right to know and participate voluntarily in medical research remained central issues in clinical research. At the same time, the Trial

---

[1] Quoted from the narrative account in [Sands 2016], p. 288.

[2] Cf. our mention of Bentham in the opening paragraphs of Chapter 1.

> revealed much about the structures and values attached to research on both the German and Allied sides. The German defence counter-attacked by challenging the ethical standards and practices in Allied medical research.[3]

Subsequent trials, while much debated, widened the breach between scientific and ethical values. The sixth of them was the 1947–1948 Nuremberg "IG Farben" trial of leading German industrial chemists who were accused, among other things, of having provided the chemicals for the gas chambers in the Nazi extermination camps. Reflecting on this trial turned out to be an unexpected challenge for the 1928 Nobel Prize winner Adolf Windaus (1876–1959), himself a well-known opponent of the Nazi regime, because of the high scientific esteem he had for some of the accused colleagues.[4] Even if not shared by everyone, a new critical sentiment with respect to science began to spread at the time and reverberated in Science International after World War II: ethics. The exclusion politics against the central powers after the First World War had also been carried by admonishing moral declarations. But that moral reasoning had dealt with Nations, and insofar as the moral rules were to be applied to individuals, to scientists, the dominant criterion was the nationality of the person in question. This changed after World War II. The tribunals and the Declaration of Human Rights vindicated genuinely supranational ethical norms and opened up the possibility of condemning individuals on a would-be global juridical basis.

This new ethical dimension of science would meet with growing concern about the possible atomic self-destruction of mankind, giving pacifism a novel twist and urgency.

> After the Second World War, scientists in many Western nations attempted to reestablish a scientific internationalism that, as they understood it, had been only suspended during two world wars. Bringing the scientific ethos to bear on geopolitics, they mobilized for world government and international control of nuclear weapons. Eventually, however, Cold War tensions demanded that Western scientists view science from within a Cold War paradigm. In one scholar's words, the Cold War produced a "bipolar scientific internationalism" that united the scientific community under an anticommunist and pro-Western ideology. With Western governments dispensing larger and larger amounts of funding and influence, scientists had professional reasons to oppose the Soviets.[5]

I am not aware of any documented general survey of mathematicians' reactions to this new ethical dimension. The book [Booß-Bavnbeck & Høyrup 2003] is in itself such a reaction. We ended Chapter 7 with an anecdote involving Pontryagin and Grothendieck at the Nice ICM in 1970, quoted from this book. I personally remember Hans Grauert (1930–2011) telling me proudly during a conversation in the 1970s that he had once refused to carry out a calculation involving complex functions of several variables that an American colleague had asked him to do in order to solve a problem related to the H-bomb. Grauert immediately added that the colleague had solved the problem himself shortly afterwards using more elementary methods.

---

[3] See [Weindling 2004], p. 3.

[4] See [Schauz 2021], pp. 332–333.

[5] See [Rubinson 2012], p. 247.

Following the timeline of the past 70 years in this chapter we will have several occasions to mention human rights issues that affected ICMs and the IMU.

8.1.2 UNESCO and ICSU

Recall that the first IMU, founded in 1920, belonged to the International Research Council (IRC). The latter would transform into ICSU in 1932, which allowed international scientific unions as full members, on a par with nations—see Section 4.4.3.1. Independently of IRC/ICSU and their scientific unions, there was also ICIC, the International Committee on Intellectual Cooperation. It answered directly to the League of Nations, but would occasionally meet with IRC institutions to discuss, for instance, questions of bibliography—see Section 4.2 above.

In the new world after 1945, the League of Nations was quickly replaced by the United Nations (UN), and the analog of ICIC would be greatly expanded into a specialized agency of the UN, an Intergovernmental Organization of a kind that had never existed before: the *United Nations Educational, Scientific and Cultural Organization*, UNESCO. Very potent and active for about half a century it would, among all its activities, also leave its mark in the domain of Science International, including programs of the IMU.

> The twenty Member States that founded UNESCO in London on 16 November 1945 wished to resume the work of the defunct International Committee on Intellectual Cooperation in a considerably broadened scope. Meeting amidst the smoking ruins of the British capital, they asserted that, in the words of American poet Archibald MacLeish, 'it is in the minds of men that the defences of peace must be constructed'. In the second half of the twentieth century, UNESCO carried out a considerable number of tasks: promoting the right to education; effectively contributing to the rescue, safeguarding and enhancement of humanity's cultural and natural heritage; providing support for artistic creativity, so often stifled and endangered by the new technological and economic environment; mobilizing political leaders to increase and share scientific knowledge; promoting the free flow of words and images; and attempting to reduce the flagrant imbalance in access to information and means of communication available to industrialized and developing countries.[6]

The fact that scientific teaching and research would play such an important part in the program of the new organization was, to a large extent, the fruit of two years of work by Joseph Needham (1900–1995), the British biochemist and historian of science of socialist convictions, who would later edit, among many other things, a much discussed history of mathematics in China.[7]

> Born in London in 1900, Needham studied medicine and biochemistry at Cambridge but he also had a keen interest in religion and philosophy. His political commitment was forged during the Great Depression. The massive unemployment resulting from the economic crisis that began in 1929, led many people to criticize the role of science and its applications to industry. It also brought about a reduction in both finances and employment within the field of scientific research. Needham joined the International Council of Scientific Unions

[6] See [Gopal et al. 2008], p. 8.

[7] Cf. [Werskey 1978] as well as [Jami 1996].

> (ICSU) and, throughout the 1930s, benefited from his experiences with "movements for social relations in science." Needham was part of an idealistic generation of scientists who wanted to use discoveries and their applications to improve living conditions for all and to develop democracy.
>
> The war did not interrupt this commitment. Quite the contrary. Needham, like most of his peers, was horrified by the way the Nazis deformed and used science to justify the racist ideology that led to the Holocaust. Most scientists participated directly in the struggle against Nazism. Even during the war, several conferences were organized in London by the British Association for the Advancement of Science (BAAS) and the British Association of Scientific Workers (AScW) to discuss the post-war role of science. Participants were determined that science and its applications be used for the well-being of all. The importance of international scientific cooperation would be paramount. In February 1945, several foreign delegations took part in the "Science for Peace" Conference during which the creation of international scientific associations was notably discussed. From 1946 onward, these same scientists quite naturally met up again at UNESCO, ICSU or the World Federation of Scientific Workers to put into practice their ideas and projects.[8]

Thus it was Needham who, in connivance with Julian Huxley (1987–1975), the first director of UNESCO, succeeded in effectively squeezing the letter 'S' into that acronym. Looking back from the mid-1980s the following could be said:

> By the insertion of the S in Unesco not only did the Preparatory Conference introduce "science" to Unesco, it also started a process that is still a major factor in international scientific cooperation today forty years later by the adoption of the following resolution:
>
> "that the Preparatory Commission of the United Nations Educational, Scientific and Cultural Organization be requested by this Conference to instruct its Executive Committee to consult with the International Council of Scientific Unions on methods of collaboration to strengthen the programmes of both bodies in the area of their common concern, and that the plans thus formulated be reported to the first Conference of Unesco, with recommendations for a suitable working arrangement with the International Council of Scientific Unions".
>
> This resolution was discussed by the ICSU Executive Committee at its meeting on 4 December 1945, which set up a small group that met Dr. Joseph Needham, Head of the Natural Sciences Division of the Preparatory Commission of Unesco, on 29 May 1946, when the possibilities of collaboration were discussed. Subsequently, a statement and draft agreement were prepared for submission to the ICSU General Assembly in July 1946, which discussed the possible agreement, and to the inaugural meeting of the Permanent Organization of Unesco that met at the end of 1946. One point that caused some controversy and a change in the initial draft was the degree to which other U.N. bodies had scientific activities. . . .
>
> At the first General Conference of Unesco in Paris in November 1946 a Sub-Commission on Natural Sciences was established. The Chairman was H[omi Jehangir] Bhabha [1909–1966] who, twenty years later, as Director of the Tata Institute, was to act as host to the ICSU 11th General Assembly in Bombay.[9]

Homi J. Bhabha was killed in a plane crash at Mont Blanc days after this 1966 General Assembly. TIFR, the Tata Institute for Fundamental Research, was founded in June 1945—cf. Section 8.3 below.

---

[8] From the introduction of [Petitjean 2006]; references given there.

[9] See [Baker 1986], p. 1.

The considerable contributions that UNESCO has provided to ICSU over several decades added a new tier to the institutional diagram of Science International. Back in the 1920s and 1930s there had only been occasional contacts between IRC/ICSU and ICIC. Now, however, in dealing with the intergovernmental UNESCO, the non-governmental organization ICSU was collaborating with the top agency of the United Nations in charge of international scientific relations. Not surprisingly it could, and did, happen that UNESCO would consider pursuing its work for Science International in other ways than with ICSU.[10]

> Although ICSU has never been dependent on the existence of any other international body (League of Nations, United Nations, UNESCO, or other), and, as we have seen, originated and had established its major features before either UN or UNESCO was created, these two are the most important external international structures affecting the life of ICSU. Without UNESCO and the other UN bodies, ICSU would be very different, although it would still exist and be very effective.[11]

We will not go into the impressive list of projects that fleshed out the collaboration between UNESCO and ICSU. Suffice it to just name one successful example: the *International Hydrologic Decade* 1962–1972. Several UNESCO grants for concrete projects of the IMU will be mentioned in the consecutive subsections of Section 8.2 below. Given the nature of UNESCO and the fact that its first decades coincided with the era of decolonization, concern about Science and Technology in what was then called the Third World would seem like an obvious domain of activities bringing UNESCO, ICSU, and other agencies together. However, work in this direction took a while to get organized; here is what Baker had to say about it in 1986:

> It was not ... until 1966 at the General Assembly in Bombay that ICSU set up a formal structure to be specifically responsible for helping scientists and technologists in developing countries: the Committee on Science and Technology in Developing Countries (COSTED).
>
> COSTED prepared within the framework of the U.N. Conference on Science and Technology for Development a Symposium on Science Technology and Development at which the views of scientists and technologists from the developing world were expressed. This Symposium was held in Kuala Lumpur in April 1979 and, with the Symposium organized in Singapore in January 1979 by ICSU and a number of other non-governmental organizations, formed part of a series of symposia which provided inputs to the Colloquium of the Advisory Committee on the Application of Science and Technology (ACAST) held in Vienna the week preceding the U.N. Conference.
>
> After an initially slow start the Committee is now developing into an effective organization for assisting scientists and technologists in developing countries.[12]

As far as the IMU is concerned, we will address this concern in Section 10.1.3 below.

---

[10] Such a threat occurred—and was quickly overcome—for instance in 1949, when the Executive Board of UNESCO decided it "should not continue such grants to [scientific] Unions for their normal activities but should make these grants to new bodies, in order to start them and keep them going for a few years; in particular for projects of special interest to Unesco"—see [Baker 1986], p. 4.

[11] See [Greenaway 1996], p. 183, from the beginning of the Chapter "ICSU and UNESCO".

[12] See [Baker 1986], p. 14–15.

Somewhat analogous comments apply to the question of science education. The influence of UNESCO seems to have been crucial in putting this subject on the ICSU agenda. However, it was not at all a new theme for Mathematics International.

> The seventh General Conference [of UNESCO] in 1952 authorized the Director-General "to stimulate and facilitate the improvement of natural science teaching, with particular reference to methods, manuals, teaching equipment and audiovisual aids". This began a movement that added education and training of scientists to the much more general problem of educating the general public about science. One of the travelling exhibitions, on the Construction of Laboratory Apparatus for Schools, shown in Cairo in 1955, brought together the problems of disseminating and teaching science.[13]

Baker then continues to acknowledge that these concerns had been attended to in the domain of mathematics early on by the International Commission for Mathematical Instruction (ICMI)—see Chapter 9 below.

We end our sketch of the times when the partnership between ICSU and UNESCO was going strong with a general appraisal, which is worth quoting:

> It would be an exaggeration to suggest that relations between ICSU and Unesco have always been smooth. There have been some periods when ICSU felt that its activities were not fully appreciated nor was sufficient notice taken of its advice. For example, in 1966 at the 11th General Assembly the President of ICSU, Sir Harold [Warris] Thompson [1908–1983], in his address said: "At the Vienna Assembly, it was agreed that ICSU should accept the invitation of Unesco to become its principal scientific advisor, and this relationship is now quoted in many Unesco publications. While I realise, of course, that scientists within ICSU are advising Unesco on special matters, and some of the ICSU Programmes are receiving financial help, which we much appreciate, I am uneasy about the position and do not feel that our attempt to coordinate activities and to ensure a planned distribution of funds is satisfactory, and I hope that in the discussions between Unesco and the Officers of ICSU our advice may be found more acceptable".
>
> Since then relations have greatly improved and it would appear that 1966 was a turning point for at the next General Assembly in 1968 the Assistant Director-General of Unesco for Science, Prof. A[lexei] Matveyev , said: "It has been said that ICSU and Unesco may be regarded as two sides of the same coin, and in one sense I agree. I do not intend to toss this coin to see which side will come down on top, but I do feel that ICSU and Unesco are complementary and both essential for the promotion of science and for international cooperation in scientific research: their cooperation in complete confidence is indispensable—and is, indeed, progressively being achieved". After praising the friendly cooperation between the two organizations he ended as follows: "Are Unesco and ICSU two sides of the same coin? Perhaps. But I prefer another metaphor. I prefer to think of them as two sides of a Möbius [strip]".[14]

At least since the 1980s, UNESCO has been directly involved in burning issues of world politics. The US withdrew from UNESCO for the first time between 1984 and 2003, and then again in 2017, followed by Israel.

[13] See [Baker 1986], p. 16.

[14] See [Baker 1986], p. 7.

As for UNESCO's contributions to global mathematics, they are currently still reflected in the work of local institutions such as CIMPA in Nice, France. Also, with a view to broader outreach, the IMU took the initiative and secured national support from many countries to have UNESCO proclaim 14 March as the International Day of Mathematics.[15]

8.1.3 The New IMU

In Section 6.2.1 about mathematical conferences in the 1930s, we had occasion to quote the Swiss mathematician Rolin Wavre. A first version of his piece had possibly already been penned before or at the beginning of World War II, but when the long planned book [Lionnais 1948] was finally ready to go to press, the mathematician from Geneva added to his chapter a remark on the possible future of the IMU:

> As we said before, the IMU does not in fact exist any more. But it would be advantageous to reconstruct it. It could contact the United Nations and respond to all consultations, on behalf of the United Nations or on behalf of other international organisations. No need to insist further on the interest of reaching first an agreement between the mathematicians and within ICSU, so that the latter could present our requests and needs before UNESCO. Already this March [1947] a meeting has taken place to resuscitate the IMU; in the Spring in Paris, on the occasion of two international conferences sponsored by the Rockefeller Foundation, this question is going to be taken up again in a meeting of mathematicians from different countries.[16]

Here the motivation to revive the IMU after World War II is clearly linked to the newly transformed scene of international agencies under the umbrella of the United Nations, with the new potent donor UNESCO, as well as ICSU on its continuing mission. Even for those who remembered how mathematicians had fared quite well in the past with regular ICMs organized in an *ad hoc* manner, without the help of an IMU, this refurbished scene of world organizations would be an attractive factor. It may well have been the key to explaining why the IMU could be mounted in about five years after the end of World War II.

On the other hand, even though the Paris meeting that Wavre alludes to was held as an appendix to top international mathematical conferences steeped in the very best tradition of the 1930s,[17] it could awaken dubious memories of how the IRC and IMU had been conceived in the wake World War I. Indeed, that Paris meeting was eyed rather sceptically by American mathematicians who had already taken the lead in the double project of organizing the next ICM and building a new IMU.[18] At the

[15] For more information on the International Day of Mathematics, see [URL 10].

[16] See [Wavre 1948], pp. 302–303; my translation.

[17] We are alluding here to Section 6.1.2 above. Cf. the part "The Springboard: the Colloque d'Analyse Harmonique, Nancy 1947" of the paper [Barany et al. 2017].

[18] See [Barany 2016a], pp. 170–176.

1936 ICM in Oslo, US mathematicians had offered to host the next ICM in 1940 in Cambridge, Massachusetts, and the offer had been accepted. When the Second World War began, the congress was canceled for 1940.

> [A]n Emergency Executive Committee was set up "to take the initiative for resumption of activity." This the Committee did in 1946, an important year because decisions about the international policy to be followed in science, in general, and in mathematics, in particular, were then made. . . .
>
> In April 1946, the Emergency Executive Committee for the ICM, headed by Marston Morse, reported to the American Mathematical Society (AMS) that it was interested in the revival of plans for the Congress only if it could be an open meeting to which all mathematicians would be invited irrespective of national allegiance. The Council of the AMS agreed with this principle, which set the tone for international mathematical cooperation after 1945.
>
> . . .
>
> The mathematical community of the United States felt that, as an organizer of the ICM, it could study the possibilities of re-forming an International Mathematical Union. In the summer of 1948, the responsibility for all preparations concerning the planned Union was delegated to a three-man committee, consisting of Marshall Stone (Chair), John Robert Kline , and Marston Morse.[19]

So what motivated the Americans around Marshall Stone to try and recreate the IMU? The question arises because Stone's public rhetoric tended to rather stress the successful mathematical tradition of *informal* international contacts through *ad hoc* meetings. In a short letter to the editor of *Science Magazine*, for instance, Stone wrote in 1941:

> I have always understood that [the old IMU] lost the effectiveness it might have had and in the end went out of existence (in 1936, I believe) chiefly because the majority of mathematicians did not approve the political origins and development of the Union. However that may be, there is no question that the mathematicians of the world had every reason to be pleased with the effectiveness of that rather informal but close cooperation which, among other things, made possible their successful and important quadrennial international congresses.[20]

Stone knew exactly what he was talking about because he had participated in the ICMs in 1932 and 1936, as well as the International Topological Congress in Moscow in 1935. However—as Michael Barany has acutely observed, and then dismissed—Stone would use the same rhetoric in 1947, now in favor of resuscitating the IMU:

> Beyond such a union's practical motivations, like facilitating East-West exchanges and drumming up financial and diplomatic backing in the international arena, Stone insisted on a higher purpose "of a psychological rather than a practical order." Namely, a new union "would give concrete expression to the deep-felt desire for international scientific cooperation and would be a step of incalculable importance in restoring to mathematics the international character it enjoyed before the war." With this, Stone did not of course have in mind the factious patchwork of actors and institutions of the interwar mathematics community that had failed to sustain the previous International Mathematical Union. Rather, the "international

[19] From the epilogue of Olli Lehto's chapter in [Parshall & Rice 2002], pp. 393–394.

[20] See *Science*, Vol. 94, No. 2432 (August 8, 1941), p. 138.

> character" Stone and his committee sought to "restore" was a fiction vigorously touted by Stone and his American colleagues after the Americans took the International Congress's mantle in 1936.[21]

Well, Stone's personal congress experiences of the 1930s were no fiction, and contrary to Barany, I do think he appreciated their international character. Stone's enthusiasm for international congresses also showed in his presidency of the Subcommittee on Conferences planned to be held in connection with the 1940 ICM.[22] It is remarkable, though, that he used them as arguments in favor, not just of the next ICM, but of a new IMU. Was it maybe natural for the son of the Chief Justice of the United States to look for a solid legal structure, especially at a time when international agencies were *en vogue*? Or was there a peculiar American twist to the project: building a new IMU on the basis of the new global primacy of the US after World War II? Discussing these issues in the narrow context of Mathematics International is complicated by the fact that both the actors of the 1940s[23] and some historical notetakers tended to reduce the question of the political dimension of the IMU to the exclusion policy that had dominated the old IMU.

In his book on the history of the IMU, Olli Lehto chose to call the post-World-War-II axiom of Mathematics International, according to which national allegiance must not be a reason for exclusion, the "American Declaration of Universality."[24] Indeed, in the wake of World War II, the opposition of US mathematicians against the IRC exclusion policy of the 1920s—see in particular Section 4.4.2 above—undoubtedly helped to steer clear of excluding Germany and Japan from the upcoming ICM and the new IMU. Thus the organizers of the ICM

> hoped that German and Japanese mathematicians would be represented, and that those they termed "decent" would receive the needed subsidies for travel. The requirement for permission to enter the United States made them confident, meanwhile, that "notorious Nazis who attempt to attend the Congress" would be unable to do so. It is not clear whether they had particular "notorious Nazis" in mind, nor what kind of problem they imagined the attendance of such figures would present.[25]

However, Lehto's solemn section title: "American Declaration of Universality," invites starry-eyed idealizations of what was going on, all the more so as Lehto opens that section of his book with a quote from a 1946 declaration of ICSU's president to the effect that "[w]e are inclined to keep politics as far from science as possible, for we know how much the International Research Council, the predecessor of ICSU, suffered after the First World War by not discriminating sufficiently in this respect; the development of international scientific cooperation then was prevented for at least ten years."[26] Obviously, the decision against the politics of exclusion as practiced after World War I does not purge the procedure of the American committees of all

[21] See [Barany 2016a], p. 165.

[22] See [Hollings & Siegmund-Schultze 2020], p. 129.

[23] See [Barany 2016a], p. 167.

[24] See [Lehto 1998], title of Section 4.1, pp. 74–77.

[25] See [Barany 2016a], p. 159.

[26] See [Lehto 1998], p. 74.

political aspects. After all, Cold War was clearly in the air at least since the Potsdam Conference between Churchill, Truman, and Stalin in the Summer of 1945. This rendered the political reflexes of the 1920s increasingly obsolete, and it gave the ideal of "universality" a different twist.

> By 1948, [the organizers of the ICM] had to worry about whether U.S. or Soviet governments would allow travel to the Congress for even those Soviet mathematicians who were informed and willing to attend. Mathematicians interested in fostering a single, global disciplinary community had more and more reasons to fear a mathematical world cleaved in two.
>
> It was for this reason that accessing Russian writings and mathematicians (including via "governments, national academies, and other international bodies") was near the front of Marshall Stone's rationale for reconstituting an International Mathematical Union as quickly as possible.[27]

In fact, no Soviet mathematician would participate at the 1950 ICM. Then Stalin died in 1953. The 1954 ICM in Amsterdam saw a delegation of four Soviet mathematicians. The USSR would officially join the new IMU in 1958.[28]

At any rate, being neither apolitical nor falling back on the exclusion reflexes of the 1920s, after years of hard work, the Congress did take place in 1950, and the new IMU was indeed founded, although several years may compete for the honour of being its year of birth:

> After careful preparations by worldwide correspondence under Stone's direction, draft statutes were presented to the "Union Conference" in New York in August 1950. This conference was attended by the delegates of the National Committees for Mathematics of twenty-two countries. Consensus about the statutes and by-laws was reached, and it was decided that the new IMU would come into existence as soon as it had ten member countries. This quota was reached in September 1951. Among the first ten to join were Germany and Japan.
>
> The activities of the IMU began after the First General Assembly held in Rome in March 1952. By the end of the 1950s, several important targets had been reached: the IMU was readmitted to the ICSU; a subcommission, the International Commission on Mathematical Instruction [ICMI], was established to continue the work of the old Commission on the Teaching of Mathematics; the Soviet Union and other socialist countries of Europe became members; the first *World Directory of Mathematicians* appeared; and the development was initiated that gave the IMU sole responsibility for the mathematical program of the International Congresses.[29]

8.1.4 Gathering "a Very Large Part of the Mathematical World"

Having looked at the rhetoric surrounding the foundation of the new IMU, we are well prepared to read in J.R. Kline's Secretary's Report about the 1950 ICM at Harvard University a stylistic exercise about apolitical politics facing the 'Iron Curtain':

[27] See [Barany 2016a], p. 164.

[28] See the diagrams in Section 10.4.1 for an overview of Soviet participation in the ICMs.

[29] See [Parshall & Rice 2002], p. 394. More details on the founding of the new IMU are presented in [Lehto 1998], Chapter 4.

> In attempting to maintain the non-political nature of the Congress, many serious difficulties had to be overcome. In the solution of these problems, officers of the Congress found the various officials of the Department of State most sympathetic and helpful. As a part of the effort to keep the Congress apolitical, they tried to secure a visa for every mathematician who notified them about any visa difficulties before cancelling his passage. As far as they know only one mathematician from any independent nation was prevented from attending the Congress because he failed to pass a political test and this man did not notify the officers of the Congress about his difficulties. Only two mathematicians from occupied countries failed to secure visas. Mathematicians from behind the Iron Curtain were uniformly prevented from attending the Congress by their own governments which generally refused to issue passports to them for the trip to the Congress. Their non-attendance was not due to any action of the United States Government.[30]

Five years had passed since the end of the war. The Cold War mapped out the image of the world. The Soviet Union had already tested its own atomic bomb. The Korean War had started two months before the opening of the congress. In this war the US intervened on behalf of the United Nations. They in turn were boycotted by the USSR and only recognized Taiwan as China. McCarthyism had begun haunting the US.

The resulting visa problems were a major challenge for the organizers, which the quote only indicates politely. Michael Barany has investigated two cases in particular: the well-known expert on stability theory for differential equations José Luis Massera (1915–2002) from Uruguay, who had excellent connections in the U.S., in particular Princeton, and published in journals like the *Annals of Mathematics*; and Laurent Schwartz, one of the 1950 Fields Medalists, the other one being Atle Selberg (1917–2007).

> Beneath Kline's official accounting of who could or could not attend for whatever reason, there is no way to know with certainty how many mathematicians would have made the trip under different political circumstances but were dissuaded at one point or another. Kline did not record the name of the neutral mathematician who failed a political test, but if he had Uruguayan Communist mathematician José Luis Massera in mind then the facts on the ground give a different picture from Kline's implication that he simply neglected to inform the Congress's officers of his troubles. If it was not Massera, then Massera's difficulties stand in evidence of just how substantial a political barrier Kline was prepared not to blame on U.S. authorities. Either way, Massera's political tests help foreground the tangible consequences of American anti-communism in a critical period for intercontinental mathematics.[31]

Even though Massera was barred from participating in the congress, the 1950 ICM did a lot to upgrade South America's place on the mathematical map of the world, through relations with North America, to which some of the principal organizers of the Cambridge ICM had personally contributed.

> In terms of setting a mold for postwar connections between mathematicians across continents, however, the 1950 Congress's most important legacy may well have been the new ties it reflected between mathematicians of North and South America. For Stone, in particular, such North-South connections had a triple importance for his efforts at re-establishing an

[30] Proceedings ICM 1950, vol. 1, p. 122.

[31] See [Barany 2016a], pp. 215–216. On the subsequent pages up to p. 228, Barany presents the archival evidence he has managed to collect on Massera. See also [Barany 2016b].

> International Mathematical Union. First, they offered an important pretext for his UNESCO-centered organizational approach. Second, they furnished a valuable resource to help him carry out that approach. And third, in view of his struggles to enroll European mathematicians his Latin American connections furnished his undertaking's most notable and unequivocal dividend.[32]

Turning to the case of the Fields Medalist Laurent Schwartz, we should first note that in 1950 the Fields Medal did not yet have the nimbus it enjoys today. Apparently neither Schwartz nor Selberg had ever heard of this prize before they learned that it was going to be awarded to them.[33] But the worldwide success story of Schwartz's theory of distributions, from Nancy, via Copenhagen—Harald Bohr presided over the Fields Committee—to the Congress and the Medal, increased the visibility of the medal.[34]

Schwartz had already experienced visa problems in the US more than a year before the ICM at Harvard, when he was denied entry to the US from Canada. The apparent reason was his earlier Trotzkyite political activity.[35] As Schwartz was one of several French scientists encountering visa problems, André Weil, Henri Cartan, and other French mathematicians prepared to protest vigorously against the lack of international openness of the host country for the upcoming ICM.

> What we have to seriously envisage (and on this point Stone authorizes us to make it known that he shares our view) is the possibility of relocating the Congress, either to Canada or to Europe. In the latter case, one would need to find a country where the invitation of German colleagues would not create comparable difficulties. If France could guarantee this, one could think of France; otherwise it would be better to think of Denmark. In any event, we would have to organize letters—either personal letters (essentially from colleagues having received an invitation to attend the congress), or collective letters (from Mathematical Societies etc.)—asking that the possibility to relocate be considered urgently, given that it becomes more and more obvious that a Congress held in America could not have the international character that one would like it to have. The choice of Cambridge [Mass.] was made in 1936 when nobody could imagine that such a situation would arise ... [36]

The affair gathered a certain momentum. It allows us to read the above quote from Kline's report, and also Lehto's praise of the 'American Declaration of Universality,' with the necessary grains of salt. The Congress, as we know, did take place as planned. Laurent Schwartz, and also Jacques Hadamard, did obtain their visas in the end.[37]

> The 1950 International Congress of Mathematics decisively shaped the discipline's international stature, both in the personal and intellectual connections it created and reshaped and in the institutional arrangements (foremost the International Mathematical Union) forged

---

[32] See [Barany 2016a], p. 183.

[33] See [Barany 2016a], p. 250, footnote 1.

[34] See [Barany et al. 2017] for a masterly description of this story.

[35] The American Consul at Strasbourg, in a report sent to Washington, had apparently called Schwartz a 'Stalinist' instead. This is was what Henri Cartan had heard, who therefore called the consul an "idiot" in his letter to Weil of 21 July 1949; see [Audin 2011], p. 265.

[36] From Weil's letter to H. Cartan dated 15 July 1949; see [Audin 2011], p. 264-265; my translation.

[37] For details and references, see [Barany 2016a], pp. 228–245.

> around it. It represented mathematicians' first postwar effort to grapple with the entirety of their discipline: its theories, people, institutions, nations, politics, and practicalities. Its universalism, such as it was, was necessarily that of non-exclusion. For those same practicalities, politics, nations, institutions, people, and even theories made universal inclusion impossible.[38]

Laurent Schwartz's political engagement continued, particularly in the French context of the Algerian War, which lead to his dismissal from *Ecole Polytechnique* 1961–1963. I intended to illustrate Schwartz's political presence by reproducing the title page of the French magazine *L'Express* No. 343, 16 January 1958, which shows a portrait of Laurent Schwartz. Since *L'Express* never answered my request for the reproduction rights, I suggest that the reader visit [URL 69].

The visa problems and threats of boycott of the Harvard Congress invite us to return to a question we have asked earlier (see Section 8.1.3): what motivated Marshall Stone to invest so heavily in the preparation of a new IMU? As Barany suggests, the mounting political pressure could jeopardize the Cambridge ICM, but it was less likely to sabotage the concept of the IMU. Betting on the IMU could be a way to reap a lasting result from the enormous organizational effort.

8.2 IMU Time Intervals

Both for the current overview of the past seventy years of Mathematics International and for our subsequent data analysis of ICM-related excellence—see Sections 10.3 and 10.4—it seems adequate to organize these 70 years into five consecutive periods, each containing three or four ICMs, whose years are used to name the period. This division structures the current section.

8.2.1 Gearing up to Run Mathematics International: The New IMU 1950–1962

There are various ways to recount the story of how the new IMU got down to work and how it increasingly asserted itself in a situation that was a priori not favorable to a worldwide approach. Olli Lehto for instance writes about the first General Assembly of the IMU in March 1952, in Rome:

> In the historic Villa Farnesina the delegates were to breathe life into the Union. The Statutes constituted the framework but allowed much leeway for the Union's activities. In fact, the old Union, which had been a failure, had had statutes very similar to those of the new IMU. In Rome no reference was made to the prewar General Assemblies. It was rather the tradition

[38] See [Barany 2016a], pp. 248.

> stemming from the first International Congress of Mathematicians in 1897 in Zurich that guided many discussions. The decisions on how and in what concrete ways the IMU should implement its broadly defined objectives had many similarities to those recorded in 1897.[39]

These observations are justified by the agenda—see the end of Section 8.1.3 above—and probably also with respect to the atmosphere in which the Assembly was held. This, however, must not obscure fundamental differences, in particular the fact that the new IMU would command financial resources provided via UNESCO and ICSU. This put the young union in an enviable position for Mathematics International, which had not existed like this before. It could hope to play a more active part in the organization of the ICMs—see Section 8.2.1.1 below. Furthermore, the IMU could begin to co-sponsor smaller international meetings of the kind that had proved their worth as accelerators of international research cooperation in the 1930s—see Section 8.2.1.2 below.

Finally, the new IMU was interested in adapting for its own objectives the model of philanthropic grants that had done so much for scientific exchange in the 1920s and 1930s. Thus Albert Châtelet (1883–1960), Harold Davenport (1907–1969), Børge Jessen (1907–1993), and Kinjiro Kunugi (1903–1975), as well as the Secretary of the IMU *ex officio*, were elected in 1952 into a committee "to study all methods of facilitating the exchange of mathematicians, both Professors and students, between nations . . . "[40]

> This Commission on Exchange was in existence until 1979. It was then replaced by the Commission on Development and Exchange (CDE), with the main objective of promoting mathematics in developing countries. As long as the emphasis was just on organizing the exchange of mathematicians, the results were disappointing, in spite of competent management of the Commission. A world organization was not much needed to steer and coordinate such exchange, which grew rapidly anyhow and was largely carried on through individual contacts.[41]

Lehto's dismissal—"in spite of competent management"—of the work of the original commission underestimates how much times had changed during the 27 years of the existence of that commission. In the early fifties, remembering the effects that philanthropic grants had had on the international integration in the 1920s and 1930s, the idea to float a similar program on account of the IMU was eminently reasonable—all the more so as the IMU then was still a long way from being truly global. Twenty years later, however, there had been such a dramatic explosion of opportunities for young researchers from many countries to travel for scientific reasons that the right reflex was to focus more specifically on those countries which had the greatest difficulties in getting their share. Hence the restructuring of the *Commission on Development and Exchange*. The latter would in turn transform into today's *Commission for Developing Countries*—see Section 10.1.3 below.

---

[39] See [Lehto 1998], p. 95.

[40] See Bompiani to the colleagues just listed, 22 March 1952, in [Arch. IMU], SF 12 Ser 1 digital, Box 62(3)-3.

[41] See [Lehto 1998], p. 97.

8.2.1.1 The IMU and the ICMs

The first of our five periods comprises four ICMs, of which we have already discussed the initial one organized after World War II, in 1950 in Cambridge, Massachusetts, in a parallel effort with the founding of the new IMU. The other three ICMs of this first interval all took place in Europe: in Amsterdam in 1954, in Edinburgh in 1958, and in Stockholm in 1962. A recurring theme in the opening lectures of these congresses was the concern about the growing size of the ICMs. Both Veblen in 1950 and J.A. Schouten in 1954 warned that

> there is a limit to congresses of this kind. This limit will perhaps be reached very soon if the number of mathematicians goes on increasing as rapidly as it does now and if in the future, as I fervently hope, big countries with a great number of good mathematicians will break with the system of sending a very small delegation, the extent of which is in no way proportional to the mathematical importance of the country involved. This system of sending a small delegation only is entirely wrong, the chief aims of a mathematical congress being, as Professor [Carl] Størmer [1874–1957] pointed out in his presidential address at Oslo, to enable the direct exchange of ideas from man to man and to give a great number of younger people the opportunity to get the personal contacts they need for orientation and stimulation. The average age of participants at our congress is $40\frac{1}{2}$ and that is too old.[42]

In 1958, William Vallance Douglas Hodge (1903–1975) linked the ever growing size of the ICMs to the expansion of mathematics through increasing specialization, a tendency that called for more and more specialized conferences. His statement alludes to a new field of activity of the young IMU:

> In recent years there has been a steady growth in the number of symposia held, many with the support of the International Mathematical Union. These symposia have done excellent work in advancing research in special fields. But this is not enough. It is essential for the well-being of mathematics that there should be periodic gatherings attended by representatives of all branches of the subject, and this for several reasons: in my personal opinion, the most important reason is that gatherings such as this serve as an invaluable safeguard against the dangers of excessive specialization.[43]

We will discuss the co-sponsoring of specialized international 'symposia' below. As for the reflections about the nature and objective of the ICMs, they would quickly result in a greater involvement of the IMU in the organization of these Congresses, which has steadily grown ever since. Here is how Otto Frostman (1907–1977) put it in his opening address of the 1962 ICM in Stockholm:

> To be able to present a scientific programme worthy of an international congress it was therefore decided at an early stage to seek the assistance of the International Mathematical Union, and at a meeting in Zurich in November 1960 a small Consultative Committee was appointed with Professor [Georges] de Rham [1903–1990], Lausanne, as chairman. The wide experience and knowledge represented in the Consultative Committee itself and strengthened by contacts with experts from all over the world, made it possible to choose the subjects and speakers for the one-hour addresses and to appoint chairmen of the international

[42] See Proceedings ICM 1954, Vol. 1, p. XXXVIII. Note that Schouten in 1954 apparently expected no women to participate at ICMs. See also Heinz Hopf's contribution to the debate about the usefulness of ICMs in his closing address, Proceedings ICM 1954, Vol. 1, p. 154.

[43] See Proceedings ICM 1958, p. L.

> panels which have proposed the half-hour speakers. At subsequent meetings the Consultative Committee brought the information gathered to the Swedish representatives and all decisions were made in agreement. It must be clearly stated that the Swedish Committee takes the full responsibility for the organization of the congress, but without the invaluable help of the panels and the Consultative Committee the scientific programme would not have been adequate.
>
> The part performed by the International Mathematical Union in preparing the scientific programme of this congress is a leading one, and is well suited to act as a precedent for any future international congress. It seems therefore quite natural that the President of the International Mathematical Union should preside over the general sessions of the Congress. . . [44]

It was thus within the first decade of its existence that the (second) IMU could invite itself to a new level of involvement with the regular sequence of ICMs. The history of the International Congresses of Mathematicians, which spans roughly the last 120 years, thus splits into two halves of about sixty years each:

- Between 1897 and 1958, thirteen ICMs could be organized. As a rule, each of these ICMs decided in favor of an offer made to hold the following one. The organizers were then essentially autonomous. As described in Section 4.3, the 1920 ICM in Strasbourg was exceptional because it was part of a much more general, political scheme to redesign Science International after World War I. The old IMU was only founded a day before the 1920 ICM began and thus did not play any role in its organization; its foundation was but another piece of that bigger scheme. For the 1924 ICM, which was relocated to Toronto, the chief role of the IMU was to impose its regulations, particularly the exclusion of German participants. At the Bologna ICM in 1928, even though its organizer Pincherle was also President of the IMU, the Union was effectively excluded from the very efficient organization of the Congress—see Section 4.4 above. The subsequent ICMs then returned, equally successfully, to the pre-World-War-I model. Once the new IMU was founded in 1952, it would contribute to the expenses of organizing the ICMs—something the old IMU had never been able to do. But the local Organizing Committees of the ICMs remained essentially autonomous until the Edinburgh ICM in 1958.
- As we saw in the quote from Frostman's Stockholm address, a new "milestone"[45] of the IMU's implication in the ICMs was reached in 1962. Between 1962 and 2018, fifteen ICMs have been held with increasing participation of the IMU. We will discuss the subsequent evolution of the IMU's role in organizing the ICMs in Section 10.2 below.

[44] From Otto Frostman's (1907–1977) opening address in Proceedings ICM 1962, pp. XXXVIII–XXXIX.

[45] See [Lehto 1998], p. 139; see also the following Chapter 7 of [Lehto 1998] for details.

8.2.1.2 The IMU and Specialized Conferences

In 1953, the IMU started to co-sponsor high-level international symposia in active areas of mathematical research, profiting from ICSU and UNESCO funds. The first two such conferences were

> a Symposium on Differential Geometry ... held in Padua, Bologna, Pisa, 21–26 September 1953, under the joint auspices of these universities and the IMU; and a Symposium on Topological Groups and Their Representations (in Banach Spaces), ... held in the United States under the joint auspices of the National Research Council (USA) and the IMU.
>
> These two symposia, in Italy and the USA, opened a long series of IMU sponsored conferences. The Italian Symposium was truly international. Of the ninety-six participants, the Italians formed the majority, fifty-one, but the other forty-five came from fifteen countries on four continents. Two of them were from the USSR, as a first indication of Soviet interest in cooperation with the IMU. The American Symposium, at Columbia University, New York, was different: twenty participants, eighteen from the USA, two from Germany.[46]

The series continued in 1954 with three satellite conferences of the Amsterdam ICM organized by the "Wiskundig Genootschap", i.e., the Dutch Mathematical Society, with the moral and financial aid of UNESCO, ICSU, and the IMU: a *Symposium on Stochastic Processes*, a *Symposium on Algebraic Geometry*, and a *Symposium on Mathematical Interpretation of Formal Systems*.[47]

The following year, in September 1955, the memorable *International Symposium on Algebraic Number Theory* was held in Tokyo & Nikko. It was organized by The Science Council of Japan under the joint sponsorship with The International Mathematical Union. At a time when the first ICM held in Asia (in Kyoto in 1990) was still 35 years in the future, this meeting in Japan was in itself an exceptional event, as was the mathematically related *International Colloquium on Zeta Functions* organized in Bombay, India, a year later.[48]

The involvement of the IMU in such conferences was a new feature, facilitated by new financial resources, and contributed to making the IMU more visible. It was also an active vindication by the young international union of the trend that had begun to establish itself in the 1930s. We have presented this trend in Section 6.2.1, illustrating it in particular by the outstanding example of the International Congress on Topology at Moscow in 1935. Let us now take a closer look at the *International Symposium on Algebraic Number Theory* in Tokyo & Nikko twenty years later. It turns out that it was not only mathematically significant, but the problems encountered in its organization teach us something about the 1950s and the young IMU.

> An International Symposium on Algebraic Number Theory was held in Tokyo and Nikko, Japan on September 8–13, 1955. It was attended by 64 mathematicians, of whom 10 from foreign countries: France, Germany, India and the United States of America. Professor T.

[46] See [Lehto 1998], pp. 106–107.

[47] See Proceedings ICM 1954, Vol. 1, p. 159. Cf. Proceedings ICM 1954, Vol. 3.

[48] An overview of all the Symposia co-sponsored by IMU in the 1950s and 1960s can be gleaned from the *Bulletin of the International Mathematical Union*, which at the time was edited by the Austrian Mathematical Society under the trilingual title *Nouvelles mathématiques internationales / Internationale Mathematische Nachrichten / International Mathematical News*.

Takagi, the founder of the class field theory, attended it on September 9 as Honorary Chairman of the Symposium. It was organized by an Organizing Committee under the Science Council of Japan, with Professor S. Iyanaga as Chairman, Professor Y. Akizuki as Secretary and with three foreign members nominated by the International Mathematical Union: Professors K[omaravolu] Chandrasekharan [1920–2017], C. Chevalley and S[aunders] Mac Lane [1909–2005]. It was co-sponsored by the International Mathematical Union, whose Executive Committee approved the proposal of the Science Council of Japan to hold this Symposium, endorsed by the decision of the Japanese Government at its Cabinet meeting on October 22, 1955. Thus a financial aid was given by UNESCO through the International Council of Scientific Unions and the International Mathematical Union; it was also aided by a Society for Supporting the Symposium formed principally with representative people in the financial and industrial circles of Japan, as well as by the foreign governments and institutions concerned, which contributed towards the travel expenses of the foreign participants.

. . . . . .

As Japan is situated in a remote corner of the world distant from the western countries, and as this was the first symposium of its kind to be held here, this was considered by the Japanese public interested in mathematics as a particular good occasion to have contact with the ranking mathematicians from abroad. A Public Lecture Meeting by three participant mathematicians, Professors E. Artin, A. Weil and C. Chevalley was held on September 8 in response to the wish of the interested general public. The contents of these lectures were translated into Japanese and published in Japanese periodicals. Moreover, the foreign participants were invited to deliver lectures and to participate in seminars in universities in various parts of Japan before their going home.[49]

The foreign participants, whose number was limited by the available funds for travel costs, were[50] André Néron (1922–1985) and Jean-Pierre Serre[51] (b. 1926) from France; Max Deuring from Germany; Kollagunta Gopalaiyer Ramanathan (1920–1992) from India; Emil Artin, Richard Brauer (1901–1977), Claude Chevalley[52], Kenkichi Iwasawa (1917–1998), André Weil, and Daniel Zelinsky[53] (1922–2015) from the USA.

The meeting was a tremendous success. Mathematically, it reshaped the theory of complex multiplication of abelian varieties thanks to the interaction of Yutaka Taniyama (1927–1958), Goro Shimura (1930–2019), and Weil.

This was the most beautiful, the most joyful and the most seminal mathematical gathering that it was ever given to me to attend. Only the Zürich ICM in 1932 left with me a somewhat comparable memory. Is this just a result of age? It seems to me that, as this sort of meetings is getting more and more frequent . . . , they inevitably turn stodgy. In the country of andhra, they say, that old men bemoan the fact that the peppers no longer taste as strong as they used to . . . [54]

---

[49] See [Proceedings Tokyo & Nikko 1956], p. I.

[50] In the order of the list given in [Proceedings Tokyo & Nikko 1956], p. VII. Goro Shimura counted nine foreign participants, supposedly because he would not consider Iwasawa a foreigner to Japan; see for instance [Shimura 2008], p. 105.

[51] Who had been awarded the Fields Medal the year before.

[52] He was the only participant who was not from Japan but had visited the country before, in 1953, invited by Iyanaga.

[53] At the time he was Guggenheim Fellow at the IAS.

[54] See [Weil 1979], Vol. II, p. 541. In the same note Weil recalls his previous cordial relations with a number of mathematicians from Japan, some of whom were present at the Symposium.

Today this symposium is particularly remembered for a list of problems that Taniyama circulated at the meeting, and which contained a conjectural precursor of the Modularity Theorem according to which every elliptic curve over the rationals is modular, one of the outstanding achievements of the second half of the twentieth century in pure mathematics. The proof of this theorem was completed in 2001 by Christophe Breuil (b. 1968), Brian Conrad (b. 1970), Fred Diamond (b. 1964), and Richard Taylor (b. 1962); a weaker version of it had been the crucial point in Andrew Wiles's (b. 1953) proof of Fermat's Last Theorem in 1995.[55]

However, in preparing the conference, the organizers and the IMU had to overcome financial as well as diplomatic problems that provide an interesting marginal note to the unperturbed narrative of the new IMU's separation of mathematics from politics. Indeed, the Symposium's organizer Shôkichi Iyanaga—who had been a member of the IMU Executive Committee 1952–1954—, the IMU Secretary Enrico Bompiani, the IMU President Heinz Hopf, and Saunders Mac Lane, the member of the IMU Executive Committee who was appointed to handle the financial affairs of the meeting, spent the better part of their 1955 correspondence coping with "the German problem" of this conference.[56]

Here is, in a nutshell, what happened. According to the rules of the IMU for co-sponsored specialized conferences, which had been updated at the IMU General Assembly in 1954, the national members could make proposals for participants of the IMU co-sponsored symposia. Thus Erich Kamke (1890–1961), chair of the German National Committee of Mathematics and former second Vice-President of the IMU (1952–1954), transmitted in February 1955 a proposal of participants from Germany to be invited to the Symposium on Algebraic Number Theory: Helmut Hasse, Max Deuring, and Ernst Witt (1911–1991), in that order. While the scientific quality of these number theorists was undisputed, Hasse's name met with serious reservations from intended participants working in the US because of his former affinity with the Nazi regime (cf. Section 6.3.3). It seemed likely that some of the colleagues from the USA would refuse the invitation if Hasse was going to be present. On the other hand, inviting no mathematician from Germany at all could be seen as an exclusion of this country from an IMU co-sponsored conference on a subject in which Germany has a distinguished tradition.[57] In May 1955, Iyanaga decided to invite Deuring, who graciously accepted on the condition that his trip could be

[55] We refer the reader to [Harris 2020] for more details, and a number of reflections, on the evolution of the conjectures and arguments that led up to the Modularity Theorem. See also the first four appendices of the autobiography [Shimura 2008], as well as Section 8.2.3.2 below.

[56] This correspondence is kept at [Arch. IMU], SF1 Ser 12.1. The two letters from which we quote in this section are in the folder [Arch. IMU], SF1 Ser 12.1 F7.2. My interest in it was kindled by Antina Scholz from Wuppertal. She is preparing a thesis on the international relations of German mathematicians after World War II; a preliminary version of her chapter on the "Hasse – Tokyo case" was very helpful in preparing the much shorter account I am giving here.

[57] In 1954, during the early phase of planning the Tokyo conference, the German mathematicians Carl Ludwig Siegel (who had returned from the US to Göttingen) and Martin Eichler (1921–1992) were on the organizers' list of prospective participants. They were subsequently dropped as being closer to the orientation of the Symposium to be organized at Bombay in 1956.

Fig. 8.1 On the train to Nikko for the 1955 conference on algebraic number theory. Shown are (left to right) T. Tamagawa , J.-P. Serre, Y. Taniyama and A. Weil. Source: [Shimura 1989]. Credit: John Wiley & Sons, Inc.

defrayed. This would require more effort; the IMU President Heinz Hopf himself contacted German authorities about this. Iyanaga personally wrote to Hasse and to Kamke taking responsibility for inviting Deuring.

> But it was very difficult for me that we had to forgo inviting Professor Hasse this time.
>
> When I write this I am afraid I might create the impression that this decision was forced upon us by the IMU. This is not at all the case, as you can see from the copy of the enclosed excerpt from a letter of President Hopf to Professor Mac Lane. As you can see, the IMU is unaffected by the political division of the world, and remains faithful to the principles of tolerance. I am solely responsible for our decision, which is motivated by our desire to hold a colloquium in an atmosphere as harmonious as possible, while we are completely inexperienced and without confidence in matters of diplomacy. Of course, we have the greatest respect for Professor Hasse, and intend to find another occasion to invite him here. May I ask you once more to trust in my assurance that neither we nor the IMU feel the slightest malevolence towards Professor Hasse.[58]

Hasse was furious. He tried, in vain, to identify a person among the organizers who had provoked his not being invited, and threatened to turn the matter into a public affair among the German mathematical community. Both Kamke and the then President of the German Mathematical Society D.M.-V., Georg Nöbeling managed to avoid this.

[58] Iyanaga to Kamke 6 June 1955. My translation from Iyanaga's original German.

Even though it was only about a specialized conference, this affair tested both the IMU in its self-proclaimed engagement to separate mathematics from politics, and the mathematical community in West Germany. Knowing about the history of the first IMU—see for instance the rift about German participation at the Bologna ICM in 1928; see Section 4.4.3—the colleagues in charge on both sides took care to frame the problem as an issue pertaining to one person, which was best to be solved by personal intervention. In a long letter to Mac Lane, Hopf stated his take on the problem. The following clippings from the letter convey an impression of the discussions within the IMU Executive Committee.

> I shall now formulate and explain what, in my opinion, should be the standpoint of IMU; this opinion is a personal one, confirmed, however, in the essential points by a conversation with Bompiani, Chandrasekharan and Koksma. . .
>
> (1) The people to be invited to the symposium shall always be chosen, of course, in first line from purely scientific points of view. However, if IMU co-sponsors a Symposium it is, in order to strengthen the international significance of the Symposium, desirable that the list of invitees be as international as possible; that means that this list should not be too homogenous from the geographic point of view. – Now, if one looks on the list of invitees for Tokyo then, among the nine or ten who probably will be able to accept, the first six . . . are professors at Universities in USA. . . . I take it that this homogenity is due, in first line, to scientific reasons . . . Anyway, what I wish to point out is this: from the point of view of internationality, which must be one of IMU's main points of view, it would be desirable to increase the international character . . .
>
> (2) The German National committee for mathematics, according to IMU's Rules for Colloquia, has called the attention of the Organizing Committee to some German mathematicians whose participation it considers desirable. This demand must be seriously considered by the Organizing Committee; and just with respect to No. (1) above, it should be considered with a maximum of goodwill. . . .
>
> (3) However, the above considerations have a rather theoretical character and we have to face the Nazi-problem. Here I wish to insert, once more, two general and theoretical statements:
>
> (a) In connection with the question "How shall we (=IMU) deal with Nazis?", I think that, for many reasons, we must act as liberally and large-minded as possible. (If necessary, I can try to explain the reasons for this principle at another occasion.)
>
> (b) For the success of a Symposium or a Congress, not only the scientific level is relevant, but it is also important that the whole atmosphere and the personal relations between the participants be friendly; and for a small Symposium, this atmosphere depends much more on the single personalities than for a big Congress.
>
> What I call "our general Nazi-problem" consists in the question how far the points of view (a) and (b) can be combined. My opinion is that "in general" this combination will be possible, with the help of enough tactfulness on the side of the guests and enough diplomatic skill on the side of the hosts. However, besides the "general" cases, there exist the "exceptional" ones, and the difficulty will be to determine the degree of "exceptionality" in a given special case. . . . . .[59]

Hopf goes on to give his own opinion about Hasse, concluding that he "would be glad" if Hasse could be invited nonetheless. However, he clearly leaves the decision to the organizing committee: "I can assure you, . . . I certainly shall not criticize or grumble afterwards. (But I should appreciate your giving some thoughts to my remarks in (1).)"

[59] Hopf to Mac Lane 24 April 1955; we reproduce Hopf's original English.

Nowhere in the IMU correspondence around the Tokyo & Nikko Symposium have I found an attempt to explicitly justify Hasse's exclusion as a kind of punishment for his actions in the 1930s and 1940s. The discussion always focusses on personal incompatibilities between participants.[60]

The address of H.R.H. The Prince Philip, Duke of Edinburgh, to the opening of the 1958 ICM contains the line: "Friendship between nations grows from personal friendship between individuals."[61] This may sound like a noncommittal echo of Hurwitz's words at the opening of the first ICM in 1897 at Zürich, which we have mentioned in Section 1.4.1.2. However, whereas Hurwitz tried to avoid elaborating on the political dimension of the event, the Duke of Edinburgh actually stressed the political significance of personal contact. The handling of the "German problem" at the Tokyo-Nikko Symposium by a strictly person-oriented approach, avoiding as far as possible political implications for the IMU, may indicate that the Royal motto did capture a relevant aspect of the 1950s.

8.2.1.3 New Members

With or without support from the IMU, the very first years of its existence were marked by active mathematical exchange across the Western World, including continually increasing contacts between different parts of the Americas. The travel-intensive life of today's researchers, especially young researchers, which had begun to be seen between the World Wars, took hold of the part of the world which called itself free.

> Political developments in the world were reflected in the membership. Of the twenty-two member countries in 1952, fourteen had been members of the Union in 1932. Eight new members [Argentina, Austria, Cuba, Denmark, Finland, Germany, Pakistan, and Peru] replaced the nine old ones that had not joined. However, already in 1958, the members of 1932 except for two [Egypt and South Africa] had become members of the new Union.[62]

As the IMU asserted itself, the attitude of the Eastern block changed. Poland had sent two observers already to the very first General Assembly in Rome in 1952.[63]

> The membership of Poland in the IMU in July 1956 signaled a new policy of the Socialist countries of Europe towards the Union. On 1 March 1957 the USSR became a member. Bulgaria, Czechoslovakia, and Hungary were admitted in May 1957 and Romania in March 1958. The applications of these countries were unanimously accepted by all the Union member countries that took part in the voting. These new memberships represented an essential enlarging of the sphere of the Union.[64]

---

[60] Ten years later, in 1965, Carl Ludwig Siegel's attitude would be different when he started a series of motions to make Hasse's exclusion from the Göttingen Academy of Sciences official—see [Schappacher 2015b].

[61] See Proceedings ICM 1958, p. XLIX.

[62] See [Lehto 1998], pp. 95; 305–306.

[63] The list of participants is reproduced in [Lehto 1998], p. 94.

[64] See [Lehto 1998], p. 122.

One may add to this list that the GDR (East Germany) joined the IMU in 1966, after having adhered to ICSU in 1961.

The USSR had been the first nation to deposit its ratification of the United Nations Charter in 1945. A year after Stalin's death, in April 1954, the USSR also joined UNESCO. Before World War II, the USSR had not been a member of ICSU, although it did belong to certain scientific unions; for example, to the International Astronomical Union (IAU). In 1954, the USSR also joined ICSU itself. "During the so-called Cold War the interaction with the Russian member was fully maintained and proved to be of special significance."[65]

Indeed, Lehto's presentation in the above quotes could be slightly twisted by asking why the Soviet Union *only* joined the IMU in 1957. Beyond formalities,[66] the main reason for this seems to have been that the IMU was a new organization that could not build on administrative continuities from before World War II. To see the difference, here is what happened in the International Astronomical Union: The Executive Committee of the IAU had its first meeting after the war in March 1946 at Copenhagen and brought itself up to date by replacing the retiring Vice-Presidents. On this renewed Committee, the history of the IAU dryly remarks:

> With the United States and the USSR having emerged from the War as the dominating world powers, it would become customary henceforth to have both of these powers represented on the Executive Committee.[67]

This is indeed what was also done in the Executive Committee of the IMU as of 1959. For instance, the two Vice-Presidents serving from 1959 to 1962 were Pavel Alexandrov and Marston Morse. Thus also the new IMU, whose foundation was based in the USA and surrounded by intense anti-communism, adapted to the peculiar rules of Science International in times of Cold War. Interpreting this instead à la Lehto as a victory of the 'apolitical' agenda of the IMU[68] strikes us as rather idiosyncratic. It actually seems not to have prepared Lehto very well for some of the internal clashes in the Executive Committee in the years to come, which he relates in later chapters of his book.

---

[65] See [Greenaway 1996], p. 88.

[66] Lehto points out that the procedure "was delayed because the composition of the Soviet Committee for Mathematics was made known only in December 1956"; see [Lehto 1998], p. 122.

[67] See [Blaauw 1994], p. 143.

[68] See the lower half of page 122 in [Lehto 1998] as well as the description of the applause he obtained for his speech at Warsaw, pp. 236–237.

8.2.2 From Moscow to Helsinki: 1966–1978

In due course, once the USSR had joined the IMU, an ICM was called to Moscow in 1966. The other ICMs of the second time span were held in Nice, France, in 1970; in Vancouver, Canada, in 1974; and the second period ends with the Helsinki ICM in 1978.

8.2.2.1 Mathematics in Moscow

Looking back forty years after the event, the historian of mathematics Sergei Sergeevich Demidov had this to say about the era of Soviet mathematics which was marked by the Moscow ICM.

> An important event in the life of the Soviet mathematical community was the 1966 International Congress of Mathematicians in Moscow, which hosted a record number of participants (more than five and a half thousand!). At this congress our country declared itself one of the leading mathematical powers of the world, and, especially important, our mathematicians felt themselves to be competent and respected members of the world mathematical community. An awakened spirit of freedom found its expression in the growth of free thinking and even dissidence among Soviet mathematicians. Let us just recall the famous "letter of the ninety-nine", which was a response to [Alexander Sergeyevich] Esenin-Vol'pin [1924–2016] being placed in a psychiatric hospital in 1968, or, what would have been quite impossible before, the election in 1970 of [Igor Rostislavovich] Shafarevich [1923–2017], then an active defender of human rights, as president of the Moscow Mathematical Society.
>
> Of course, one should not look for information about this on the pages of *Uspekhi Matematicheskikh Nauk* (to print any such thing in that time of absolute censorship would have been simply impossible). However, and it is especially important to stress this, the dissident persons remained authors in the journal: the significance of the research material to be published remained as the one and only decisive criterion. Despite the repression brought down on the head of [Sergei Vasilyevich] Fomin [1917–1975] as one of the signers of the aforementioned "letter of the ninety-nine", his name as deputy editor was not removed from the title page of the journal.[69]

It is less clear—and may depend on varying personal experiences—how long this upbeat period lasted. Vladimir Mikhailovich Tikhomirov (b. 1934) speaks of "the great outburst of mathematical achievements during the period . . . 1950–1975," and adds that "[t]he beginning of the 1970s is often referred to as the 'stagnation period.' Emigration took root in that period along with the persecution of scholars."[70]

The historic origins that made the tremendous success of Soviet mathematics possible go back as far as the second decade of the twentieth century—see Section 2.1.1 above. Then the seminal years of the 1920s and 1930s with their strong presence of Russian mathematicians in the international networks emerge, with hindsight, as another root of the later success story—see Chapter 6, in particular Sections 6.1.3 and 6.2.1. Two iconic figures stand out among the founders of the Moscow mathematical center, because of the amazing breadth of their mathematical interests

[69] See [Demidov 2006], p. 796.

[70] See [Tikhomirov 2000], pp. 1115 and 1120.

and production: Kolmogorov, and his student Israel Moiseevich Gelfand (1913–2009). This not to say that all the mathematicians that constituted the 'Russian school' in the second half of the twentieth century were connected to Kolmogorov by a sequence of PhD advisorships, however impressive Kolmogorov's list of more than 80 thesis students is. At the Stockholm ICM in 1962, for example, three Russian mathematicians had been invited as plenary speakers: Eugene Borisovich Dynkin (1924–2014)—who was not allowed to come himself, so his talk was presented by his former PhD advisor Kolmogorov—, Gelfand, and Shafarevich, who was mentioned in the above quote and who had been a student of Boris Nikolayevich Delone's (1890–1980).

When things came to pass, it was the Cold War that stood at the cradle of the explosive development of Soviet mathematics, and shaped to no small extent the specific conditions in which this hub of mathematical culture and creativity would grow, bear fruit, and attract worldwide attention.

> The mid 50s are marked by the arrival of a new generation on the scientific scene.
>
> Demands of the military industrial complex also influenced the work of the main educational institution where mathematicians were taught, [the Mechanical-Mathematical Department] Mech-Math of Moscow University. This department was considerably enlarged in 1952. Many students were accepted on a special basis, and received special additional military training along with the main course of studies. This system was abolished soon after Stalin's death, but the department remained within the new boundaries, which also contributed to the development of mathematics in the 50s/60s. In 1954–55 practically all restrictions on applying to the Mech-Math Department (concerning ethnical origins and parents who had served terms in concentration camps) were removed, while the interest to sciences, to mathematics in particular, was very high. This attracted bright and creative young people to the department and they could find there everything they wanted. The lecturing and teaching staff was brilliant then. The living conditions of that time contributed to the concentration of all creative forces in large cities, especially in Moscow. And all these forces, accumulated during the previous decades, participated in the scientific seminars at the Moscow University. As it was told there were about a hundred of them, and one's scientific life began from the first year of study.[71]

Those legendary seminars at Mech-Math were large meetings, both in terms of numbers of participants and of their duration. They were facilitated by the usually limited duties required by the professional contracts that mathematicians tended to get. Ideas and news were exchanged both informally and through officially presented problems, talks, ... and counter-talks. They were also a major occasion to get access to the most recent Western literature, which was not translated yet. In his personal reminiscences,[72] Alexei Nikolaevich Parshin (b. 1942) stresses the fact that there was no compartmentalization along mathematical specialities. He also advances the hypothesis that, in spite of the possible access to foreign mathematical literature, the relative isolation due to serious travel restrictions actually created a sort of protected habitat, in which some of the most prestigious original ideas of twentieth century mathematics could grow.

[71] See [Tikhomirov 2000], p. 1114.

[72] See [Parshin 2006] and [Parshin 2010].

To say more, we would have to try and describe the roles that many of the colleagues, whose names and achievements are well known today, played in various seminars. This is what is done to a certain extent in the existing literature, often by way of anecdotes.[73] In Section 10.4 below we shall supply some quantitative analysis which impressively shows the overwhelming weight of Moscow, and its university in particular, for world mathematics at the time. For instance, during our first two periods, i.e., between 1950 and 1978, Mech-Math emerges as the most distinguished mathematical institution in the world, from the point of view the ICMs (in spite of the fact that no Soviet delegation took part in the 1950 ICM)—see Section 10.4.2.

8.2.2.2 The Politics of the Fields Medals

Michael Barany has suggested that Moscow marks a turning point in the history of the Fields Medal, which was triggered by the fact that Georges de Rham, as head of the Fields Medal committee for the 1966 Congress, codified the age limit of 40 years. As long as this rule was not formally established as a criterion for admissible candidates, the Fields Committees had to actually give some thought to the question of which of the proposed candidates promised the most extraordinary work *in the future*. After 1966, attention shifted to established achievements of mathematicians under forty.

> [H]owever flawed the processes were before 1966, they forced a committee of elite mathematicians to think hard about their discipline's future. The committees used the medal as a redistributive tool, to give a boost to those who they felt did not already have every advantage but were doing important work nonetheless.[74]

Barany's thesis cannot be properly checked on the basis of what is currently known, and it certainly cannot be read off the list of Fields Medalists. It is also not clear to what extent the recent study about the declining productivity of Fields Medalists after their award sheds light on the question.[75] Barany's idea should certainly be investigated once the archives of the Fields Committees become accessible to historians. This is not the case at the moment because the IMU has imposed an extravagant 70-year embargo on all the files of the Prize Committees.

The list of Fields Medals does, however, present an obvious discontinuity in 1966 in that, thanks to an anonymous donor, starting with the Moscow Congress, four Fields Medals, instead of two, would as a rule be awarded at each ICM.[76]

The 1966–1978 period saw the first Fields Medals awarded to mathematicians from the Soviet Union. The first one was Sergei Novikov (b. 1938) in 1970 in Nice. In Helsinki in 1978, it was Grigory A. Margulis's (b. 1946) turn. However, neither of them was allowed to travel abroad to receive their Medals. A Soviet member

[73] For a panorama in English, see the whole book [Bolibruch et al. 2006].

[74] See [Barany 2018], p. 273.

[75] See [Borjas & Doran 2015].

[76] See [Lehto 1998], p. 168.

had been appointed to each of these Fields Committees: Shafarevich for 1970, and Yuri V. Prokhorov (1929–2013) for 1978. On the other hand, "Pontryagin knew that one of the Fields medalists was G.A. Margulis, from the USSR, and he was furious about this choice."[77] In his published answer to accusations that his rejection of Margulis was based on antisemitism[78], Pontryagin squarely claimed he had "never done anything which could be considered as anti-Semitism."[79]

As a matter of fact, the absence or presence of Fields Medalists at the award ceremony had been a political issue since 1950—recall the issue of Laurent Schwartz's admission to the US discussed in Section 8.1.4. Among the four Fields Medalists of 1966 Michael Atiyah (1929–2019), Paul Joseph Cohen (1934–2007), Alexander Grothendieck and Stephen Smale (b. 1930), it was Grothendieck who did not travel to his father's country of origin to receive his award.

> His father had fought all his life for freedom and self-determination and against the powerful in this world. Grothendieck's sympathy was always with the poor, the persecuted, the oppressed, those in the shadows, and he always held leftist, liberal, and possibly even anarchist political convictions. But for many years these convictions were not expressed in political actions. In the late 1950s and early 1960s, he opposed the French war in Algeria as a matter of course, but in contrast with many of his closest colleagues such as Schwartz, Chevalley, [Pierre] Samuel [1921–2009], or [Pierre] Cartier [b. 1932], he did not participate in public protests. At least he took the matter seriously enough to consider emigrating to the United States.
>
> Grothendieck's political commitment became publicly visible in the summer of 1966, when he refused to travel to Moscow to receive the Fields Medal at the International Congress of Mathematicians (ICM). This was his protest against the persecution and imprisonment of the Russian writers Yuri Daniel and Andrei Siniavsky. This action attracted a lot of attention. Some years later it was held very much against Grothendieck by orthodox communists and socialists who played a big role in the student movement.
>
> His next political action was a trip, made at his own initiative, to Hanoi and North Vietnam during the last three weeks of November 1967 in the middle of the Vietnam War.[80]

On the other hand, Smale's presence in Moscow stirred political debates in the USA, even though the immediate reason for his trip was to receive his Fields Medal on 16 August 1966. Those Fields Medals were handed out in Moscow, not by a political representative of the Soviet Union, but by the President of the Academy of Sciences of the USSR, the applied mathematician Mstislav V. Keldysh (1900–1978).

> As a member of the mathematics faculty at the University of California at Berkeley, Smale was active on the campus's Vietnam Day Committee, which had organized efforts to block troop transports and otherwise to protest the war. The worst of the loyalty oaths and blacklists that shook many in the academic community seemed to have passed, but Cold War politics continued to stir controversy, and universities were centers for such provocation and confrontation.
>
> On August 5, 1966, the San Francisco Examiner reported that Smale had been subpoenaed to appear before the House Un-American Activities Committee for his antiwar activism. The article insinuated that rather than face the committee, Smale had fled to Moscow. The acting

[77] See [Lehto 1998], p. 205.

[78] See [Kolata 1978].

[79] See [Pontryagin 1979], p. 1083 (reply item 1).

[80] See [Scharlau 2008], p. 935.

Fig. 8.2 Two of the four Fields Medalists of 1966: Alexander Grothendieck (1970, Public domain) and Stephen Smale (1966, Caroline Abraham, Courtesy Springer Verlag).

> chairman of Smale's department, Leon Henkin [1921–2006], rushed to notify the media that Smale was on his way to Moscow not to avoid HUAC but rather to attend that summer's International Congress of Mathematicians.[81]

The upshot of this story is that the prestige of the Fields Medal—since then routinely compared to the Nobel Prize, as Barany points out—could to a certain extent outweigh political slander. In other words, the IMU—which could pride itself on gaining increasing control of the mounting of the ICMs,[82] including the nomination of the Fields Committees—commanded an international renown that echoed beyond purely mathematical recognition. Widely visible public events like ICMs naturally acquire political connotations.

This became obvious again at the Helsinki ICM in 1978. Politically, the city of Helsinki in the 1970s was known for the *Conference on Security and Cooperation in Europe*, a key element of European détente policy with a view to defusing the Cold War. Its Final Act was signed in Helsinki on August 1, 1975.

> [D]uring the 1970s, the growing détente between the United States and Soviet Union led scientists to question the bipolar internationalism that had divided U.S. scientists from their Soviet peers. Rather than fight the Cold War on a scientific level, scientists began to try

---

[81] See [Barany 2015], p. 18. A memorable Associated Press cablephoto—for the reproduction of which I tried in vain to obtain permission—taken after Smale's Moscow press conference appeared on the front page of *The New York Times* on 27 August 1966, and again in *Science*, New Series, Vol. 154, No. 3745 (Oct. 7, 1966), p. 132. Documents relating to "The Case of Stephen Smale" which evolved from that press conference were collected and published, including a letter to the editors by Serge Lang, in *Notices of the American Mathematical Society*, Vol. 14, no. 6, issue no. 100 (October, 1967), pp. 778–786. Cf. the autobiographical account [Smale 1984].

[82] See, however, [Lehto 1998], pp. 166–167, for the specific difficulties that the IMU experienced in trying to assert the role of the Consultative Committee when preparing the ICM at Moscow.

> and transcend the Cold War by embracing a transnational rather than bipolar vision of scientific internationalism—one based on the ideals of the scientific discipline itself, rather than geopolitics.
>
> Meanwhile, the concurrent surge of interest in human rights around the world created another opportunity for science to play a role in geopolitics and transnational relations. The connection between science and human rights got a boost from the Helsinki Final Act, signed on August 1, 1975. This landmark agreement between East and West, relatively dismissed by Western leaders at the time, focused mainly on social, political, and economic rights, but the agreement also included a section that described science and technology as activities that "contribute to the reinforcement of peace and security in Europe and in the world as a whole." Scientific cooperation between nations in particular "assists the effective solution of problems of common interest and the improvement of the conditions of human life."
>
> ......
>
> The American Mathematical Society created a Committee on Human Rights of Mathematicians, while the Society for Industrial and Applied Mathematics Council established its own Committee on Human Rights in the late 1970s.[83]

And thus we can read in the September 1978 issue of *Nature* in a short note entitled "Boycott of Soviet contacts is for individuals, says NAS":

> [L]ast month, when the International Congress of Mathematicians was held in the "human rights" city of Helsinki, right on the Soviets' doorstep (subsidised excursions to Leningrad were laid on as a side attraction), several Soviet mathematicians failed to arrive to deliver their papers. One of these people, a Dr. Margulis, was scheduled to receive a medal. The ceremony was carried out in his absence, and the formal "presentation" of his work was greeted by a standing ovation, pinpointing in a telling manner the persistent Soviet practice of denying visas to scientists invited to international conferences.[84]

The author then goes on to discuss the merits of boycotting collaboration with Soviet peers, and the hesitant position of the US National Academy of Sciences on this question.

The Wolf Prize. Leaving this subsection about the Fields Medal, during the time period 1966–1978, let us mention in passing that 1978 was the first year when Wolf Prizes were awarded in Israel. The prize was founded in 1975 to honour "achievements in the interest of mankind and friendly relations among people" by the inventor and philanthropist Ricardo Wolf (Ricardo Subirana Lobo; 1887–1981), who had been the Cuban ambassador to Israel until Cuba broke off diplomatic relations with Israel in 1973. Wolf Prizes are awarded, essentially annually, in six separate categories: the arts, agriculture, chemistry, mathematics, medicine, and physics.

The first two mathematicians to receive a Wolf Prize, in 1978, were Israel M. Gelfand and Carl L. Siegel. They were followed in 1979 and 1980 by Jean Leray, André Weil, Henri Cartan and Andrey N. Kolmogorov. The first Wolf Prize

[83] See [Rubinson 2012], pp. 247–249.

[84] See [Rich 1978].

winner who had previously received a Fields Medal was Lars Ahlfors (Fields Medal in 1936, Wolf Prize in 1981). Incidentally, Grigory Margulis was awarded a Wolf Prize for mathematics in 2005.[85]

8.2.2.3 The ICSU and SCFCS / CFRS as Authorities of Reference

There was yet another issue of political boycott that surfaced at the Helsinki ICM, this time unrelated to the Fields Medals. Lehto's memories of it indicate the reason why the IMU did not—and still does not—have an instance of its own to deal with such matters:

> During the Congress in Vancouver, as soon as the Site Committee had made their decision [in favor of holding the next ICM at Helsinki], I was summoned and had to give a solemn promise before the Committee that the Helsinki Congress would observe the ICSU principle of free circulation of scientists. Even though it is not in the power of mathematicians to see that the principle is upheld, such a pledge is not without importance. Knowing what lay ahead, I had been in contact with the Ministry for Foreign Affairs in Helsinki. They were aware of the ICSU principle and authorized me to give the agreement. I thought that in the case of Finland, where the cornerstone of foreign policy is to maintain friendly relations with all countries of the world, my promise would be a sheer formality. But unexpectedly, problems arose with the Republic of South Africa, and the ICSU principle was put to use ...[86]

What was unexpected for Olli Lehto was that the Finnish Minister of Education, having failed to oblige the IMU to ban South Africans from the ICM in view of the anti-apartheid boycott, refused to greet the Congress at the opening session.[87]

Historically, an interesting aspect of this story is the appearance of ICSU as the reference institution for what it means to be 'apolitical.' The reader will recall that the old IMU, created in 1920 under the auspices of the IRC, had to espouse the politics of exclusion of the central powers imposed by that Council after World War I. Analogously, but going in the opposite direction, the new IMU, as a member of ICSU, followed the principle of free circulation of scientists as outlined by ICSU.

The handling of this principle inside ICSU has its own history. It is discussed in some detail until the early 1990s under the title "The free conduct of science" in what may be the most elaborate chapter of Greenaway's 1996 history of the IRC and ICSU. After recalling a Presidential statement from 1934 and quoting a policy resolution on political non-discrimination accepted by the General Assembly in 1958, Greenaway continues:

> A new discussion was triggered by the attitude of the North Atlantic Treaty Organization (NATO) which had been set up in 1949. It imposed a virtual ban on the provision of visas for entry into Nato countries for scientists who were East German nationals. At its meeting in Prague in October 1962 the ICSU Executive Board decided to ask its officers to take action

[85] See [URL 11].

[86] See [Lehto 1998], pp. 194–195.

[87] See [Lehto 1998], pp. 196 and 206.

in the general matter of visas. By 1963 a Working Party was able to present a report and recommendations which were the foundation of an ICSU mechanism which has survived, with modifications of constitution and mandate, to the present day. Successive resolutions became more bold and positive. The language of the 1963 resolution forming the Standing Committee on the Free Circulation of Scientists (SCFCS) is different from that of the 1958 resolution. It reads:

"The Assembly *considering* that ICSU has a declared policy supporting free international collaboration among scientists, but *noting* with regret that there are still parts of the world where difficulties exist in the free passage of scientists, *reaffirms* the declaration of 'political non-discrimination' adopted by the VIII General Assembly of ICSU in 1958, *resolves* that, in holding ICSU meetings, and meetings of ICSU Scientific and Special Committees and Inter-Union Commissions, the Council shall take all measures within its powers to ensure the fundamental right of participation, without any political discrimination, of the representatives of every member of ICSU concerned and of invited observers, *recommends* that this policy be adopted also by the Unions adhering lo ICSU for all their activities, *invites* the ICSU national members also to follow this policy, and *requests* the Executive Committee to create a standing working group to assist the Officers' Committee to find solutions to various specific problems associated with the implementation of this resolution."[88]

For more details about the work of the *Standing Committee on the Free Circulation of Scientists* (SCFCS), and the rich archive of cases that it assembled in the process, we refer the reader to the further developments in Greenaway's book. To sum up the work of this committee briefly, we quote from a more recent report:

In the 1960s up to the early 1990s the difficulties encountered were based mostly on the nationalities of the visa applicants. Nationals of NATO countries were refused visas to Warsaw pact countries and vice versa. Individuals from China:Taipei were refused visas for China:Beijing and vice versa. During the time of apartheid, South African scientists were unwelcome in a large number of countries around the world. Israeli nationals also encountered difficulties obtaining entry to certain countries and these difficulties are still apparent today. A similar ongoing situation exists for Cuban scientists wishing to attend conferences in the USA. In all these instances, the discrimination was not against individuals as scientists but against nationalities irrespective of the specific reason a person wanted to travel. Reference by the local academy to the principle of Universality of Science was often used, and accepted, as a reason to make an exception for scientists.

As a general rule only cases of visa refusals to conferences of Members of the ICSU family were considered by the ICSU committees. Cases of persecution of scientists within a country were usually referred to the local academy or similar organizations. It was only in the early 21st century that typical human rights cases began to be taken up by the committee and acted upon.[89]

Between 1996 and 2002, an additional *Standing Committee on Responsibility and Ethics in Science* (SCRES) existed and the abbreviation SCFCS was redefined as *Standing Committee on Freedom in the Conduct of Science*. At the ICSU General Assembly in Suzhou/Shanghai in 2005, a unique *Committee on Freedom and Responsibility in the Conduct of Science* (CFRS) was set up with a new remit,

[88] See [Greenaway 1996], pp. 94–95.

[89] See [Schindler 2017].

which covers not only the special rights of scientists but also the special responsibilities that are concomitant to those rights. The new Terms of Reference also reflect the fact that, on the one hand, visa refusals are less frequent than in the 1980s and 1990s but, on the other hand, problems of scientific misconduct, plagiarism and conflicts of interest are becoming more and more apparent.

... In 2008 a completely new booklet (with a violet cover) was published by the new committee, with the title *Freedom, Responsibility and Universality of Science*. This highlights not only freedoms of scientists, including freedom of movement, freedom of association, freedom of expression and communication, and access to data, information and research materials, but also scientists' responsibilities for the conduct of science and their responsibilities to society.[90]

The term 'Universality of Science', which was felt by many to be ambiguous and which was dropped from the committee's name already in 2005, was finally also removed from the ICSU Statutes at the 2017 General Assembly in Taipei, which prepared for the merger between ICSU and the International Social Science Council (ISSC) into the larger NGO *International Science Council* (ISC).[91]

Many scientific organizations nowadays have ethics committees, with remits that vary considerably. The AMS was already mentioned in a quote towards the end of Section 8.2.2.2 above. Another example is the European Mathematical Society (EMS). One reason to insert the present section was to indicate the historical constellation which brought about the fact that the IMU does not have its own committee dealing with ethical questions—ranging from the deontology of the publication process all the way to human rights issues—and relies instead on the ISC line of action on Ethics, Freedom and Responsibility in Science.

8.2.2.4 The Union Lectures

We have seen in Section 8.2.1.2 that the newly founded IMU conquered new territory by co-sponsoring specialized conferences. Another bid to broaden its activities was initiated in 1971: the Union Lectures. The idea was to

invite, from time to time, a distinguished and active mathematician, of international standing, to give a set of four to six lectures, on important new developments in mathematics ... The lectures are intended to be on about the same level as the one-hour survey lectures given at the International Congress of Mathematicians, but fuller, in greater detail, and more often than once in four years, though perhaps not in the year of the International Congress.[92]

---

[90] See [Schindler 2017].

[91] I thank John Ball for sharing documents with me that reflect discussions about the principle of 'Universality of Science' inside ICSU after 2002.

[92] See *Bulletin of the International Mathematical Union* No. 2, 2 September 1971, p. 5. Cf. [Lehto 1998], pp. 178–179.

The Executive Committee of the IMU was very careful to frame these events appropriately:

> IMU's role is
>
> (i) to ensure such high standards in the choice of the lecturers as will lead people to consider it a distinction to be chosen a lecturer;
>
> (ii) to keep the whole of mathematics in view, to the extent that it is humanly possible, and to watch for important new developments which warrant a survey;
>
> (iii) to keep the lectures on an international plane free from nationalistic or political consideration;
>
> (iv) to give each lecturer an honorarium to be determined by the Executive Committee;
>
> (v) to ensure that the choice of centres for the delivery of such lectures is made with due regard not only to their excellence but to their geographical location (so that, for instance, the lectures are given at centres both in the East and in the West);
>
> (vi) and generally to further the advance of mathematical research and the dissemination of the results thereof (especially among all member countries of the Union).[93]

All in all only twelve of these lectures were realized, between 1971 and 1988; most of them were published afterwards in the journal *L'Enseignement mathématique*.[94] The visibility thus attained was finally judged poor compared to the scientific quality; so the series was disbanded.

A single one of these twelve events took place "in the East"—as it says in item (v) above—: Jacques-Louis Lions's (1928–2001) Moscow lectures in November 1972. Another Union Lecture was given in Tokyo, by Friedrich Hirzebruch (1927–2012). Two were hosted by the IHES at Bures-sur-Yvette. Three took place in the US; the first two Union Lectures were hosted by the IAS in Princeton. Four were presented at various places in Switzerland.

The very last Union Lecture was given by Vladimir Igorevich Arnold (1937–2010) at Oxford in 1988, when he was finally allowed to leave the USSR in order to honor this invitation, which dated from 1976. One other Russian mathematician gave a Union Lecture: Anatoli Georgievich Vitushkin (1931–2004), at UCLA, in the Spring of 1978. He had been invited there anyway, and Pontryagin raised no objection of accepting this as a Union Lecture. The other members of the IMU Executive Committee at the time insisted that accepting Vitushkin did not mean forgetting about Arnold's invitation.[95]

After 1988, the program was reoriented exclusively towards developing countries, where local institutions were asked to apply as hosts of such lectures.

---

[93] See *Bulletin of the International Mathematical Union* No. 2, 2 September 1971, pp. 5–6.

[94] The complete list of the Union lectures is reproduced, with minor mistakes in the references, in [Lehto 1998], pp. 324–325.

[95] In [Arch. IMU], SF1 Ser 11, Box 11 A, numerous documents show that Russian colleagues continued to be reminded of Arnold's standing invitation over the years. Once Arnold had delivered his lectures, the problem of how to pay him took almost a year to be sorted out.

8.2.3 New Horizons: 1982–1990

Let us move on to our third period. It is the first of two intermediary periods comprising only three ICMs each.

In 1978, the choice for the site of the next ICM was soon narrowed down to two competing cities: Warsaw and Jerusalem. It seems that fear of local political instability played a role in the decision against the Israeli offer.[96] However, fundamental political opposition whipped by social protest and catholic ardour, and intensified by the economic crisis, had marked the Polish scene for a number of years at the time. In December 1981, Jaruzelski's government declared martial law and battered public opposition. Among those sent to jail were of course also mathematicians. Reacting to this situation, the IMU Executive Committee meeting in Paris in April 1982,

> after considering the scientific prospects for a Congress at the present time, decided to postpone the Congress to the later part of August 1983, with the following understanding: a discussion of the situation will take place at the meeting of the General Assembly in Warsaw in August 1982. On the basis of this discussion and in light of the scientific outlook for a congress in August 1983, the E[xecutive] C[ommittee] will reconfirm or cancel the Congress in November 1982. No alternative site will be considered.[97]

The IMU General Assembly did take place in Warsaw on 8–9 August 1982 as planned. In the Fall of 1982, the ICM was rescheduled for the following year. It is only with hindsight that these Polish events appear today like a prelude to the end of the Soviet Union. Similarly, the flow of Russian Jewish immigrants to Israel since 1971 may also be seen as an early precursor of the brain drain in the 1990s.

The other two ICMs of the third period returned to the original quadrennial rhythm: the ICM in Berkeley, California, took place in 1986, and the very first ICM that was organized neither in Europe, nor in North America, was held in Kyoto, Japan, in 1990. A picture of the opening session of the Kyoto ICM has been chosen for the cover of this book.

8.2.3.1 Politics

There were moments when the responsibility of the IMU for running the ICMs, which had increasingly asserted itself since the Stockholm ICM in 1962, seemed to be challenged on political grounds. Olli Lehto, in the section of his book entitled "Politics interferes with the IMU," narrates such a moment in 1979 during the preparation of the Warsaw Congress. Certain Soviet mathematicians were unhappy about the way their community was treated by the Consultative Committee (as the program committee was still called at the time). They tried to retaliate by political allegations

---

[96] See [Lehto 1998], p. 202.

[97] See Introduction to the Special Issue on occasion of the Ninth General Assembly of the IMU in 1982, of the *IMU Bulletin*.

against the role of the IMU. Olli Lehto translates at length the memorandum penned after a meeting of the Polish organizer of the ICM Czesław Olech with Vinogradov and Pontryagin in Moscow in November 1979. Let us just quote this much:

> 1. The categorical dissatisfaction of the Soviet Committee of Mathematicians with the procedure by which invited speakers were selected was noted. Further, Vinogradov and Pontryagin noted that the Consultative Committees have systematically discriminated against Soviet candidates, rejecting strong candidates proposed by the Soviet Committee and including in the program candidates well known to be weak. In this behavior of the Consultative Committees, an important role was played by the openly racist propaganda of the Zionists, widely advertised by the Western press.
>
> 2. Western mathematicians with Zionist ideology have taken advantage of the ICMs for anti-Soviet political activity, which has nothing to do with and is detrimental to the scientific work of the Congress.
>
> ......
>
> 4. The Soviet Committee is of the opinion that the procedure in force before the Stockholm ICM of 1962 should be restored; i.e., invited speakers should be elected by the national Organizing Committee on the recommendation of the participating countries.[98]

It turned out the next day that what was voiced here was not a general opinion of all the leading Russian mathematicians, nor a government requirement, and could thus be resolved through personal interventions by Moscow Academicians.[99]

Another difficult issue with obvious political connotations was whether it was such a good idea to decide in November 1982 to go ahead and reschedule the Congress in Warsaw for 16–24 August 1983, at a time when the martial law meant to crush *Solidarność* was still in force. Indeed, martial law was lifted only on 22 July 1983, hardly a month before the ICM. Lehto relates this story under the heroic title "Mathematics Above Politics",[100] naturally insisting on the point of view of the organizers.

I have occasionally tried to get a clearer picture of ICMs, and of the IMU, by looking at roughly simultaneous events within the international community of historians—see Section 1.4.1.1, and the beginning of Chapter 7. The problems around the Warsaw ICM may be compared with tough debates among historians about meeting in Moscow. In 1957, the General Assembly of the International Committee of Historical Sciences (ICHS) was to meet in Moscow. This meeting was called off, following a personal decision of the IHCS President, in reaction to the Soviet quelling of the Hungarian uprising in 1956: "both for reasons of political ethics and to prevent a split between the members of the ICHS."[101]

Years passed, and an International Congress of Historians was planned to be held in Moscow in August 1970, when a similar obstacle arose, this time from the Soviet military intervention against the Czech reform communists in the Summer of 1968.

---

[98] See [Lehto 1998], pp. 215–216.

[99] See [Lehto 1998], pp. 217–218.

[100] See [Lehto 1998], Section 10.4, pp. 229–237.

[101] See [Erdmann 2005], p. 255.

> The situation was reminiscent of 1956. ... Should one, and could one now prepare an International Congress in the Soviet capital as if nothing had happened? Opinions within the ICHS were divided.
>
> The British position was unambiguous. The national committee thought that it was not right to hold the Congress in the Soviet Union, and passed a resolution requesting that the Bureau locate it in another country, or cancel it if that should not be possible. The British national committee would not participate in a Congress in the Soviet Union. This resolution was presented to the other forty-one national committees and the press, together with a statement that denounced the Soviet action as an 'unprovoked aggression,' as an 'attack particularly aimed at the freedom of speech and writing,' but at the same time declared that the British committee did not intend to call for a general boycott of the Congress and that it did not claim to speak for all British historians since there were also arguments against staying away. The American Historical Association also came out against Moscow. The Dutch likewise made an 'official démarche,' stating that the principles of the ICHS prohibited holding the Congress in Moscow. The national committees of other countries, including France, expressed reservations without raising such objections. By contrast, Italian and Austrian historians warned against the consequences of preventing the Congress. The West German historians refrained from taking a position because they anticipated the admission of the GDR as a new member, and believed that their absence from the Moscow Congress would not be in the German interest, 'since in this way the GDR would practically be made the sole representative of German historical scholarship.'[102]

At long last, the 1970 Moscow Congress of Historians was maintained thanks to personal diplomacy. It did see a few strong personal political declarations by some of the guests, though. The historian's account of the historians' Moscow Congress concludes:

> The prudent behavior of the Soviet hosts, who attempted not to exacerbate the conflict, had a specific political background. A few days before the Congress's opening, a German-Soviet treaty had been signed in Moscow inaugurating the new 'Eastern policy' (*Ostpolitik*) of the Federal Republic of Germany ... It was an element of the general détente policy of the 1970s, which was based on de facto recognition of the political situation and zones of influence created by the war. This influenced the Congress's atmosphere. There was a deliberate endeavor to establish contact despite the existing tensions.[103]

Rather than flatly declaring it to be "Above Politics," the way in which the Warsaw ICM could take place in 1983 shortly after the end of the period of martial law in Poland calls for a careful historical analysis of the events. Lacking the necessary sources, this cannot be done here.[104] It remains a task for the future. Looking back from today, the Warsaw Congress does mark, within the evolution of Mathematics International, the beginning of the end of the Cold War, and the dawning of a new age.

> The basic political configuration of international relations ... remained constant. The antagonism between the two competing political systems, which nonetheless rely on cooperation ..., is an element of the 'long- term' perspective. It has had its impact on the Congresses. But at the same time, they have been influenced by surface events fluctuating between the

[102] See [Erdmann 2005], p. 252.

[103] See [Erdmann 2005], p. 255.

[104] Cf. the summary account in [Curbera 2009], pp. 207–217.

> behavioral patterns of the Cold War and de-escalation. A further political parameter modifying the basic configuration of polarity and convergence has been the gradual emergence of a pentarchy—the United States, the Soviet Union, Japan, China, and Europe—overlying the political world dualism of the two hegemonic powers.[105]

This is also reflected by the fact that in 1985, ending years of subtle discussions, "China" joined the IMU, with two Adhering Organizations, the Chinese Mathematical Society and the Mathematical Society located in Taipei, China.[106] Furthermore, while the Democratic People's Republic of [North] Korea had been a member of the IMU since 1966[107], the Republic of [South] Korea joined the IMU in 1982.

We introduced Section 8.2.2.3 above with a reference to ICSU's policy of the free circulation of scientists and the way it came up in connection with the Helsinki ICM. The same issue arose again ten years later, potentially jeopardizing the 1990 ICM in Kyoto.

> In accordance with the sanctions enacted by the United Nations, Japan had prohibited citizens of South Africa from entry into Japan for the purpose of cultural exchange. However, an exception could be made for meetings associated with ICSU if the scientists signed their application for a visa including the statement: "I do not hold any racial prejudice nor do I belong to any racially discriminatorial organization."[108]

This practice of the Japanese authorities—and similarly by India—was considered by ICSU as being incompatible with its free circulation policy. As a consequence, ICSU changed plans for its 1988 General Assembly, moving it away from Tokyo, to Beijing. The IMU, however, decided to tolerate Japan's practice, and preparations for the ICM could proceed.

Concerning such issues as would fall under the remit of the SCFCS, we have noted above—see Section 8.2.2.3—that this Committee of ICSU initially dealt predominantly with visa denials. However, over the years humanitarian cases grew more frequent.

The files on Human Rights issues 1975–1990 in the IMU Archive show a great variety of cases[109]: Complaints about NSF refusing to finance US mathematicians to travel to the Warsaw ICM; Cuban mathematicians not admitted to the US to participate in the Berkeley ICM; mathematicians left out of the World Directory of Mathematics by National agencies; 'Refuseniks'; and much more.

Let us single out just one individual example. We have already mentioned, in Section 8.1.4, the case of the leading Uruguayan mathematician and communist politician José Luis Massera, who could not participate in the 1950 ICM for lack

---

[105] These general remarks in [Erdmann 2005], p. 244, concern the long period 1960–1985.

[106] See Section 10.6 in [Lehto 1998], pp. 242–250.

[107] North Korea would change to observer status as of 1 January 2003, and has not been a member of the IMU since then.

[108] See [Lehto 2005], p. 254.

[109] See [Arch. IMU], SF1 Ser 21.

of a US visa. Under the civilian-military regime in Uruguay 1973–1985, Massera's life became a nightmare, which prompted Henri Cartan in Paris and Israel Halperin (1911–2007) in Toronto to launch an international campaign in 1982:

> As is well known, the professor of mathematics J.L. Massera has been tortured and then held a prisoner in Uruguay in spite of actions to obtain his freedom by thousands of individual scientists and scientific societies over six years. Some months ago, an international Campaign was started by Professors Henri Cartan of France and Israel Halperin of Canada to persuade the Government of Uruguay that their image in the world would suffer more damage by continued imprisonment of Prof. Massera than by anything he might say if he were released.
>
> This International Campaign now has the formal support of the Mathematical Societies of Canada, France, Yugoslavia, Italy, Denmark and Czechoslovakia. We anticipate that this list will grow as other Mathematical Societies can arrange to put the question to their memberships.
>
> International Campaign-Massera would like to ask the IMU to take one or more of the following actions:
>
> 1. Issue a public statement expressing the wish that Prof. Massera be allowed to go immediately to France or Italy, in both of which countries he has standing invitations.
>
> 2. Recommend to adhering National Organizations and National Committees of Mathematics that they in turn take such action as they find appropriate to obtain the release of Professor MasseraProf. Massera .
>
> I appreciate that some voices will be raised in opposition to this request on the grounds that the IMU is not authorised to get involved in politics. But it should be clear to all that this is a question of simple humanity and does not involve political attitudes or influence. I imagine that we would all agree that if the IMU had at a certain point in time protested the inhuman treatment of Banach and many other scientists that no one to-day would criticise that action. ...[110]

Many mathematicians pleaded in favour of Massera.[111] The IMU has consistently abstained from throwing its considerable reputation behind such causes or campaigns. The attitude of its Executive Committee is concisely summed up, for instance, in a letter from 1987:

> The E.C. felt that even though as individuals the members are very much concerned with human rights IMU as an organization should refer such matters to ICSU.[112]

This way of deferring humanitarian issues seems to be by and large what IMU is still doing today—in spite of the considerable position of influence that the IMU has gained, and continues to develop, in particular through contacts with political instances in the context of the International Congresses.

---

[110] From Israel Halperin to the Secretary of IMU, J-L. Lions, 7 May 1982, [Arch. IMU], SF1 Ser 21. Massera's case is mentioned in this book only with a view to the conundrum it posed for the IMU. For more information on Massera's life, mathematics and politics, see [Markarian & Mordecki 2010] and [URL 66].

[111] To name but one, slightly later example: Robert P. Langlands's letter to the editors of the *New York Times*, which was published on 11 January 1983 in Section A, p. 18—see [URL 67].

[112] From Lehto to Rosenzweig, 26 May 1987, [Arch. IMU], SF1 Ser 21.

Fig. 8.3 José Luis Massera and his wife Marta Valentini after his liberation, March 1984. Source: [Markarian & Mordecki 2010].

8.2.3.2 Architectural Conjectures and Unifying Research Programs

Both in Part I and in Part II of this book, I have included glimpses of certain developments in mathematics that in my view capture a characteristic aspect of the era considered: various takes on the foundational crisis are discussed in Chapter 2, and Section 6.1 is dedicated to Emmy Noether's Legacy. For the current Part III, which deals with the past seventy years, a section on 'Computers and Mathematics' could pursue the diverging developments of two fields and professional communities, compare the Fields Medal and the Turing Award, follow the history of what used to be called the Nevanlinna Prize and will now be the Abacus Award, etc. However, I felt I knew too little about this vast subject.

Therefore, I shall briefly present instead a peculiar new feature that came to the fore in the development of pure mathematics of the last fifty years, building on the broad consolidation of the 1930s (Chapter 6). It was Barry Mazur (b. 1937) who identified this as a phenomenon in its own right, and christened it *Architectural Conjectures*.[113]

That mathematicians often find themselves trying to solve problems which have been formulated by others, sometimes quite a while ago, is no news. Some of these problems can be formulated as a precise statement, which experts have their reasons

[113] See [Mazur 1997].

to expect to be true, but for which no accepted proof is known. Such a statement is today called a *conjecture*, although colleagues may be more or less fussy about granting this label to a given statement. This word is relatively new, though. The most famous conjecture of all, which is still unproven, is called the Riemann *Hypothesis*. Let us recall a few examples of famous conjectures that have been proved recently.

- A 1922 article by Louis Mordell ends with a list of five statements, of which he writes that "the preceding work suggests to me the truth of the following statements ..., none of which, however, I can prove."[114] The last one of these statements—to the effect that any smooth curve of genus at least 2 defined over the rational numbers has only finitely many rational points—became known as *Mordell's Conjecture*. It was proven by Gerd Faltings (b. 1954) in 1983, which earned him a Fields Medal in 1986.
- A statement claimed, and not proved, by Pierre de Fermat (160?–1665) in the margin of the edition of Diophantus's *Arithmetica* that he used became known as *Fermat's Last Theorem*. It was finally established by Andrew Wiles, as a corollary of the much more general result to the effect that every elliptic curve defined over the rational numbers is modular. Just slightly too old then for the Fields Medal, Wiles received a special IMU silver plaque at the Berlin ICM in 1998.
- A question about three-dimensional compact manifolds, on which every closed curve can be contracted, asked by Poincaré in 1904, which became known as the *Poincaré Conjecture*. Grigori Y. Perelman (b. 1966) was offered the Fields Medal in 2006 for his proof of it. Analogs and generalizations of this conjecture in higher dimensions had been established earlier; Stephen Smale and Michael Freedman (b. 1951) were awarded Fields Medals in 1966, resp. 1986, for those theorems.

These examples set the tone for the traditional model of what conjectures are and how they command mathematical attention. Ideally, they are as simple to state as they are hard to prove. The proof, once found, will typically be conceptually very far from the original statement. It is precisely the tension between the elaborate theory needed for the proof, and the elementary succinctness of the claim, which may stimulate the working mathematician, and invariably impresses the public.

In contrast to this, the *Architectural Conjectures* singled out by Mazur are of a different kind. They are

> mathematically precise assertions, as well-milled as minted coins, provisionally usable in the commerce of logical arguments; less than 'coins' and more aptly, promissory notes to be paid in full by some future demonstration, or to be contradicted. These conjectures are expected to turn out to be true, as, of course, are all conjectures; their formulation is often a way of "formally"packaging, or at least acknowledging, an otherwise shapeless body of mathematical experience that points to their truth. From these conjectures, implications may be perfectly rigorously made. Best, if the conjectures are, loosely speaking, "testable", or "falsifiable" in the sense that they imply a stream of particular, numerical perhaps, predictions many of which may be directly checked. But these conjectures are architectural in that they

[114] See *Proceedings Cambridge Philosophical Society* 21 (1922), p. 191.

play the role of "joists" and "supporting beams" for some larger mathematical structure yet to be made. These conjectures sometimes round out a field by being clear, general (but not yet proved) statements enabling one to understand where a certain amount of on-going, perhaps fragmentary, specialized work is headed; they provide a focus. Their formulation sometimes serve to "allow the field to proceed": a research program may continue, conditional on the truth of these statements, in order to see what lies further down the road. One effect of the formalization of Conjecture is to give concrete language—"a local habitation and a name"—to expectations, analogies, hoped-for constructions, etc., long before the methods needed for their elucidation are available, giving us a rich source of palpable "historical artifacts" about ideas at an early stage in their development.[115]

Mazur's inspiration to focus on this new class of conjectures came from Alexander Beilinson's (b. 1957) Conjectures.[116] Starting from special cases known since the nineteenth century—specifically, Dirichlet's analytic class number formula—Beilinson built an amazing hypothetical framework of cohomology-like theories to theoretically predict in extreme generality the transcendental part of special values of generalized zeta-functions in terms of 'regulators' afforded by the framework. The required cohomological tools are based on algebraic K-theory and Deligne-cohomology. K-theory was famous, and infamous, for combining a perfectly general scope with the fact that very little could be proved about concrete K-groups, even in fairly elementary situations. One of the first (sectional) ICM talks on this circle of ideas was given by Christophe Soulé (b. 1951) in Warsaw.[117]

Developing such an architectural conjecture, Beilinson has taken Emmy Noether's approach to another level. As we have explained in Sections 6.1.1 and 6.1.2, Noether was always eager to turn analogies into logical deductions from a more general framework. Even in the absence of a general unified theory, she was tempted to speak and reason as if such a general theoretical frame was available. This being said, the degree of generality and its connection to established theories, which Beilinson boldly postulated, was simply unheard of in the 1930s.

In the 1980s, however, it seemed less extravagant, in spite of the sequences of thorny translations it forced on all those who wanted to sort out special cases. The reason for this was that other largely conjectural research programs, sometimes verging on fairy-tales, had begun to gather excited followers. Robert Langlands's (b. 1936) Program, whose origins date back to the late 1960s and 1970s, is the best-known example marking this new era of pure mathematics. It predicts profound correspondences between number theory and geometry, between Galois groups and automorphic forms, involving the representation theory of algebraic groups. It continues to be one of the most far-reaching mathematical research programs. But it is not an Architectural Conjecture in the specific sense indicated above, because every single case study that appeared to fall under the Langlands Program had to be spelled out in its own way before it could be worked on.

Even when one is not in a position to prove Langlands's conjectures, a standard procedure is to reason on one side of the correspondence to deduce interesting and surprising consequences on the other side and then to prove these consequences directly. Specialists call this a "guide

[115] See [Mazur 1997], pp. 198–199.

[116] Mazur mentions the volume [Rapoport et al. 1988].

[117] See Proceedings ICM 1983, pp. 437–445.

> to intuition," as if to confirm that a "non-logical cognitive phenomenon" is involved. The most spectacular example was undoubtedly the series of deductions that led to the discovery by Gerhard Frey [b. 1944] and Ken Ribet [b. 1948], in the mid-1980s, that Fermat's Last Theorem could be proved by establishing one part of the Langlands correspondence. Since the Langlands program had already shown such exceptional explanatory power that it had grown "too big to fail," this dramatically intensified the belief among number theorists in the truth of Fermat's Last Theorem and gave Andrew Wiles the confidence he needed to spend seven years in an ultimately successful effort to prove as much of the Langlands correspondence as he needed.
>
> The Langlands program, meanwhile, flourishes as never before, in large part thanks to the new ideas Wiles brought to bear on a problem that had motivated number theorists for centuries but that is now well advanced along Weil's cycle of knowledge and indifference. One of the most fruitful approaches to proving Langlands's conjectures was developed by the Soviet-born mathematician Vladimir G. Drinfeld [b. 1954], using the full range of techniques developed by ... Grothendieck ..., which is fitting, since the Galois representation side of the Langlands correspondence is naturally interpreted as an avatar of Grothendieck's motives, and vice versa. Having absorbed earlier work of Goro Shimura, [Pierre] Deligne [b. 1944], and Langlands, Drinfeld defined several new (Grothendieck-style) geometries to bridge the two apparently unrelated "structures" that matched Langlands' predictions and, in so doing, launched the geometric Langlands program and incidentally began the process of aligning these structures into categories. Drinfeld and Alexander Beilinson, and later Ed[ward V.] Frenkel [b. 1968] and his collaborator Dennis Gaitsgory [b. 1973], were among the first to construct a Langlands correspondence as a relation between categories.
>
> Drinfeld's geometries looked extremely strange when he defined them; did they preexist his definition? Are they avatars of a platonist Langlands correspondence or do they bring the Langlands correspondence into being? In designing his geometries, Drinfeld was clearly guided by the hope of applying Weil's topological insight; his definition's merit was that it provided just the fixed points he needed. In 1990 I was sitting in Manin's seminar in Moscow when Beilinson stood up and interrupted the speaker to explain that the objects studied in algebraic geometry were an illusion, maya, to hold fixed points together. Of course he said no such thing, but that's what I heard... [118]

Laurent Lafforgue (b. 1966), who had been a student of Christophe Soulé's and wrote his thesis under the direction of Gérard Laumon (b. 1952) on one of Drinfeld's key notions, received a Fields Medal at the Beijing ICM in 2002 for proving the Langlands Correspondence in the function field case.

8.2.4 Mathematics Without Borders? 1994–2002

This fourth and penultimate of our five periods, like the preceding one, counts only three ICMs. In 1994 it was the third time, after 1897 and 1932, that the ICM returned to Zürich. Neutral Switzerland continued to hold its own as the center of gravity of mathematical Europe.

[118] See [Harris 2015], pp. 201–202. The reader who may be slightly mystified by the final twist of this quote may want to look up that book.

8.2.4.1 A New Global Dimension

The 1998 ICM did not go very far from there. It convened in Berlin, the capital of the newly reunited Germany. This was the second ICM to be held in Germany; 94 years had passed since the Heidelberg ICM, which had been organized at a time when even the first IMU had not been conceived yet. The Berlin ICM could proudly present itself as the first one to take advantage of the would-be borderless World Wide Web. Never before had email been the dominant means of communication for mounting an ICM. In the same vein, the 1998 IMU General Assembly held in Dresden, before the Congress, initiated the IMU Committee for Electronic Information and Communication (CEIC)—see Section 10.1.1 below.

The last ICM of the short fourth period was also the first ICM of the twenty-first century. It took place in 2002 at Beijing. This twenty-fourth ICM organized since 1897 was only the second to be held outside of Europe and North America, the first being the Kyoto Congress of 1990. Even more could be said about it. As the President of the Chinese Mathematical Society, Kung Chin Chang put it in his Berlin announcement of the upcoming ICM:

> All the past congresses were held in developed countries. Now, the next congress, the first in the new century, will be held for the first time in a developing country. This will add a new chapter to Prof. Olli Lehto's book *Mathematics Without Borders*.[119]

To be sure, China was rapidly catching up with the developed nations. One may also add the observation that the 2002 Congress took place while the President of the IMU 1999–2002 himself represented the global reach of the organization: Jacob Palis (b. 1940) from IMPA[120] in Rio de Janeiro was the first South American to preside over the IMU. He was the second IMU President who was neither North American nor European, after K. Chandrasekharan's presidency 1971–1974. The Honorary President of the Beijing ICM was Shiing-Shen Chern, whom we have met in Section 6.4.

The border between the two German States had been officially abolished, but quite a few new independent nations, each one with its own border, sprang up during those years around the boundary of the former Soviet Union, and in the Balkans.

It is therefore not as surprising to see the total number of countries represented at the ICMs—according to the official lists published in the Proceedings—jumping up from the 40 countries (including two Germanies) counted in Kyoto in 1990, to 92 countries present in 1994 in Zürich. In Berlin in 1998 the count went up to 98 countries. It attained 104 in Beijing in 2002.

These numbers actually reflect, in the crystalline world of mathematics, a general international trend of the period considered. Without elaborating on it in detail, one should recall that the World Trade Organization (WTO), based in Geneva, Switzerland, began its work on the first of January 1995, following the 1994 Marrakesh

[119] See Proceedings ICM 1998, Vol. I, p. 58. The book alluded to is of course [Lehto 1998].

[120] Cf. Section 8.3 below.

Agreement. The goal of the WTO is to frame and foster worldwide trade, in particular also of intellectual property. This may remind us of the Goethe quote which we have placed at the very beginning of Part I. All but a handful of countries are either members or at least observers of the WTO today. The creation of the WTO marked the transition into a new world, replacing the General Agreement on Tariffs and Trade (GATT) that had been enacted in 1948.

Furthermore, the period we are focussing on here coincides with the very time slot where the creation of the International Criminal Court (ICC) could succeed. This unique international instance, which can complement national jurisdiction in particular when it comes to crimes against humanity, has been in operation since July 2002. It had been proposed after World War I. After the Second Word War, allied *ad hoc* tribunals in Germany and Japan had been instituted based on analogous charters—see Section 8.1.1. Today the ICC exists. However, almost twenty years after its foundation, ICC membership is much less global than that of the WTO, and its procedures are occasionally likened to a kind of neo-colonialism.

But let us leave world politics in order to focus again on the more specific realm of Science International. We have noted in Section 8.2.2.2 that after the 1975 signing of the Helsinki Final Act of the Conference on Security and Cooperation in Europe, one could observe a "symbiotic relationship between science and human rights" in certain quarters.[121] At the turn of the millennium, however, the constellation looked quite different.

> [I]t became all too easy to set human rights aside in pursuit of scientific internationalism. Scientific internationalism, after all, had been a core scientific value longer than human rights. By the late 1990s, with the Cold War over and the Soviet Union disintegrated, the journal *Science* detected a sense of apathy among scientists, declaring in a headline, "Human Rights Fades as a Cause for Scientists." Such had not always been the case. . . . . .
>
> Having transcended bipolar internationalism, human rights gave scientists a way to criticize the Soviet government while sympathizing with those who lived under its brutal regime. Human rights also gave U.S. scientists a way to avoid the partisan domestic bickering over nuclear weapons yet still claim relevance to geopolitics. The human rights of the West—including freedom of speech, intellectual freedom, and the right to dissent—would also be the values of the international scientific community, and both would flourish. Human rights served not just as a platitude but as a force that shaped the *mores* of a discipline that, on the surface, appeared to have little to do with human rights.
>
> At the same time, . . . human rights could regress to being simply a noble idea. Scientists . . . embraced and then lost interest in human rights. . . . Earl Callen [had] argued for a discrete movement with limited goals. He demanded of Paul Flory: "We better ask what our goals are, and at what point we are willing to resume exchange. The SOS pledge (and the very name of that organization) focuses on the maltreatment of particular individuals." If the boycott and human rights campaign was to be a broad, indefinite campaign, where should it end? What about reform of the whole Soviet system? Callen wondered. South Africa? Argentina? What about a boycott of the United States for not having signed the SALT agreements? Should the human rights campaign ever end or become a dominant feature of scientific life? He needn't have worried about this slippery slope. The same American Physical Society that Shawlow had declared "actively working" for human rights in 1981 changed noticeably: at the annual

[121] See [Rubinson 2012], p. 250.

APS meeting in 1998, despite attempts to raise concerns over the human rights of scientists in China, only two participants out of several thousand registered for a workshop on human rights. Science would be international, but noncontroversial.[122]

Let us leave these general reflections by quoting from László Lovásc's (b. 1948) presidential address to the IMU General Assembly at Bangalore in 2010.

There are some events which bring the IMU in contact with politics. The most dramatic ones are when a mathematician is kidnapped or unjustly imprisoned. Needless to say, we do our best to help, but this is not always straightforward. For example, when a mathematician was kidnapped, we made confidential contact with our local colleagues, and were advised not to make any public move, because this would make hostage negotiations more difficult by increasing the stakes for the kidnappers. Luckily, our colleague was eventually freed. In another case, when massive protest seemed the best route, we joined this protest (unfortunately, our colleague was probably murdered by that time).

...... Visas for scientists and treatment of foreigners is becoming, unfortunately, an increasing concern. Under threats of terrorism, governments are often tightening their visa policies to irrational levels. IMU stands firmly by the principle that no scientist should be punished for actions of his or her government. The EC has joined the protest of ICSU against US visa policies by moving our EC meeting from the US to another country. Unfortunately, unexpected events like terrorist attacks can create difficult situations like we experienced with Indian visas for this meeting. It took enormous efforts on the part of the Indian organizers to make sure that the delegates and Congress participants get visas, for which I would like to express our gratitude.[123]

8.2.4.2 Fields Medals, and Prizes for the New Millennium

We have seen in Section 8.2.2.2 above that, thanks to an anonymous donation, the habitual number of Fields Medals awarded at each Congress had risen to four since the Moscow ICM in 1966. But there were exceptions to this rule. The Fields Medal Committee for the 1974 ICM at Vancouver, for instance,

decided, at the outset, and not without discussion, to confine the award to mathematicians under forty, as in the past, The names of some who have done brilliant work in recent years, but who are now on the wrong side of forty, have had regrettably to be omitted. Even so, more than a score of names figured on our first list. The task of reducing that number was by no means easy. There was a great deal of consultation, deliberation, and reflection. The Committee elected finally to select two names for the award. That decision was reached as unanimously as one could reasonably expect. We are aware of the very strong claims of many of those not selected, some of them so young that many Congresses will meet before they are forty. Nevertheless, we are convinced that the two selected are mathematicians of exceptional merit, whose work has advanced the development of important branches of our science.[124]

Those two winners in 1974 were Enrico Bombieri (b. 1940) and David Mumford. Shortly afterwards the number of Fields Medals awarded was reduced to three both in 1983 at Warsaw and at the Berkeley ICM in 1986. In 2002 at Beijing it dropped to

[122] See [Rubinson 2012], pp. 259–260. The quote from *Science* refers to issue no. 282 (October 9, 1998), p. 216.

[123] See *IMU Bulletin* 59 (October 2010), p. 14.

[124] See Proceedings ICM 1974, Vol. I, p. xvii.

two once more: Laurent Lafforgue and Vladimir A. Voevodsky (1966–2017) were handed their Medals by the highest ranking politician who had ever served at an ICM until then, the President of the People's Republic of China Jiang Zeming. All three prize winners at Beijing: L. Lafforgue, Voevodsky, and the winner of the 2002 Nevanlinna Prize Madhu Sudan, were born in the same year 1966.

With the benefit of hindsight a superficial, yet suggestive connection emerges between Voevodsky and Sudan; at some point of their careers, both investigated the use of computers to check mathematical proofs. Sudan did so early on. He used the standard logical formalization of mathematical statements, and was chiefly interested in probabilistic results. Voevodsky on the other hand started to explore the help of computers in checking complicated proofs after he discovered a gap in earlier work of his own. Ten years after the Beijing Congress he would organize a special research year at the IAS in Princeton about computer verified proofs, but based on a very different, type-theoretic formalization of mathematics.

The period we focus on in this section saw the breaking of the new millennium, and with it a new type of mathematical challenge combining difficulty, prestige and philanthropic money.

> The Clay Mathematics Institute (CMI) grew out of the longstanding belief of its founder, Mr. Landon T. Clay, in the value of mathematical knowledge and its centrality to human progress, culture, and intellectual life. Discussions over some years with Professor Arthur Jaffe [b. 1937] helped shape Mr. Clay's ideas of how the advancement of mathematics could best be supported. These discussions resulted in the incorporation of the Institute on September 25, 1998, under Professor Jaffe's leadership. The primary objectives and purposes of the Clay Mathematics Institute are "to increase and disseminate mathematical knowledge; to educate mathematicians and other scientists about new discoveries in the field of mathematics; to encourage gifted students to pursue mathematical careers; and to recognize extraordinary achievements and advances in mathematical research." CMI seeks to "further the beauty, power and universality of mathematical thinking."
>
> Very early on, the Institute, led by its founding scientific board—Alain Connes [b. 1947], Arthur Jaffe, Edward Witten [b. 1951], and Andrew Wiles—decided to establish a small set of prize problems. The aim was not to define new challenges, as Hilbert had done a century earlier when he announced his list of twenty-three problems at the International Congress of Mathematicians in Paris in the summer of 1900. Rather, it was to record some of the most difficult issues with which mathematicians were struggling at the turn of the second millennium; to recognize achievement in mathematics of historical dimension; to elevate in the consciousness of the general public the fact that, in mathematics, the frontier is still open and abounds in important unsolved problems; and to emphasize the importance of working toward solutions of the deepest, most difficult problems.
>
> After consulting with leading members of the mathematical community, a final list of seven problems was agreed upon: the Birch and Swinnerton-Dyer Conjecture, the Hodge Conjecture, the Existence and Uniqueness Problem for the Navier–Stokes Equations, the Poincaré Conjecture, the P versus NP problem, the Riemann Hypothesis, and the Mass Gap problem for Quantum Yang-Mills Theory. A set of rules was established, and a prize fund of US$7 million was set up, this sum to be allocated in equal parts to the seven problems. No time limit exists for their solution.[125]

---

[125] See [Carlson et al. 2006], p. vii.

Incidentally, the lavishly illustrated volume that "sets forth the official description of each of the seven problems and the rules governing the prizes"[126] contains exactly 105 portraits (each showing a single person), including a picture of Landon T. Clay. Several famous mathematicians are shown more than once, at various places in the book. Of these 105 pictures, exactly 3 show a woman: Sophie Germain and Sofia Kovalevskaia are portrayed in Jeremy Gray's historical introduction, and Olga Ladyzhenskaya (1922–2004)—who had been on the shortlist of the Fields Medal Committee for 1958[127]—is given a place in the gallery for the Navier–Stokes Equation.

As explained in the long quote above, the problems were well known before the Clay Foundation went about singling them out. One may reasonably doubt that the substantial amount of money newly attached to them can effectively speed up their resolution. Precisely one of the Millennium Problems has been settled so far: the Poincaré Conjecture, which we have briefly mentioned in Section 8.2.3.2.

Apart from the prizes bound to the Millennium Problems, the Clay Mathematics Institute also attributes its yearly *Clay Research Award*. The illustrious list of awardees, which begins with Andrew Wiles in 1999, contains several Fields Medalists.[128]

8.2.5 Global Reach from a New Homebase: 2006–2018

The list of the four ICMs that make up our last period speaks for itself: 2006 Madrid, Spain; 2010 Hyderabad, India; 2014 Seoul, South Korea; and 2018 Rio de Janeiro, Brazil. For each of these four countries it was the first ICM they hosted, and the four metropolises chosen for these congresses belong to four different (sub-)continents. This in itself suggests global reach, an appeal to worldwide communication, which is equally evident in fact that the IMU proudly counts today more than 80 member countries.

The global picture is further vindicated by the fact that during this period a new standard was reached for travel subsidies helping mathematicians from developing countries to attend ICMs. Specifically, the Korean Mathematical Society offered to pay for 1,000 participants from developing countries, thus introducing a new level of responsibility, which is now expected from organizers of ICMs.

All the ICM Proceedings since 1897 contain opening texts that evoke the formidable task of preparing the event, and recall the flow of the opening and closing ceremonies. In the case of the 2014 ICM at Seoul, these pages conjure up the momentous awakening of a mathematical nation. Let us quote the beginning of the story unfolded there.

[126] See [Carlson et al. 2006], back cover.

[127] See [Barany 2018].

[128] See [URL 12].

> The first mathematician from Korea to attend the Congress made it to Helsinki International Congress of Mathematicians (ICM) in 1978 with a help of an IMU travel grant program for developing countries. It could be said that the mathematical research in Korea until the early 1980s was rather isolated and sporadic at best. Korean mathematical community began its globalization efforts by joining IMU in 1981 as a Group I member. In the mid-1980s, Korea endeavoured to jump start its mathematical research inviting renowned mathematicians from abroad to deliver lectures and providing young Korean mathematicians with opportunities to glimpse at mainstream mathematics. Visible spinoffs occurred, including modernization of academic curriculums and diversification of research centers. In light of such advancement, Korea became a Group II member of IMU in 1993. In the 1990s, Korea made quite a progress in improving its mathematical research both in the quantity and quality, partly aided by the influx of talented Korean mathematicians educated abroad who returned to Korea after obtaining their degrees. The establishment of the first research institute in Korea devoted to mathematics and theoretical physics, Korea Institute for Advanced Study (KIAS), in 1996 provided the infrastructure for further development.
>
> A dramatic display of the progress of Korean mathematics was made at Madrid ICM in 2006, to where three mathematicians, Jun-Muk Wang, Jeong Han Kim and Yong-Geun Oh, were invited as first Korean ICM invited speakers. This ignited a festivity among the Korean mathematical community. Inspired by the evidence of the momentum in Korean mathematics, Korean Mathematical Society (KMS) applied for a raise in its IMU group level and was announced as a Group IV member in 2007. In the history of IMU, this still remains the only instance in which the group level of a member country was raised by two in one shot. Highly motivated by this series of developments and to continue the momentum, KMS decided to place a bid to host ICM 2014.[129]

Instead of going into detail about the four ICMs of this period—two of which have also marked my personal memories—we shall highlight developments that have marked the inner life of the IMU during these last years. Before doing this, though, let us briefly take note of some recent changes in the prize landscape.

8.2.5.1 New Prizes

We started this book with the Fields Medal; we mentioned the creation in 1978 of the Wolf Prize at the end of Section 8.2.2.2, and talked about the prizes offered by the Clay Mathematics Institute—both for the solution of the "millennium problems" and the yearly Clay Research Awards for other outstanding mathematical achievements—in Section 8.2.4.2. The most recent period that we are addressing now saw the establishment of a number of further international prizes related to mathematics, both inside and outside of the realm of the IMU. The overall landscape of prizes concerning mathematical achievements is large, variegated and evolving.[130] A serious cartography of this whole landscape would have to look deeper into the structures of current scientific philanthropy. This is beyond the scope of this book. The present section simply illustrates recently established international prizes that

[129] See Proceedings ICM 2014, Vol. 1, p. 4.

[130] See for instance the attempt to list and order mathematics awards proposed in [URL 13].

are in principle open to women and men from all nations, independently of membership in an association.[131] We start with prizes associated to the IMU.

The *Gauss Prize* of the IMU has been awarded since 2006 to honor scientists whose mathematical research has had an impact outside of mathematics, be it in technology, in business, or in people's everyday lives. The IMU's *Chern Medal Award* was granted for the first time in 2010; it honors an individual whose accomplishments warrant the highest level of recognition for outstanding achievements in the field of mathematics. This new Medal, awarded once every four years by the IMU and the Chern Medal Foundation, is by far the most highly endowed prize managed by the IMU. It comprises a cash award of $250,000, plus the same amount for the funding of organizations chosen by the Medalist to foster mathematical research, education, or other outreach programs. Also given away for the first time at the Hyderabad ICM in 2010, the *Leelavati Prize* honors major accomplishments for increasing public awareness of mathematics.[132]

Outside of the realm of the ICMs, the annual *Abel Prize* has been awarded every year since 2003 to one or two mathematicians by the King of Norway. The IMU does take part in nominating the Abel Committee, which is elected by the Norwegian Academy of Science and Letters and chooses the Abel Prize winners. The Abel Prize was consciously modeled to be as close to a Nobel Prize for mathematics as possible. In particular, looking at the list of prize winners, which starts with Jean-Pierre Serre in 2003, we see older mathematicians being honoured for their lives' contributions to mathematics. This fact and the generous cash award of 7,5 million Norwegian Kroner mark the principal differences with the Fields Medal.

The Hungarian Academy of Sciences resuscitated for the new millennium the *International János Bolyai Prize of Mathematics*, which is today awarded every five years to a mathematician for a monograph with important new results published in the preceding 10 years. It had been bestowed on Henri Poincaré in 1905, and in 1910 on David Hilbert. After this it lay dormant for ninety years. The prize winners since 2000 were Saharon Shelah (b. 1945), Mikhail Gromov (b. 1943), Yuri Manin (b. 1937), Barry M. Simon (b. 1946), and Terence Tao (b. 1975).

In reminiscence of the well-known mathematician in Zürich—we have mentioned him in Section 6.1.3, and seen him acting in a peculiar affair as President of the IMU in Section 8.2.1.2—the *Heinz Hopf Prize* has been awarded in Zürich once every two years since 2009. It is confined to honor achievements in pure mathematics. All awardees of the prize give a public *Heinz Hopf Lecture*. However, the series of Heinz Hopf Lectures started earlier and not all of them are linked to the prize.[133]

---

[131] The well-known Cole Prize in Algebra or in Number Theory, for instance, requires awardees to be members of the American Mathematical Society AMS.

[132] For more information on these prizes, see the IMU's website. For the ICM Emmy Noether Lecture, see Section 10.1.2.2 below.

[133] See [URL 14].

Finally, we mention two recent philanthropic prizes sponsored by entrepreneurs:

The *Shaw Prize* was founded in 2002 by the Hong Kong media businessman and philanthropist Run Run Shaw (1907–2014), in the first place to honor individuals whose significant works have recently been achieved and who are currently active in their respective fields. Prizes are awarded every year in the three branches: astronomy, the life and medical sciences, and the mathematical sciences. The first two prizes awarded in the latter category in 2004, resp. 2005, went to Shiing-Shen Chern and Andrew Wiles.[134]

A very recent addition to the list of international prizes that carry a specific section for mathematics is the *Breakthrough Prize*. It has been awarded since 2015 in three categories: physics, the life sciences, and mathematics. It was founded by Sergei Brin, Priscilla Chan & Mark Zuckerberg, Yuri & Julia Milner, and Anne E. Wojcicki. Between 2015 and 2021, twelve mathematicians have been awarded the prize. The list suggests a preference for seminal work in pure mathematics, especially geometry and arithmetic.[135]

Worth three million dollars, the Breakthrough Prize has the highest endowment among all the prizes we have mentioned. Apart from this major award, the same foundation also provides valuable encouragements: the *New Horizons in Mathematics Prize* ($100,000) aiming at early-career researchers, and the *Maryam Mirzakhani New Frontiers Prize* ($50,000) presented to women mathematicians having recently obtained their PhDs.[136]

Maryam Mirzakhani (1977–2017), the first female Fields Medalist (2014), and Peter Scholze (b. 1987) both received Clay Research Awards in 2014. In 2016, Scholze refused the New Horizons in Mathematics Prize that was offered to him. In spite of his young age, he was already past the early stages of his career, and would be awarded a Fields Medal in 2018. Twenty or more years from now it should be interesting to compare the awardees of the various prizes, and to see to what extent all these medals have shaped their own profile or acted as mutual multipliers.

8.2.5.2 The New Homebase

Leaving the ICMs and the ever more plentiful prizes for mathematicians, we now take a brief look at the inner circle of the IMU under the presidencies since 2003 of John M. Ball, Lásló Lovász, Ingrid Daubechies, Shigefumi Mori, and Carlos E. Kenig. They and the very active IMU Secretaries General Martin Grötschel (2007–2014)

---

[134] For more information, see [URL 15].

[135] See [URL 16] as well as [URL 17]. Most of the media coverage related to the Breakthrough Prize is via Facebook.

[136] See [URL 18].

and Helge Holden (since 2015) have realized important innovations in the way that the IMU functions. These include the timely updating of the work of IMU committees in view of current challenges, which will be discussed in Section 10.1 below.

Another novelty highlights the impressive ability of the leading personnel running this big organization to reflect on the obvious danger of self-perpetuating structures and personal constellations. I am alluding here in particular to the recent creation of the *Structure Committee* (SC), whose mission it is to scrutinize the general setup followed by the ICM program committees. The SC will be presented in the last part of Section 10.2 below.

As far as the concrete, daily functioning of the IMU is concerned, the most visible and consequential change in the last decade was the creation of a new stable homebase for the IMU Secretariat. Since 1950, the home of the IMU secretariat had moved through eight countries and three continents, because

> IMU conducted its business at the institution of the IMU Secretary General which usually also served as the legal domicile of IMU. At the General Assembly 2010 in Bangalore, India, the Weierstrass Institute in Berlin, Germany (WIAS) was elected as the host institution of the permanent secretariat.[137]

Fig. 8.4 East front of Hausvogteiplatz, Berlin, with the new IMU headquarters indicated (Courtesy IMU).

[137] See [URL 19].

A bid to host the Stable Office of the IMU had been launched in October 2007. From among ten institutions that showed initial interest, the *Stable Office Committee* in 2009 selected a shortlist of three: the Fields Institute (Toronto), IMPA (Rio de Janeiro), and the WIAS (Berlin). The WIAS received the absolute majority in the final vote at Bangalore.[138] As a result of Grötschel's adroit use of good contacts, the German Federal Government and the City of Berlin agreed to jointly grant a yearly allowance that enabled the International Mathematical Union to set up its permanent headquarters in this city. One of the arguments Grötschel used was the remark that by welcoming the IMU, Germany would finally become host to an international scientific union.

Thus, while the legal domicile of the IMU had already been transferred from Princeton, NJ, to Berlin on 1 January 2007, the IMU Secretariat physically moved to Markgrafenstrasse 32 in Berlin-Mitte, in the historic center of town, in January 2011.

> An opening ceremony celebrated the inauguration of the secretariat, guests from home and abroad enjoyed the festive event. The secretariat staff started work, it runs IMU's daily business and provides support for many IMU operations. Another highlight was the opening of the IMU Archive that moved from the University of Helsinki to its new home in the IMU Secretariat.
>
> On the occasion of the establishment of the permanent IMU Secretariat in Berlin, the Einstein Stiftung Berlin (ESB) gave a grant to the Berlin Mathematical School that initiated the IMU Berlin Einstein Foundation Program with a view to increasing interactions of young mathematicians from developing and economically disadvantaged countries with the lively mathematical environment in Berlin.[139]

In 2017 the headquarters were relocated within Berlin-Mitte, from Markgrafenstrasse to the spacious fourth floor of a new office building at Hausvogteiplatz 11a—see Fig. 8.4. Apart from the offices for the staff and a meeting room, the IMU floor also includes the rooms for the IMU Archives. The permanent office in Berlin allows for easier solutions to daily administrative problems that had haunted former IMU administrations, such as setting up a stable legal structure for the worldwide money transfers.

8.2.5.3 Back to Sicily

At the beginning of the period we are considering here, during the opening ceremony of the International Congress of Mathematicians (ICM 2006) in Madrid on 22 August 2006, the International Mathematical Union presented its new logo, which had won the international competition launched by the IMU in 2004. The design is due to John M. Sullivan (b. 1963) and Nancy Wrinkle.[140]

---

[138] See *IMU Bulletin* 59 (October 2010), pp. 18–26; 35.

[139] See *IMU Bulletin* 61 (December 2011), p. 3. Cf. Section 10.1.3 below on the IMU Commission for Developing Countries (CDC).

[140] See [URL 20] for more detailed information on the logo.

Fig. 8.5 The Borromean link in the mosaic floor of Room 18 of the Roman Villa del Casale near Piazza Armerina, Sicily. On the left, overview of Room 18; photographer Tyler Bell. On the right, detail; photographer Thomas Delzant. Below, the current IMU logo.

The logo is a particularly neat presentation of the so-called Borromean rings. This is the simplest example of what knot theorists call a Brunnian link, i.e., a non-trivial link that can be undone by cutting just one of the components. The latter name alludes to Hermann Brunn (1862–1939), a geometer and arabist from Munich, who thought about knots between 1887 and 1897, and published four texts, the last one being the talk he gave at the very first ICM in Zürich. These papers contain little more than various suggestions to measure the complexity of a given knot.[141] Brunn's enthusiasm for knots may have been intensified by memories of his father, who was an archaeologist—a fact that incidentally explains why Brunn was born in Rome.

Having started this book with an allusion to Archimedes , it is a pleasure to return to antiquity, and to Sicily, almost coming around full circle in the chronological part of my account. Indeed,[142] a beautiful Brunnian link is preserved in the marvellous mosaic floors that were in all likelihood executed by North African craftsmen from

[141] See the discussion of Brunn's work in [Epple 1999], pp. 180–182.

[142] I am most grateful to Thomas Delzant, Strasbourg, for pointing this out to me, and for letting me use the pictures that he took.

the Carthage region between 320 and 330 CE, in the magnificent Roman villa situated in the Casale district near Piazza-Armerina, Sicily[143]—see Fig. 8.5. The mosaic it situated in Room No. 18 of the Villa.[144]

This may be one of the earliest known representations of the so-called Borromean rings. In a discussion with Roger J.A. Wilson in the Fall of 2021, he pointed to three similar mosaics from Jordan, which date from the sixth, resp. fourth century CE.[145] These four mosaic representations of our pattern from Sicily and Jordan predate both the triangular Viking version of the knot from Lärbro, Sweden, and the Christian usage of a ring arrangement to represent Holy Trinity.[146]

What these beautiful mosaics were to represent or mean for a contemplator of the fourth century seems to be open to speculation only.

When this mosaic was composed in Sicily, more than half a millennium had gone by since Archimedes had lived and worked some 90 km East of this villa, in Syracuse. And more than one and a half millennia would pass before Giovanni Battista Guccia launched his international mathematical circle from Palermo, thereby contributing to what is today the cause of the International Mathematical Union.

8.3 A World Wide Web of Institutes

Picking up a historical thread from the 1930s, which is as crucial for current international relations in the sciences as it is *a priori* independent of International Scientific Unions or the ISC, let us briefly remind ourselves of the current constellation of *Locally-grounded Transnational Research Sites* in the domain of mathematics. This notion of LGTRS has been introduced and explained, starting from the initial model provided by the IAS Princeton, in Section 5.2 above. Below we present a short list of well-known, typical examples of such institutes, which are functioning today and whose core research domains include mathematics.

Before presenting this selection, it should be stressed that the list far from exhausts the institutional network of internationally entangled mathematical research. Next to their teaching missions, many distinguished University Mathematics Departments fulfill roles that are very analogous to those of an LGTRS that is exclusively dedicated to research.

---

[143] For general information about the villa, we refer the reader to the introductory chapter: "Background" in [Wilson 2016], pp. 1–25; see also [Wilson 2021].

[144] We stick to the numbering proposed by Gino Vinicio Gentili and followed in [Wilson 1983]. The numbering used in [Steger 2017] is different.

[145] See [Piccirillo et al. 1993], Fig. 136 (p. 125) from the crypt of Saint Elianus in Madaba, Jordan; Fig. 566 (p. 295) from the Church of Bishop Isaiah at Jerash, Jordan; and Fig. 684 (p. 328) from the Byzantine baths at Gadara, Jordan: the one among these three mosaics that dates from the fourth century.

[146] For these later examples, see [URL 21]. Cf. [Sansoni 1998], Fig. 143 (p. 119) and Fig. 196 (p. 164).

Furthermore, there are borderline cases of independent institutes that do not match a typical LGTRS sufficiently closely to be sampled for our short list. A first such instance has already been mentioned in both sections of Chapter 5: the *Institut Henri Poincaré* (IHP) in Paris. Today, apart from housing various regular seminars and conferences, it organizes both short-term scientific visits for small groups of researchers working on a common research project, and longer thematic programs. A somewhat analogous case is that of the Institut Mittag-Leffler in Djursholm (near Stockholm), Sweden, which was finally transformed into an independent institute along the lines of Mittag-Leffler's original idea in 1969. Today it also houses conferences as well as longer thematic activities.

This being said, here are a dozen institutes that perfectly illustrate the notion of a mathematical LGTRS. We list them in the order of their year of foundation.

1930 Institute for Advanced Study (IAS), Princeton, USA

1945 Tata Institute for Fundamental Research (TIFR), Mumbai, India

1952 Instituto Nacional de Matematica Pura e Aplicada (IMPA), Rio de Janeiro, Brazil

1958 Institut des Hautes Études Scientifiques (IHES), Bures-sur-Yvette (near Paris), France

1963 Research Institute for Mathematical Sciences (RIMS), Kyoto, Japan

1980 Max-Planck-Institute for Mathematics (MPIM), Bonn, Germany

1982 Mathematical Sciences Research Institute (MSRI), Berkeley, California, USA

1992 Isaac Newton Institute for Mathematical Sciences, Cambridge, UK

1992 Fields Institute for Research in Mathematical Sciences, Toronto, Ontario, Canada

1996 Morningside Center of Mathematics, Chinese Academy of Sciences, Beijing, China

1996 Korea Institute for Advanced Study (KIAS), Seoul, Korea

1996 Max-Planck-Institute for Mathematics in the Natural Sciences (MPIMN), Leipzig, Germany

An in-depth study of the increasingly global web of research structures is an obvious desideratum.[147] Indeed, the creation and development of each LGTRS in mathematics tells a fascinating story about how the local and national frame of mathematics was brought to interact with the worldwide dimension of mathematical research. A comparative history, say, of the twelve LGTRSs listed above can thus provide decentralized stories of Mathematics International. This will in turn produce insights which a centralized focus on a unified organization like the IMU may tend to hide. A typical problem when studying LGTRSs in a comparative perspective is to

[147] My original plan for this book included forays in this direction, helped by on-site studies—which all had to be cancelled for the time being.

strike the best balance between focussing on founding fathers and key personalities, vs. the ambient scientific and political conditions. First preliminary studies exist for some of the institutes in the above list. For example, in her lecture at the 2018 ICM at Rio de Janeiro, Tatiana Roque tried to situate the first decades of IMPA between Cold War politics, the mathematical interests of Leopoldo Nachbin (1922–1993), and the Brazilian legacy of Bourbaki.[148]

A curious reaction to the daily work at a mathematical LGTRS in the late 20th century has been preserved in a book by the Swiss sociologist Bettina Heintz.[149] Against the background of a survey of changing scientific practices over the past 200 years, Heintz's observation of the mathematicians at the Max-Planck-Institut für Mathematik in Bonn led her to conclude that "modern mathematics is characterized by peculiar features which actually leave almost no possibility for sociological analysis." She was apparently particularly impressed by the lack of fundamental quarrels about method and by the absence of long disputes after the lectures she observed at the MPIM. The mathematicians' surprisingly consensual behavior, which differed so radically from what Heintz experienced in her own field, was attributed in her analysis to the fact that mathematics has the generally acknowledged and universally respected notion of formal *proof*, which brings about a unique "coherence and rationality of arguments."

Heintz's radical reaction to dump the sociology of mathematics may sound like pleasant news for the mathematician, but it somehow misses the point. While the acceptance of thoroughly checked proofs is indeed amazingly stable, the values surrounding problems, approaches, ways to write arguments, dress up theories and so on, vary considerably over time. Such issues are crucial for the role and fate of an LGTRS in the domain of mathematics; the influences and negotiations about what kind of mathematical developments are considered interesting and promising, for the field and for the institute, and who on the international scene of mathematics can be invited to represent these choices most effectively and most visibly. In Section 10.4.2 below, we shall touch upon these questions from the peculiar vantage point of mathematicians chosen for distinguished roles at ICMs.

Apart from the typical LGTRSs presented above, there also exist institutes of a different kind that play an essential role for the international coherence of mathematical challenges and achievements: the internationally visible *Conference Centers*. The first example of this kind of institution has already been mentioned in Section 7.2.3, because it was actually a product of World War II: the Mathematical Research Institute at Oberwolfach, in the Black Forest, in South-West Germany. Although it was conceived as a mathematical research institute for the German war effort, it

---

[148] See Proceedings ICM 2018, Vol. 4, pp. 4093–4112. For the MPIM at Bonn, I have tried a first sketch of its prehistory and the first years of its existence in [Schappacher 1985].

[149] See [Heintz 2000]; both quotes in this paragraph are from p. 274; my translation.

not only managed to survive after the end of the war, but by the end of the 1960s had established itself as the first center welcoming about fifty weeklong specialized mathematical conferences of moderate size every year.[150]

Oberwolfach subsequently served as a blueprint for other mathematical conference centers attracting international attention, such as the *Centre International de Rencontres Mathématiques* (CIRM) at Luminy, near Marseille, France, which was established there at the beginning of the 1980s. Also the *Banff International Research Station* (BIRS) for Mathematical Innovation and Discovery founded in 2003 followed the same idea, with the Black Forest replaced by the Alberta Rocky Mountains in Canada. Today BIRS is a Canadian–US–Mexican joint venture with various outlets, such as the *Casa Matemática Oaxaca* under construction in Mexico.[151] An even more recent creation along these lines is the *Tsinghua Sanya International Mathematics Forum* (TSIMF) on the Island of Hainan, China. It was officially opened in December 2013.

Yet another type of international research institute, this time geared towards welcoming conference participants from developing countries, is exemplified by CIMPA at Nice, France, and the ICTP at Trieste, Italy. These institutes will be mentioned again in Section 10.1.3 below.

[150] On the early history of the Oberwolfach Institute see [Remmert 2020].

[151] See [URL 22].

Chapter 9
ICMI, The Resilient Nucleus of the IMU

Today the *International Commission for Mathematical Instruction* (ICMI) is by far the most prolific of the three commissions of the IMU. But treating ICMI just as a commission of the current IMU misses its peculiar historic significance. Rather than being a daughter of the mother union IMU, ICMI is the IMU's elder sister. An elder sister opens doors; she calls to reason; her presence makes you feel that she was there before.

ICMI celebrated the centennial of its creation in 2008.[1] As we have seen in Part II of this book, the first IMU was founded only in 1920. Since it did not survive the 1930s, a second IMU had to follow suit, but there "has been some confusion over whether the Union came again into being in 1950, 1951, or 1952."[2] The history of ICMI has its own discontinuities and uncertainties as well. Most of them are intimately connected with the fate of the first IMU. Roughly speaking, ICMI was hibernating when the first IMU was alive. This alternating activity of the first IMU and ICMI was a corollary of the politics of exclusion of the old IMU and the fact that German participation had strongly marked the work of ICMI until World War I.

In spite of these discontinuities of ICMI's history, some chroniclers of that commission have presented aspects of its history by way of variations on the theme of *longue durée*. The latter approach to history, developed by the historians of the *Annales* school, insists on the long-term evolution of structural historical patterns. In contrast, those reporting on the activity of ICMI tend to orient their accounts according to influential mathematicians who have left their footprints both in the history of Mathematics International in general and in ICMI. At the beginning of Section 1.3, for example, we have already quoted Hyman Bass's generous division of ICMI's first 100 years into the 'Klein Era', from 1908 to World War II, and the 'Freudenthal Era', post World War II until 2008, thereby blissfully skipping over the fact that both Felix Klein and Hans Freudenthal died some 20 years before the end of 'their' period.

---

[1] See the proceedings of the splendid centennial conference in Rome, [Menghini et al. 2009].

[2] See [Lehto 1998], p. 88.

N. Schappacher, *Framing Global Mathematics*,
https://doi.org/10.1007/978-3-030-95683-7_9

Trying to do justice to the history of ICMI in a brief survey, there is thus the dual difficulty of finding a fitting periodization and of doing justice to individual actors, whose activities often span a broad variety of fields: as research mathematicians, as organizers of international enterprises, or as colleagues interested in mathematical instruction at various levels. Highly distinguished researchers may have strongly divergent ideas about the teaching of mathematics. For instance, Henri Fehr, Heinrich Behnke, Marshall Stone, Hans Freudenthal, and Hyman Bass, to name but them, have all played crucial roles for ICMI, but we cannot do justice to any of these personalities in this book. If framing mathematical excellence is one of the given duties of the IMU, many key mathematicians of the past century and a half have also taken questions of mathematical teaching to heart, both in their own countries and on a global scale. This concern has time and again connected with the grassroots activities of teachers and experts on the teaching of mathematics.

The modest goal of this short chapter is thus to present a condensed overview of the development of ICMI. This task is both facilitated and rendered more difficult by the overwhelming amount of literature about ICMI that has been published over the years. In what follows we will quote from just a few of these texts. Another excellent way to get into the history of ICMI is by exploring the timeline webpage prepared on the occasion of the centennial of ICMI by Livia Giacardi.[3]

From the Early Start in 1908 to World War I. ICMI is the only international association that was founded at an ICM before World War I. This happened at the 1908 ICM in Rome.

> The idea of an International Commission to enquire into mathematical education was first suggested in 1905 by the American David Eugene Smith [1860–1944], in *L'Enseignement Mathématique*, the revue founded in 1899 by Henri Fehr and Charles Laisant [1841–1920]. A formal proposal was considered at the Fourth International Congress of Mathematicians held in Rome in April 1908 and it was resolved to establish the *Commission internationale de l'enseignement mathématique* (CIEM or, as its anglicized form now is, ICMI).[4] The first president was the great German mathematician, Felix Klein, and the first Secretary-General, Henri Fehr. ......
>
> The reasons for the formation of ICMI at that particular period are not hard to perceive. The educational systems of the major countries of Western Europe and North America had expanded during the early years of the century, new technologies set new demands, and innovators had attempted to carry out significant reforms of the (grammar) school mathematical curriculum. In Germany, Klein gave the lectures now known to us as *Elementary Mathematics from an Advanced Standpoint*, in France, a government decree of July 1905 invited 'teachers to follow a method entirely new in geometry', and in England, as a result of the efforts of John Perry [1850–1920] and others, Euclid's rule came to an end (not the spirit of Euclid, which Dieudonné was later to deplore, but the use of his *Elements*). Perry, indeed, wanted far more than just the reform of geometry. He laid stress on making mathematics useful and on linking its teaching with that of science and engineering: he argued for 'utility'

[3] See [URL 23].

[4] Added by N.Sch.: At the time, the commission would be known in the English speaking world as the *International Commission on the Teaching of Mathematics* (ICTM); in Germany as the *Internationale Mathematische Unterrichtskommission*, or IMUK for short; in Italy as the *Commissione Internazionale dell'insegnamento matematico*.

> rather than 'rigour', laboratory-based experience rather than abstraction. His influence was worldwide, ranging from the U.S.A. to Japan, whilst German educators coined the term *Perryismus*.[5]

The Central Committee of ICMI consisted of Klein, Fehr, and George Greenhill as Vice-President.

> Why these three persons? Klein was an entirely understandable choice given his reputation as a mathematician and his active involvement in the German reform movement. The Swiss Henri Fehr was likewise obvious: As editor of *L'Enseignement Mathématique*, he was well informed about national developments in mathematics teaching and about persons active in this field. But why George Greenhill? He was an applied mathematician at the Royal Artillery Institution in Woolwich, but was retired and had not been known to be involved in questions of school teaching. ... He seems to have been nominated for purely political reasons: Smith gave as a reason that Britain would host the next ICM.[6]

At a first get-together of the Central Committee in September 1908 in Cologne, to which Klein also invited Walther Lietzmann (1880–1959) as his assistant, a work plan was drawn up. In the early stage the Commission counted on 18 member countries. The core outcomes of the Commission were detailed reports about the educational system and the teaching of mathematics practiced in the member nations:

> The main work of the Commission at this time was ... the preparation of a vast survey of teaching practices in member countries. Each participating country appointed a sub-commission to prepare national reports, often in many volumes, and the result was outstanding both in terms of quantity and quality. Thus, for example, the French report ran to five volumes and that of the U.S.A. to eleven. The British contributed only two volumes, but the first of these had over 600 pages! Certainly, nothing on the same scale had been attempted before, or has been attempted since. Moreover, not only did countries comment on their own systems: but, for example, as part of the German contribution G. Wolff ... wrote a fascinating account of secondary education in England, ... which still remains a model of a successful comparative case study. That its delayed publication should have taken place in 1915 when the two countries were locked in battle is just one further bewildering and poignant fact to be recorded from those years.[7]

The flow of national reports was held together by regular international meetings.

> [T]he proper work of the Comité Central and of the Commission as a whole progressed remarkably well. The CC met every year, and from 1910 on, there were general meetings of the entire Commission. ... thematic reports were prepared, presented, and discussed in Brussels (1910), Milan (1911), Cambridge (1912), and Paris (1914), the last report having been prepared at a 1913 meeting of the CC in Heidelberg ... The climax of these activities was the Paris meeting. It was prepared in the most intense manner; it had the best participation and the most vivid discussions. There was even a satellite congress on philosophy of mathematics. It is legitimate, therefore, to speak of the work of the Commission before the First World War as a success story.[8]

---

[5] See [Howson 1984], pp. 75–76.

[6] From G. Schubring's account of the early years of ICMI in [Menghini et al. 2009], p. 115.

[7] See [Howson 1984], pp. 77–78.

[8] From G. Schubring's account in [Menghini et al. 2009], p. 119.

The 1911 meeting in Milan was the first plenary meeting of the commission. The subsequent congress in 1912 was held on the occasion of the Cambridge ICM.[9] By April 1914, the following countries were represented in ICMI: Australia, Austria, Belgium, Brazil, Bulgaria, Cape Colony (South Africa), Denmark, Egypt, France, Germany, Greece, Holland, Hungary, Italy, Japan, Mexico, Norway, Portugal, Rumania, Russia, Serbia, Spain, Sweden, Switzerland, the UK, and the USA.[10]

From One World War to the Other. We have seen in Section 3.1 above how Felix Klein, the president of the commission, compromised himself very early on in the Great War by being the only mathematician to sign the infamous German *Aufruf* "To the Civilized World." And in Section 3.2 we have seen examples from France of the intellectual warfare in the academic realm. It is therefore hardly surprising[11] that Henri Fehr published the following note about the commission already in the 1914 volume of *L'Enseignement mathématique*, in which he explicitly alludes to the academic mindset at the beginning of the war.

> The European war affects the international institutions severely. In the countries at war and their neutral neighbors all of the nation's capable men are mobilized. This makes it factually impossible to continue working as before, with the help of many collaborators. The works of peace like ours have to stand back. Besides, as they follow a common, freely chosen ideal, they require a goodwill for union that would be impossible to ask from scholars in a period as troubled as the one we are going through. Our work will inevitably be put on hold. We hope this will not be for long.
>
> Under these circumstances it is understood that a meeting of the *Commission internationale de l'enseignement mathématique* in 1915 is out of the question, and the *Comité central* finds itself obliged to adjourn its projects.
>
> H. Fehr, Secrétaire général de la Commission.[12]

By the end of the war, as we know, the scene for Science International had changed completely, and brought about the creation of the *International Research Council* in 1919 under inter-allied rule, based on the exclusion of colleagues from the central powers. Under this new regime, the (first) IMU was created and the Strasbourg ICM was held in September 1920—see Chapter 4 above. Recognizing this newly imposed politics of exclusion, Fehr announced the dissolution of ICMI in July 1920, even before the official founding of the first IMU.[13] As a result, the commission was already defunct when the first IMU was created, and no mention of earlier achievements of ICMI can be found in the Proceedings of the Strasbourg ICM.

---

[9] For more information on this meeting, see the ICMI History website [URL 24].

[10] See *L'Enseignement mathématique* 16 (1914), p. 166

[11] I thus beg to differ from Gert Schubring's account, which scolds Fehr for his "one-sided actions"—see [Menghini et al. 2009], pp. 120–124, as well as [Schubring 2008].

[12] See *L'Enseignement mathématique* 16 (1914), pp. 477–478; my translation from the French.

[13] See *L'Enseignement mathématique* 21 (1920), pp. 137–138.

Yet, several national subcommittees of ICMI kept working. Since the original ICMI was essentially based on the coordination of input prepared by the various national subcommittees, what was lacking in the early 1920s was primarily the international networking between them.

In fact the elder sister of the IMU was just hiding in the closet during those years, waiting to step out once the exclusionist wave had passed. As we know—see Section 4.4.3 above—this happened in 1928 at the Bologna ICM. Henri Fehr was there, of course; he had participated at every single ICM since the first one in 1897. In 1924 in Toronto, he had been elected one of the Vice-Presidents of the IMU, and in Bologna he acted as secretary of the unofficial IMU General Assembly that congratulated Pincherle on having ended the exclusionist politics, against the resistance of the IMU Secretary General.[14]

And at the same ICM in Bologna, on Wednesday, 5 September 1928, Henri Fehr presented before Section VI of the Congress a fairly detailed report on the activities of ICMI since its foundation in 1908.[15] The discussion following his presentation pleaded for a revival of the commission. As if this was already done, the report bestowed on Fehr the title of "Segretario generale della Commission Internationale de l'enseignement mathématique."[16] At the same ICM, Nilos Sakellariou (1882–1955) from Athens presented a *Projet pour la constitution d'une commission internationale pour l'enseignement des mathématiques*.[17]

Since the first IMU was liquidated at the 1932 ICM in Zürich, it was only during the four years 1928–1932 that the first IMU and ICMI coexisted. However, the *Commission internationale de l'enseignement mathématique* was never a commission of the first IMU. It worked as autonomously as in the first years of its existence. However, there were no yearly congresses devoted to the teaching of mathematics any more, as had been the case before World War I. There were only the corresponding sections at the ICMs in 1932 and 1936. At these ICMs, both in Zürich and in Oslo, the mandate of ICMI and its Central Committee were renewed; a renewal that had to be understood to be valid until the next Congress. This is why Henri Fehr could claim in 1952 that the commission was still in existence. Fehr had been the Secretary-General of ICMI (when it existed), ever since its creation in 1908. After Klein's presidency, which may be said to have officially ended with the dissolution

---

[14] See Proceedings ICM 1928, Vol. I, p. 83. The assembly was unofficial because it had not been convened by Gabriel Kœnigs, the Secretary General of the IMU. Cf. Section 4.4.3.2 above.

[15] See Proceedings ICM 1928, Vol. I, pp. 106–113. The impressive count of the publications of the various countries for ICMI can be seen at [URL 25].

[16] See Proceedings ICM 1928, Vol. I, p. 113.

[17] See Proceedings ICM 1928, Vol. III, pp. 157–158.

of ICMI in 1920, D.E. Smith stepped in as president in 1928, followed by Hadamard as of 1932.[18]

Starting Afresh in 1952. At the first General Assembly of the newly created IMU, which was held in Rome in March 1952, ICMI was reborn as a permanent subcommission of the IMU. Precise terms of reference for ICMI were adopted at the 1954 General Assembly in The Hague. We have described in Sections 7.2.3 and 8.1 the global constellation that stood at the cradle of the new IMU after World War II. Among other factors, we mentioned the new branches of applied mathematics which had emerged from war-related research and continued to mark the image of the sciences during the Cold War. That this general background also marked the timely reincarnation of ICMI can be felt, for instance, in a remark made by Marshall Stone, the first president of the new IMU, in his report before the 1952 General Assembly:

> The problem of determining the place of mathematics [in society] cannot be divorced from technical considerations concerning teaching methods. If we judge by the results, we must find it difficult to escape from the conclusion that our attempts to teach mathematics as part of a program of mass education have so far been, to put it bluntly, a colossal failure, traceable to our ignorance and complacency in respect to the art of teaching.[19]

Contrary to the work model of the old commission, national subcommittees—to the extent that they still existed at all—would no longer play a major role in the activities of the new ICMI. What shaped the new ICMI most decisively during its first two decades were the personalities and politics of its successive presidents—the only exception being the rather passive Albert Châtelet, who was president from 1952 to 1954, when he was about 70 years old. Heinrich Behnke first was ICMI Secretary-General under Châtelet and then served as president from 1955 to 1958. He was followed by Marshall Stone himself, who had been Vice-President under Behnke. From Stone, André Lichnerowicz (1915–1989) took over in 1963. Hans Freudenthal chaired ICMI from 1967 to 1970, and would continue to serve on ICMI's Executive Committee—as it was called since the commission's reincarnation in 1952—through 1974.

During their mandates, Behnke and Freudenthal had to sort out the relation between ICMI, the IMU, the ICMs, and the mathematical community at large, with a view to obtaining a reasonable autonomy for ICMI. In this endeavour Freudenthal would be more aggressive than Behnke. The latter's orientation was largely guided by his high esteem for Felix Klein, but he was well aware of the fact that he could not measure up to Klein's omnipresent role on the national and international scenes

[18] See [Lehto 1998], p. 316. For the composition of ICMI and its CC, and for the reports delivered in the 1930s, see also [URL 25], as well as [Howson 1984], pp. 79–80, and [Hollings & Siegmund-Schultze 2020], pp. 231–234.

[19] We take this quote from the first section of [Furinghetti & Giacardi 2022]. This article, which the authors have graciously allowed me to read when it was in the making, is my main source for the account given here of the first twenty years of the new ICMI.

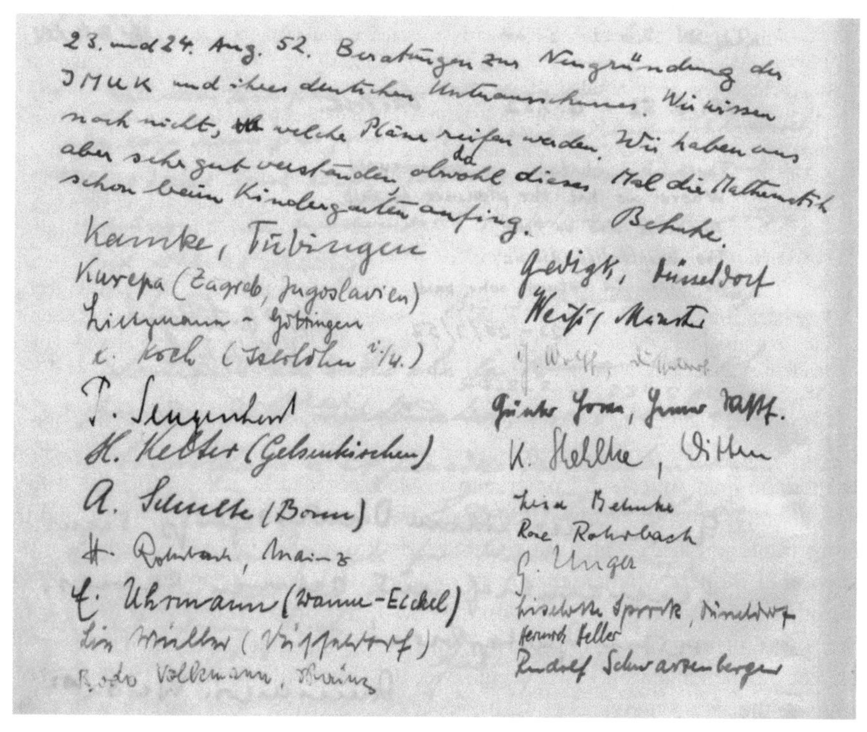
23. und 24. Aug. 52. Beratungen zur Neugründung der IMUK und ihres deutschen Unterausschusses. Wir wissen noch nicht, welche Pläne reifen werden. Wir haben uns aber sehr gut verstanden, obwohl da dieses Mal die Mathematik schon beim Kindergarten anfing. Behnke.

Fig. 9.1 Entry in the Oberwolfach Guestbook of a meeting about the renewal of (the German subcommittee of) ICMI, held in Oberwolfach, 23–24 August 1952. Behnke writes: "Debates about the refoundation of IMUK and of its German subcommittee. We do not know yet which plans will come to fruition. However, we did get along well with each other, although/because this time mathematics started as early as the *kindergarten*."

of the past. During his mandate he tried, with modest success, to render ICMI less dependent on the IMU, and he was the first to open up ICMI for educational debates of the time.[20]

> In the USA the University of Illinois Committee on School Mathematics (UICSM), headed by Max Beberman [1925–1971], was established in 1951. In the years that followed there were further initiatives in mathematics education and, eventually, the launch of the Soviet Sputnik in October 1957 persuaded the United States Congress to designate an unprecedented amount of dollars for science education. This event also led the OEEC—created to administer the funds allocated by the United States (Marshall Plan) to rebuild Western Europe after the Second World War—to deal with problems relating to the teaching of science and mathematics. . . .
>
> In 1958 the American mathematician Edward Begle was appointed director of the School Mathematics Study Group (SMSG), the largest and most influential of the so-called New Math curriculum projects in the USA. SMSG published and distributed extensive collections of books and films for teachers as well as a series of monographs for students, the New Mathematical Library. American educators feared that the Soviet Union was surpassing

[20] Cf. [Furinghetti et al. 2020].

> the United States in educational emphasis on science and mathematics, so in September 1959 a conference was held at Woods Hole on Cape Cod in the USA with the aim of improving science education in primary and secondary schools, bringing together scientists, mathematicians, educators, biologists, psychologists and other professionals. . . .
>
> In Europe, the Bourbaki group, which beginning in the 1930s had attempted to generalize, formalize, and unify all of pure mathematics, stimulated the emerging of the movement usually known as Modern Mathematics. Among the promoters were Dieudonné, Choquet, and Lichnerowicz, founding members of CIEAEM [*Commission Internationale pour l'Étude et l'Amélioration de l'Enseignement des Mathématiques*].
>
> On 28 May 1958, during the meeting of the Executive Committee of ICMI in Münster-Westfalen, Kay Piene, Member of ICMI, proposed, among the topics to be discussed by the Commission in the period 1959–1962, the study of which themes and applications of modern mathematics might find a place in the teaching programs of secondary schools.[21]

Stone and Lichnerowicz were open to projects beyond the traditional scope of ICMI or the IMU, establishing contacts with other institutions. Specifically, Stone chaired the Inter American Conference on Mathematical Education (Bogotà, 1961), which would be affiliated to ICMI as an effective regional group in 1965, and the momentous seminar at the Centre Culturel de Royaumont in 1959, where the introduction of Modern Mathematics in secondary instruction was amply discussed. On the other hand, when it came to adopting new Terms of Reference for ICMI in 1960, Stone was instrumental in mitigating Behnke's original draft, which had envisaged a greater independence for ICMI. In the end, ICMI was allowed as of 1960 to "cooperate, to the extent it considers desirable, with effective regional groups which may be formed spontaneously, within, or outside, its own structure", and it was stipulated that the "Commission may, with the approval of the Executive Committee of IMU, coopt, as members of ICMI, suitably chosen representatives of non-IMU countries, on an individual basis."[22]

These activities of ICMI reflect the general pattern of international associations that had emerged after World War II—see Sections 8.1.1 and 8.1.2 above.

> In the 1960s the action of ICMI broadened considerably: thanks to Stone and Lichnerowicz, collaborations both scientific and organizational were established with other associations such as OEEC (Organization for European Economic Cooperation) and UNESCO (United Nations Educational Scientific and Cultural Organization). These led to a greater internationalism and to the organization of numerous thematic congresses in various parts of the world.[23]

One should add here a reminder of analogous concerns with questions of science instruction, not just on the level of the IMU, but of ICSU, thus potentially pertaining to all the sciences. We have already alluded to this, and to the precocious role of ICMI in that respect, in Section 8.1.2 above.

---

[21] See [Furinghetti & Giacardi 2022], Section 5.2.

[22] The quotes are from points (f) and (g) of the 1960 Terms of Reference. See [Furinghetti & Giacardi 2022], Section 5.3.

[23] See [Furinghetti & Giacardi 2010], pp. 16. Again, these activities are discussed in detail in various sections of [Furinghetti & Giacardi 2022].

> There are few indications of major concern with science education in ICSU in its early years. The role of Unesco in stimulating ICSU's interest in education and training appears to be primordial. One of the first of Unesco's long series of publications on science education entitled *Suggestions for Science Teachers in Devastated Countries* appeared in 1948. Unesco's own interest in science education was particularly stimulated by a meeting of experts convened in 1950 to examine the place of science in general education. The fifth General Conference of Unesco in Florence in 1950 invited Member States "to develop teaching in the Natural Sciences and the dissemination among the adult public of knowledge of the methods, discoveries and applied uses of these sciences." This General Conference also drew attention to the importance of the interactions between science and society, to the application of science to the solution of urgent problems and to the need to study different methods for the popularization of science. One of the activities that developed in this field was the organization of travelling exhibitions, the first of which showed recent discoveries in physics and astronomy and was visited by almost half a million people in Latin America. . . .
>
> While these developments had been taking place in Unesco the interest in ICSU and the Scientific Unions had also been developing. Initially this was on the basis of individual science teaching commissions in the Unions, of which the . . . [ICMI] appears to have been the first and predated Unesco. . . .
>
> In 1961 ICSU created an Inter-Union Commission on Science Teaching (IUCST) which organized in Dakar in 1965 a congress on Science Teaching and Economic Progress. Its second congress, on the Integration of Science Teaching, began a series on integrated science education that continued in close association with Unesco for more than a decade. At the Dakar Congress the IUCST made cooperative agreements with the Sector for Science and Education of Unesco. The commission worked closely with Unesco during its existence and played an important role in the development of integrated science teaching.[24]

Freudenthal's presidency 1967–1970, which followed his presence as a member of the Executive Committee since 1963, marked a major turning point in the history of the commission, both from an organizational point of view and for the objectives of ICMI. It was on Freudenthal's initiative—about which the IMU was not informed ahead of time—that ICMI decided to hold congresses that were separate from the ICMs, and to create a new journal exclusively devoted to mathematics education.[25] The first issue of the new specialized international journal *Educational Studies in Mathematics* (ESM) appeared in May 1968. The traditional press of ICMI, *L'Enseignement mathématique*, continued to appear, publishing a wide range of articles, only few of which were devoted to questions of mathematical instruction. The first International Congress on Mathematical Education (ICME) was held in August 1969 in Lyon, France.

Both in the first two decades of ICMI's life as a commission of the IMU and afterwards, there were tensions between the two sisters. During Freudenthal's presidency they are particularly tangible in the correspondence that is preserved in the archives of the IMU and ICMI.[26] As usual, such disputes tend to be a mixture of personal and structural issues. For this brief overview, let us single out one structural source of potential discord between many working mathematicians and colleagues engaged

[24] See [Baker 1986], pp. 15–16.

[25] I am paraphrasing a passage from the conclusion of [Furinghetti & Giacardi 2022].

[26] See [Furinghetti & Giacardi 2022] for a few telling examples.

with ICMI: the increasing professionalization, since the late 1960s, of the domain that deals with questions of teaching.

Didactics as a Discipline. In the forthcoming paper [Furinghetti & Giacardi 2022], the authors illustrate the beginning of the new discipline by listing a few institutions created in the late 1960s and early 1970s in Europe: In 1967 the *Nordic Committee for the Modernization of School Mathematics* (Denmark, Finland, Norway, and Sweden) presented a new syllabus inspired by New Math. In 1968 the *Zentrum für Didaktik der Mathematik* was founded in Karlsruhe, Germany. In 1973 the *Institut für Didaktik der Mathematik* (IDM) was founded in Bielefeld, Germany. Starting from 1969 the first French IREMs (*Instituts de Recherche sur l'Enseignement des Mathématiques*) were established in Paris, Lyon, and Strasbourg. In 1971 Hans Freudenthal himself founded the *Institut Ontwikkeling Wiskunde Onderwijs* (IOWO), i.e., the Institute for the Development of Mathematics Instruction, at the University of Utrecht. Today it is simply called the *Freudenthal Institute*.

Focussing more on the content, the emergence of new issues in mathematics education has been described under the heading *ICMI Renaissance* in a separate chapter of the centennial proceedings, authored by Fulvia Furinghetti, Marta Menghini, Ferdinando Arzarello, and Livia Giacardi.[27] We quote a few extracts from this chapter in order to illustrate characteristic subjects discussed at the first ICMEs. The first reference, from Section 4.1 of that chapter, starts out from the *Commission Internationale pour l'Étude et l'Amélioration de l'Enseignement des Mathématiques* (CIEAEM), which has already been mentioned once above. This International Commission was founded in 1952 around a Europe-based collaboration between mathematicians, psychologists such as Jean Piaget (1896–1980), pedagogues like Caleb Gattegno (1911–1988), and secondary-school teachers. As an organization, it is independent of both ICMI and the IMU.

> [O]ne of the main features of the activity of CIEAEM ... was the use of concrete materials. This topic has a great emphasis in ICME-1 (1969) and ICME-2 (1972).
>
> In various contributions to ICME-1 we find mention of games, worksheets, films, manipulatives, even the 'modern' overhead projector, which allowed lessons to be prepared in advance, to perform movements and overlapping....
>
> We agree with Howson's opinion that educators look at the 1960s as the period of New Math, but that actually the main long-lasting idea that gathered strength in that period was the emergence of new styles of teaching and a more systemic transfer of teaching materials and ideas in the various countries.[28]

The following section of the chapter highlights the influence of psychologists.

> Piaget's theories are an important feature of the first two ICMEs. At ICME-2 the contribution of Piaget (not present) still outlines the analogy of Bourbaki's three mother structures with his structures of thinking; he identifies the causes of the failure of modern mathematics in the use of traditional teaching methods based on oral transmission....

[27] See [Menghini et al. 2009], pp. 131–147.

[28] See [Menghini et al. 2009], p. 141.

> The influence of Piaget's methods also permitted broadening the range of mathematical topics in primary school.[29]

Next, we read about the relation with the world of mathematics.

> Professional mathematicians have always been present in ICMI, first as founding members and supporters, and later as ICMI presidents. Moreover, for a long time ICMI inquiries were planned during the ICMs. In the first ICMEs the presence of mathematicians is rather high, while it decreases in the more recent congresses. Often the contributions of mathematicians consisted in examples of applications of mathematics to the real world and, as a consequence of this, in stressing the key role of mathematics and science in society. The presence of mathematicians offered support and encouragement for the various ICMI activities.
>
> ... At ICME-1 and ICME-2 New/Modern Math is still present, with contributions both for and against it. The most important of them, by René Thom (1973), stresses the contradiction of a teaching that is heuristic in principle, but is based on abstract mathematics. Thom claims that Piaget is much too confident in the potentialities of mathematical formalism: Modern Mathematics has not produced new theorems and, as far as education is concerned, does not produce new knowledge. The actual problem is not rigour, but rather the development of meaning. Modern Mathematics has eliminated Euclidean geometry in favour of algebra, but it is precisely Euclidean geometry that is the link between natural language and abstraction. Because of Thom's contribution, many authors date the end of Modern Mathematics to ICME-2. In successive ICMEs we still find echoes of the debate, in particular the discussions about the movement Back to Basics at ICME-4 in 1980.[30]

Finally, the description of the new trends of research in mathematics education comes back to the influence of CIEAEM on the Congresses held by ICMI.

> One of the key ideas in Gattegno's way of working, developed in the CIEAEM meetings of the 1950s, was to put the researcher in direct contact with the classroom. This idea was given new life at ICME-1 by the Mathematics Workshop, which included a class of children at work. ...
>
> A call for more in-depth research in mathematics education was also present at ICME-2 in the plenary of the mathematician Hassler Whitney, who had in mind the failure of New Math. In ICME-2 a WG was devoted to Research in the teaching of mathematics. Successive ICMEs would show the strengthening of research in mathematics education as a scientific discipline with new results, new theoretical frameworks, new hypotheses, and new methods of gathering and recording data.[31]

We thus see that, starting with Freudenthal's presidency, most of the colleagues working for ICMI projects tend to be "professional researchers in the teaching and learning of mathematics, i.e., didacticians." These include "significant examples of research mathematicians becoming professionally engaged with mathematics education even at the scholarly level."[32]

[29] See [Menghini et al. 2009], p. 142.

[30] See [Menghini et al. 2009], p. 143.

[31] See [Menghini et al. 2009], pp. 143–144.

[32] See [Hodgson 2015], p. 49.

Sisterly Skirmishes and Successes. Here is how one of the actors and keen observers, Bernard Hodgson, summed up some future consequences of Freudenthal's reshaping of ICMI.

> [T]he presidency of Freudenthal resulted in what might be rightly seen as "years of abundance" for ICMI, in the sense that the scope and impact of its actions expanded considerably. Not only were the newly established [journal] ESM and [the congresses] ICMEs highly successful, but also new elements were gradually added to the mission of ICMI. To name a few, ICMI introduced in the mid-1970s a notion of Affiliated Study Groups, serving specific segments of a community becoming more and more diverse. There was also a regular collaboration between ICMI and UNESCO, contributing in particular to outreach actions of ICMI towards developing countries. And later, in the mid-1980s, the very successful program of ICMI Studies was initiated. Still this deep evolution of ICMI, notably through the influence of Freudenthal himself, did not happen without some tensions with IMU, in particular as it was often the case that IMU faced decisions that were *faits accomplis*, taken without any consultation between the Executive Committees of ICMI and IMU—such had been the case for instance with the launching of the first ICME congress.
>
> Another moment of tension between IMU and ICMI happened in connection with the program of the section on the Teaching and Popularisation of mathematics at the 1998 International Congress of Mathematicians. As a consequence, the first Executive Committee of ICMI on which I served, under the presidency of Hyman Bass, had to deal with an episode of misunderstanding, and even mistrust, between the communities of mathematicians and didacticians as represented by IMU and ICMI.[33]

The President of the commission, Michèle Artigue (b. 1946), described the constellation of the two sisters in her remarkable closing address to the centenary celebration of ICMI.

> ICMI was still a structure at the interface, an interface between mathematicians and an increasing number of communities that tended to be institutionalized inside the mathematics education world. The creation of the first ICMI Affiliated Study Groups in the seventies and eighties evidences this phenomenon: PME [The International Group for the Psychology of Mathematics Education (1976)], HPM [The International Study Group on the Relations between the History and Pedagogy of Mathematics (1976)], IOWME [The International Organization of Women and Mathematics Education (1987)]. ICMI's Executive Committees (EC), whose election were controlled by IMU and its General Assembly, are insightful from this point of view: the President was always a first rank mathematician and mathematicians with an interest in education were well represented in the EC, but mathematics educators with a diversity of fields of expertise were also well represented and had officer responsibilities, being Vice-President or Secretary-General. Under the Presidency [1991–1994] of Miguel de Guzmán [1936–2004], the balance between the two communities inside ICMI executive committee progressively evolved. New tensions also arose all the more as at that time the supposed influence of mathematics educators was considered by some mathematicians as an important, if not the major, source of the observed difficulties in mathematics education, leading to such extremes as the so-called Math War in the USA. In 1998, when Hyman Bass was elected as President, Bernard Hodgson as Secretary-General, and when I entered the EC together with Nestor Aguilera as Vice-Presidents, the tension was at its maximum. At the 1998 International Congress of Mathematicians in Berlin, the project proposed by ICMI for the section on Teaching and Popularization of mathematics had been partially rejected,

[33] See [Hodgson 2015], pp. 49–50.

> and the Math War in some sense had entered the section. Voices asking ICMI to take its independence from a mother institution that expressed such mistrust were becoming stronger and stronger.[34]

The crisis which had erupted between the two sisters on the occasion of the Berlin ICM was subsequently resolved thanks to a proposal, which made the election of the Executive Committee of ICMI a procedure independent of the IMU Executive Committee.[35] This proposal was adopted at the IMU General Assembly at Santiago de Compostela in 2006.

> Retrospectively this crisis was beneficial. It obliged ICMI EC to deeply reflect about the nature of ICMI and what we wanted ICMI to be. This led us to reaffirm the strength of the epistemological links between mathematics and mathematics education ... At the same time, we were convinced that making these links productive needed combined efforts from IMU and ICMI; the relationships could not stay as they were. Since 1998, the situation has evolved very positively thanks to the joint efforts of the successive EC and especially of their presidents: Jacob Palis and John Ball for IMU, Hyman Bass for ICMI. Hyman Bass had the credibility of a[n] outstanding research mathematician, being known as the father of K-theory, but he was also someone having much more than the 'peripheric interest' in mathematics education shown by his predecessors, to use the same expression as he did in his opening lecture at this Symposium, Hyman Bass claimed that he is not a mathematics educator and he was certainly right saying this, but he knows research in mathematics education from the inside not just as an empathic observer. This makes a great difference. For me, there is no doubt that without him for pushing and guiding the evolution, ICMI would not be the structure it is today, and I would not be serving it as ICMI President.
>
> ICMI is thus entering its second century of existence still at the interface between mathematics and mathematics education but certainly stronger for playing such a role and for coping with its ambitions. For the first time next July [2008], ICMI General Assembly will elect the new ICMI EC on the basis of a list of candidates established in full consensus by a nominating committee representing the two communities. ICMI and IMU are officially collaborating on specific projects ... The two institutions are more and more coordinating their actions in the developing world thanks to the Developing Country Strategic Group (DCSG) and Commission for Development and Exchanges (CDE) structures where ICMI is represented. ... less than one third of the existing countries in the world belong to IMU, and only some fifteen more are members of ICMI without belonging to IMU. One impediment to wider membership to IMU is the requirement of independent scientific activity, which is interpreted as being some kind of sustained presence in research mathematics, but all countries are engaged in mathematics education and are thus concerned with ICMI activities. An increased collaboration between IMU and ICMI was thus considered desirable for extending their outreach and making these institutions better serve the cause of both mathematics and mathematics education worldwide.[36]

CANP. Michèle Artigue's predicament remains valid today. An impressive manifestation of these goals, which goes well beyond the continuing regular activities of ICMI indicated above, and which reflects the relentless engagement of all contributors, is the Capacity & Networking Project (CANP). It is a joint enterprise, realized over the last ten years, of the IMU and ICMI, in conjunction with UNESCO

---

[34] See [Menghini et al. 2009], p. 189.

[35] See [URL 26].

[36] See [Menghini et al. 2009], p. 189.

and the International Congress of Industrial and Applied Mathematics (ICIAM).[37] The originality of the program is to select groups of countries in given developing regions that share cultural traditions, in particular a common language, and to run long and intense workshops on the teaching of mathematics in these countries. These workshops are mounted in collaboration with regional organizations that are affiliated with ICMI.

The first program of CANP was held in Mali in September 2011. Its aim was to create a network in Sub-Saharan African countries, to face the challenges of mathematics education. Support was provided by UNESCO, CIMPA, and ICIAM. Four more regional CANP programs followed suit: CANP 2 Central America and the Caribbean was held in Costa Rica in August 2012; CANP 3 South East Asia took place in Cambodia in 2013; CANP 4 East Africa was held in Tanzania in September 2014, and CANP 5 for the Andean Region and Paraguay was held in Lima, Peru, in February 2016. Since then, the enduring challenge has been to enhance these five CANP programs and help them develop and maintain sustainable networks and activities.

This remarkable program CANP was also the activity that Jean-Luc Dorier, the Secretary-General of ICMI, chose to focus on in his presentation of ICMI at the 2021 IMU Celebration in Strasbourg.[38]

[37] See the website [URL 27] (and its appendices).

[38] See [URL 28].

Chapter 10
Framing Mathematical Excellence

Today's International Mathematical Union (IMU) derives its greatest visibility among mathematicians world wide from the International Congresses. Its very foundation was an integral part of the mounting of the first postwar ICM at Harvard in 1950. It is via the experience of the quadrennial ICMs and the published traces they leave behind that an image of mathematics continues to be framed and projected for the mathematical community at large, and for the whole world to see. In this final chapter we present a data-based study of how the most exquisite layer of this image has evolved over the past seventy years.

The hard core of this chapter—see Sections 10.3–10.5 below—presents and interprets a data-analysis realized for the occasion by Birgit Petri, Darmstadt. I am very much indebted to her for her relentless work on this project, and express my cordial gratitude.

Before focussing on this, though, let us sketch the overall structure of the IMU, and the activities of its associated bodies (apart from ICMI, which we have already considered in the preceding chapter).

10.1 The Infrastructure of the IMU

Among all the scientific unions assembled today under the umbrella of the International Science Council (ISC),[1] the IMU may well be the one with the most slender organigram. One could be tempted to explain this by the very nature of mathematics. In fact, even though stunning discoveries do exist in the world of mathematics—recall for example the *exotic spheres* uncovered by John Milnor (b. 1931) and Egbert Brieskorn (1936–2013), which stirred quite a bit of excitement in the late 1950s and 1960s—this is a far cry from naming and monitoring near-earth asteroids that might collide with our blue planet, which is one of the responsibilities that the International Astronomical Union (IAU) is involved in through its *Minor Planet Center*:

[1] See the list of Category 1 (Full Members) of ISC at [URL 29].

N. Schappacher, *Framing Global Mathematics*,
https://doi.org/10.1007/978-3-030-95683-7_10

> The Minor Planet Center (MPC) is the single worldwide location for receipt and distribution of positional measurements of minor planets, comets and outer irregular natural satellites of the major planets. The MPC is responsible for the identification, designation and orbit computation for all of these objects. This involves maintaining the master files of observations and orbits, keeping track of the discoverer of each object, and announcing discoveries to the rest of the world via electronic circulars and an extensive website. The MPC operates at the Smithsonian Astrophysical Observatory, under the auspices of Division F of the International Astronomical Union (IAU).[2]

The elusive nature of mathematical objects is not sufficient, though, to explain the slim infrastructure of the IMU. The end of the above quote contains an indication that the whole internal organization of the IAU is much more complex than that of the IMU; the Minor Planet Center belongs to *Division F* of the IAU, which is just one among nine different Divisions, each of which in turn counts several Commissions and Working Groups:[3]

- Division A Fundamental Astronomy
- Division B Facilities, Technologies and Data Science
- Division C Education, Outreach and Heritage
- Division D High Energy Phenomena and Fundamental Physics
- Division E Sun and Heliosphere
- Division F Planetary Systems and Astrobiology
- Division G Stars and Stellar Physics
- Division H Interstellar Matter and Local Universe
- Division I Galaxies and Cosmology.

This shows that the IAU has chosen—in fact, right from its beginnings, and partly building on pre-World-War-I specific networks of international collaboration—to mirror major dividing lines of the discipline in its administrative structure.[4] Several other scientific unions do the same. To name but one more example, the International Union of Geological Sciences (IUGS) counts among its constituent scientific bodies the International Commission on Stratigraphy (ICS), whose primary objective it is

> to define precisely global units (systems, series and stages) of the International Chronostratigraphic Chart that, in turn, are the basis for the units (periods, epochs and age) of the International Geological Time Scale; thus setting global standards for the fundamental scale for expressing the history of the Earth. The work of the Commission is divided between seventeen subcommissions, each responsible for a specific period of geological time. Their work is overseen and co-ordinated by an executive of five officers.[5]

Unlike these scientific unions, the IMU has never attempted to express the diversity of the mathematical sciences in its administrative structure. As far as I know, the division between pure and applied mathematics, which caused so many tense situa-

---

[2] See [URL 30].

[3] See [URL 31].

[4] Besides, as briefly mentioned in 6.2.2 above, the IAU also distinguishes itself from the IMU and most other international scientific unions by having many individual members.

[5] Quoted from the heading of [URL 32]. Note in passing that the IUGS was founded only in 1961; it is a member of the ISC alongside the International Union of Geodesy and Geophysics (IUGG), whose foundation in 1919 we have mentioned in Section 4.1.1.

tions in numerous institutions and countries during the second half of the twentieth century, seems not to have been considered a reason for an internal administrative divide of the IMU.

Meanwhile ICIAM, the *International Council for Industrial and Applied Mathematics*, came into being in 1986 in the form of a standing committee for the organization of the quadrennial *International Conferences on Industrial and Applied Mathematics*, through an understanding of the principal societies for applied mathematics: GAMM, IMA, SIAM and SMAI. Thirteen years later this committee grew into a society of societies with an increasing number of members. In contrast to the IMU, however, ICIAM does not belong to the International Science Council (ISC).

There are only three Commissions subordinated to the IMU: the *International Commission for Mathematical Instruction* (ICMI), which we have discussed in Chapter 9, the *Commission for Developing Countries* (CDC), and the *International Commission on the History of Mathematics* (ICHM). Furthermore, apart from the IMU *Executive Committee* and other purely administrative committees, three IMU Committees (i.e., structures of a possibly less perennial nature than the Commissions) are currently active: the *Committee for Electronic Information* (CEIC), the *Committee for Women in Mathematics* (CWM), and the recently instituted *ad hoc Committee on Diversity* (CoD), whose first report is expected for the 2022 ICM.[6]

Before returning in Section 10.2 to the central focus of the IMU: the ICMs, we now briefly present the substructures that have not been discussed yet.

10.1.1 The Committee for Electronic Information and Communication (CEIC)

Recall that, throughout their history, the general assemblies of the ICMs and of the IMU had repeatedly tried to add genuine issues to their agendas, particularly concerning questions relating to the reviewing and bibliography of the rapidly exploding number of publications.[7] However, not only were most of the ICMs before 1950 organized independently of the IMU, but also throughout the twentieth century, neither the first nor the second IMU played an important part in advancing those classical bibliographic projects.

There is only one exception to this general verdict, which is rightly stressed in [Lehto 1998], p. 95: the initiative—which did arise in the context of various bibliographical projects—of a global Directory or index of mathematicians (WDM). The project was decided at the first General Assembly of the IMU at Rome in 1952 and was a success for almost half a century.[8]

---

[6] See [URL 33].

[7] Cf. Sections 1.4.1.2, 4.1.1, and 4.3.2 above.

[8] Cf. [Lehto 1998], Section 6.3 for an account of the first forty years of WDM.

In this sense the creation of CEIC in 1998 was a fresh initiative. It came about in reaction to the new, electronic world of publishing, and communication in general. The Berlin ICM in 1998 congratulated itself several times on being the first to be organized and realized essentially via email.[9] The General Assembly of the IMU at Dresden that preceded the Berlin ICM thus adopted an "enabling resolution" to form a Committee on Electronic Information and Communication, which begins like this:

> 1. In the last decade, the internet has been transforming our communication and commerce. In the world of science, the internet is radically changing the modes of information transfer at all levels. Communication on hand-written and printed paper, distribution via postal mail and libraries is a system which has been stable for many centuries. We cannot foresee clearly the new system which is evolving except that it will involve electronic media and it will radically alter the economics of communication. This transformation will certainly be global and will affect mathematical research on all continents.
>
> 2. We strongly believe that the IMU can play several important roles during this transition. Among these are:
>
> i) it can provide a forum where all parties, i. e., all countries and all interest groups (individual researchers, professional societies, publishers, and libraries) can discuss the issues and it can publish proceedings to increase general understanding of all the issues involved,
>
> ii) it can recommend and promote international standards on electronic communication among mathematicians, when needed,
>
> iii) it can act as a liaison between regional, national and local groups, coordinating their initiatives and discussions.[10]

The CEIC webpage echoes this mission:

> The Internet, and the World Wide Web (WWW), have transformed mathematical communication in at least as great a way as the introduction of journals. This transformation affects all disciplines, and many of the resulting commercial pressures are beyond the control of mathematicians. Nevertheless mathematics, by its intrinsic nature and world-wide scope, has to develop a particular approach to this new situation. Changes have occurred very rapidly, and some of the habits of mathematicians—such as citation conventions, ways of building reputation, and for many mathematicians, very significant matters like promotion and working conditions—are still evolving in response to continuing changes. The IMU's Executive Committee therefore formed the Committee on Electronic Information and Communication (CEIC) in 1998 to watch these developments, to advise the EC, and through it the IMU and mathematicians generally, about these trends, and to find the best ways of evolving practice to adapt to these changes.[11]

In 2006, a specific idea was articulated:

> With the ultimate goal of creating an enduring network of digital mathematical literature, the General Assembly of the IMU endorses the new version of the "Best practices" document of its Committee on Electronic Information and Communication (CEIC), posted June 2005 ..., as well as the March 2005 draft of "Digital Mathematical Library: a vision for the Future". The digital mathematical library is a very important project that we need to do as much as we can to further.[12]

[9] See Proceedings ICM 1998, Vol. 1, pp. 27, 31, and 53.

[10] See Proceedings ICM 1998, Vol. 1, p. 54.

[11] Slightly amended clipping from [URL 34].

[12] See Proceedings ICM 2006, Vol. 1, p. 47.

This Global Digital Mathematical Library was discussed at the 2014 and 2018 ICMs. Meanwhile the whole process of scientific reviewing and publishing changed more quickly than many contemporaries had expected; but some were ready to react strongly.

> In 2012 Sir Tim[othy] Gowers [b. 1963], professor at Cambridge University, and thirty-three mathematicians from all over the world launched the movement "The Cost of Knowledge" and called to boycott Elsevier. They denounced Elsevier's lobbying for the *Research Works Act*, a bill proposed to the American Congress aimed at prohibiting *open access* mandates for federally funded research and thus reversing the policy of the National Institute of Health (NIH), which requires taxpayer-funded research to be freely accessible online. The mathematicians of *"The Cost of Knowledge"* considered it was also their duty to design alternative publishing models to recover control of the peer-reviewed journals they create and use. In June 2012, they proposed the *diamond open access* model (a terminology inspired from the *Diamond Sutra*, a treasure of the British Library that was printed in 868 in China). This model assumes that researchers should not pay to publish their articles, and should own the journals they create and peer review.[13]

I recommend Marie Farge's concise text—from which the preceding quote is taken—as a useful orientation in a debate which is far from settled. As for Diamond Access, one has to add the more recent information that in 2017, Elsevier bought Digital Commons-Bepress—which had originally been founded by researchers from Berkeley—and thus in a way also the label of Diamond Access.

The fact that Elsevier was explicitly targeted by the movement not only met with opposition from some colleagues, but potentially put the IMU into a difficult situation insofar as the official journal *Historia Mathematica* of the *International Commission on the History of Mathematics* ICHM (see 10.1.4 below) is published by Elsevier.

Marie Farge also criticizes the questionable spread of bibliometric indices. She mentions the IMU via Ingrid Daubechies's blog of 2012:

> When alternative open access models will have proven to be effective (i.e., for the quality of articles they publish, the efficiency of their dissemination and financial viability), editorial boards might be able to emancipate existing journals. Indeed it might be necessary for a community of researchers to take back control of the best, and often the oldest, journals they use to publish their results. Emancipating a journal means that its intellectual property is transferred from the publisher to the editorial board, the publisher being then paid as service provider and no more the owner of the journal's title, as proposed in 2012 by IMU (the International Mathematical Union).[14]

Rather than going into details of the ongoing debate among mathematicians, we invite the reader to consult the presentation of the *International Mathematical Knowledge Trust* (IMKT)[15], which is based on the corresponding panel at the 2018 ICM in Rio de Janeiro.

---

[13] See [Farge 2017], p. 3.

[14] See [Farge 2017], p. 5. See the blog entries at [URL 35], in particular Ingrid Daubechies's opening of this exchange.

[15] See [Ion et al. 2018].

To conclude this first mini-portrait of an IMU commission, it may not be superfluous to point out that commissions and their subject matters do not exist in separate bubbles. As for CEIC, both in view of the more than centennial history of Mathematics International and because of the immediate importance of publication for the professional life of mathematics, the IMU cannot but occupy itself and feel responsible for the best possible worldwide organization of the information stream of mathematical research. On the other hand the IMU is of course not at all the only organization which naturally has to attend to these issues. Other organizations concerned include for instance the International Science Council (ISC), and there are plenty of national or even regional, local or institutional agencies that all share in the responsibility of the mathematical publication system. That this intrinsic constellation is clearly reflected in the structure and in the concrete activities of the Committee for Electronic Information (CEIC), transpires, for instance, from the IMKT panel in 2018 that we mentioned. The same panel also shows the natural connection between CEIC and the IMU Commission for Developing Countries (CDC) that will be presented in Section 10.1.3 below.

Another natural connection which cannot currently be realized within the IMU is with deontological or ethical questions in the domain of publications, all the way from increasing problems—also in mathematics!—with plagiarism, to the slew of predatory journals and their business model of 'open access', which the individual researcher has to pay for.

10.1.2 Women in Mathematics

Emmy Noether died in 1935, shortly before her 53rd birthday. The following year, the first Fields Medals were awarded at the Oslo ICM. Even though the forty year age limit for the Fields Medal was only fixed in the 1950s,[16] speculating whether Emmy Noether could have been considered for a Fields Medal in 1936 had she lived longer is idle in the absence of documentary evidence; in fact, very little seems to be known about the work of the very first Fields Medal Committee.[17] Emmy Noether was awarded the prestigious Ackermann-Teubner Memorial Prize in 1932, jointly with Emil Artin, for their work on modern algebra; but this was a German prize, not the award of an international organization.[18]

[16] See Michael Barany's analysis of this consequence of the selection process for the 1950 Medals in [Barany 2018].

[17] See [Hollings & Siegmund-Schultze 2020], p. 225–228.

[18] See [Rowe 2021], pp. 184–185.

As we all know, it was not until the Seoul ICM in 2014 that the mathematician Maryam Mirzakhani became the first woman to be awarded a Fields Medal. In 2014 the *glass ceiling*[19] of the mathematical profession was thus, for once, pushed up all the way to the cupola framed by the IMU. In March 2015, the creation of the Committee for Women in Mathematics (CWM) was approved by the IMU Executive Committee. Its first meeting was held in Italy in September 2015, thirteen months after the Seoul Congress. The CWM thus appears as a very recent body that tries to guide and influence the IMU and Mathematics International out of a long historical burden. It does, however, build on some forty years of initiatives to improve the representation of women mathematicians, at the ICMs and elsewhere.[20]

10.1.2.1 Poor statistics

There is a fallacious but instructive argument that was once developed by Emmy Noether's most successful student Bartel L. Van der Waerden, about the possible role of women in mathematics. We present it here because, although the author tried to mobilize a certain amount of sophistication, there is just one thing he blatantly failed to see and take into account[21]: the *glass ceiling* which tended (and still tends) to block women mathematicians from distinguished professional careers. The metaphor of the invisible glass ceiling thus proves its worth one more time.

In 1967, he set out to "prove" (!) in a letter to Ms. Auguste Dick (1910–1993)[22] that only a decisive biological factor can explain why there are so few women among the famous mathematicians and theoretical physicists. For his argument he only focussed on these two domains, which according to Van der Waerden had the advantage of excluding factors like access to laboratory facilities that could be more socially selective than the purely theoretical paperwork of mathematicians and theoretical physicists. Van der Waerden draped his 'proof' of the inferior mathematical talent of women in the form of a statistical test of the null hypothesis of equally distributed talent for mathematics, even though technical details like error margins and so forth are not given.[23]

---

[19] We adapt this general concept from gender studies to the peculiar international constellation of mathematics. More generally, about the *Techo de cristal* in the world of mathematics, cf. the talks given (in Spanish) at the 2016 meeting *Women in Mathematics in Latin America: Barriers, Advancements and New Perspectives*; videos made available by the *Banff International Research Station* at [URL 36].

[20] For a condensed overview of such earlier initiatives, see [Mihaljević & Roy 2019], p. 118.

[21] We have mentioned a somewhat analogous criticism of politically charged statistics, which at the time was deconstructed by Messedaglia, in Section 1.1.5.1 above.

[22] Auguste Dick was the first biographer of Emmy Noether, see [Dick 1970].

[23] Van der Waerden is most famous for his textbook *Moderne Algebra*, and well-known among experts for his book on group theory for quantum mechanics, as well as for his contributions to algebraic geometry. However, he also developed early on a keen interest in applied mathematical statistics. This is reflected, besides his correspondence, in a number of articles as well as the textbook [Van der Waerden 1957]. In the mid-fifties he even organized a meeting on statistics at Oberwolfach.

erh. 11.3.67

Mathematisches Institut
der
Universität Zürich

CH 8006 Zürich, Rämistrasse 71
den 8.März 1967

Frau Dr.Auguste Dick
Marxergasse 18
1030 Wien

Sehr geehrte Frau Dr. Dick,

Besten Dank für die Zusendung Ihres sehr interessanten Aufsatzes "Die Mathematik und die Mädchen".

Die statistische Untersuchung, von der ich sprach, habe ich inzwischen durchgeführt. Ich habe mich dabei auf theoretische Physiker und Mathematiker beschränkt, die in den Jahren 1900 - 1950 studiert und in den Jahren 1910 - 1960 ihre grössten Leistungen vollbracht haben. Ich schätze, dass in den Jahren 1900 - 1950 mindestens 20 % der Studenten an europäischen und amerikanischen Universitäten Mädchen waren. Die"Nullhypothese", die wir prüfen wollen, ist die Hypothese, dass Knaben und Mädchen von Natur aus gleich begabt sind. Aus zwei genau gleichen Reservoirs von Knaben und Mädchen im Alter von 18 bis 20 Jahren werden also durch einen sozialen Selektionsprozess diejenigen ausgewählt, die nachher an den Universitäten studieren. Bei der Selektion spielt die Begabung eine grosse Rolle, aber auch soziale Gesichtspunkte und Vorurteile. Je stärker die Begabung, um so weniger spielen die sozialen Gesichtspunkte eine Rolle. Bei den Mädchen ist die Selektion strenger als bei den Knaben, also ist zu erwarten, dass die schliesslich zum Studium zugelassenen Mädchen im Durchschnitt etwas höher begabt sein werden als die Knaben (s.Figur).

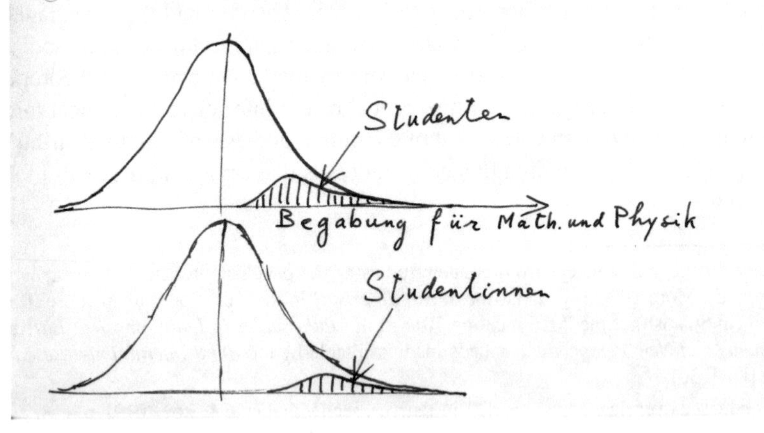

-2

Fig. 10.1 The first page of Van der Waerden's letter to Auguste Dick. Distribution of 'talent for mathematics and physics' among all students, resp. female students. Credit: [Arch. ÖAW].

I only consider theoretical physicists and mathematicians who were students between 1900 and 1950 and who realized their greatest achievements in the years 1910–1960. I estimate that in the period 1900–1950 at least 20% of the students [of mathematics or physics] at European and American universities were women. The 'null hypothesis' we want to test is

> that boys and girls are by nature equally gifted [in mathematics]. Starting from two equally big samples of boys and girls aged between 18 and 20, a process of social selection chooses those who will then study at a university. In this selection talent is a very important factor, but so are social aspects and prejudices. The stronger the talent the less effective will the social factors be. For girls this social selection is stricter than for boys; therefore it is to be expected that the girls who are finally admitted as students will on average be somewhat more talented than the boys.[24]
>
> Selecting now among the mathematicians and theoretical physicists of that period those with the most extraordinary achievements, our null hypothesis makes us expect more than 20% of them to be women.

Van der Waerden then goes on to list 35 names of excellent physicists sampled from recent source editions about the physics of the period in question, and finds not a single woman among them.

> Thus, among 35 leading physicists, there is no woman. For mathematics, there are no such editions of sources from which one may choose names. But if one would ask mathematicians from various fields to compile lists of leading mathematicians, I am sure that among the 25 or 30 best ones there would be only one woman: Emmy Noether. ... In the end we obtain a list of
>
> $$35 + 25 = 60$$
>
> top mathematicians and theoretical physicists which contains only one woman.
>
> But according to our null hypothesis one would have to expect more than 20%, i.e., more than 12 women. Such a massive deviation cannot arise by chance. The null hypothesis thus has to be rejected.

Never mind the sketchy presentation of the argument, and the casual estimates of certain numbers or percentages; after all, he is writing a letter, not a research paper. The principal reason why Van der Waerden's argument lamentably fails to establish any biological factor whatsoever is that he never deigns to wonder about what it takes for a good student to manage a successful academic career. This fallacy he shared with very many people at the time, as I vividly remember from personal discussions. It took a lot of initiatives to start to curb this widespread attitude.

Without trying to go into details about the social mechanisms that create the *glass ceiling*, we do have to indicate how strongly the image of excellent mathematics shaped by the IMU was antagonistic to the role of women mathematicians. An overall analysis of the presence of women mathematicians at the ICMs has been attempted in [Mihaljević & Roy 2019]. The authors also mention a few factors that have influenced the career possibilities of women since World War II.

> It took 60 years to reach a share of women among ICM speakers comparable to that in 1932. Among the manifold reasons for this situation are undeniably the impact of some historical and political developments. The aftermath of World War II was characterized by a rollback in society as a whole. The 1950s experienced a return to conservative gender roles, in which

[24] This and the following quotes are from Van der Waerden's letter to A. Dick of 8 March 1967, [Arch. ÖAW], Nachlass Auguste Dick. I thank R. Siegmund-Schultze for having shared this original document with me; I had only been aware of the carbon copy, without the drawing, in [Arch. ETH], Hs 632:1854 (Van der Waerden papers).

women were expected to take care of the domestic sphere, leaving the work places to the men who were coming back from the battlefields. These conceptions had impact on university education as well. During the conservative post-war era in Germany[25], for instance, the share of female students decreased significantly, and there was general agreement that men should take precedence in accessing the limited study places. However, some countries managed to overcome some of these barriers in women's university education and research faster than others. Partially, these general trends are also reflected in country-based differences regarding the presence of women speakers at postwar ICMs: in the 11 congresses between 1950 and 1990, of the 24 talks given by women, almost all delivered by speakers from the United States, France, United Kingdom, or Russia but none by speakers from Italy or Germany. By contrast, in the ten congresses before World War II of a comparable total of 27 talks by women, three of those speakers were from Germany and four from Italy.[26]

A more refined study of countries or world regions would obviously be very interesting. Overall [Mihaljević & Roy 2019] count 4,120 invited ICM contributions—all the way from 1897 to 2018—among which they determined (partly via automatic treatments, partly by hand) 202 that were presented or authored by women. This amounts to not quite 5% of the total. As indicated in the above quote, the variation over time is considerable, with no coherent trend over long periods.

In our approach, the database that will be explored in various directions in the present chapter was built to reflect the cupola of mathematical excellence framed by the IMU at the ICMs since 1950. It does not contain all the 4,120 speakers counted in [Mihaljević & Roy 2019] but is restricted to plenary speakers, prize winners as well as those who gave a laudatory address for a prize winner. On the other hand, next to these speakers and prize winners, we equally take into account those who served on a program committee or a prize committee. Among these 540 persons, we find a total of 31 women: not even 6%. And also for our criteria the distribution over time varies considerably; one third of these women first entered our database because of a function they held at the 2018 ICM in Rio.[27]

[25] The authors clearly refer to West Germany here. A careful study of the career options of women mathematicians in countries of the Eastern Block during the Cold War would be interesting.

[26] See [Mihaljević & Roy 2019], pp. 117–118.

[27] Here is the complete list, ordered by the year of the first ICM [in parenthesis] where they acted in a function which brought them into our database. Within each ICM, the order is determined by the function: Fields Medalists, their laudatory speakers, and members of the Fields Medal Committee go first; before plenary speakers and members of the Program Committee; followed by people related to the Nevanlinna Prize; and the Gauss Prize—see also 10.3 below. Mary Lucy Cartwright [1958; 1 person]; Joan S. Birman, Karen Uhlenbeck [1990; 2]; Ingrid Daubechies, Marina Ratner [1994; 2]; Dusa McDuff [1998; 1]; Frances Kirwan, Sun-Yung Alice Chang, Shafi Goldwasser, Michèle Vergne [2002; 4]; Claire Voisin, Margaret Wright [2006; 2]; Irit Dinur, Raman Parimala, Kim Plofker , Eva Tardos [2010; 4]; Maryam Mirzakhani, Vera V. Serganova, Hélène Esnault, Barbara Keyfitz [2014; 4]; Alice Guionnet, Hee Oh, Lai-Sang Young, Sylvia Serfaty, Nalini Anantharaman, Catherine Goldstein, Ulrike Tillmann, Laure Saint-Raymond, Maria Esteban, Motoko Kotani, and Bin Yu [2018; 11].

Just as in the work by Mihaljević and Roy, our data equally reflect the extreme underrepresentation of women after World War II.[28] For instance, at the four ICMs between 1966 and 1978, not a single woman met the criteria for being entered into our database. And the ICMs between 1966 and 2002 brought a total of 347 persons into the base, only 9 of whom were women, i.e., hardly 2.6%.

After Emmy Noether's plenary lecture at Zürich in 1932, the world had to wait 58 years, until 1990 in Kyôto, to see another woman give a plenary talk at an ICM: Karen Uhlenbeck (b. 1942). At that same ICM, Joan S. Birman (b. 1927) was the first woman to deliver a laudatory talk for a Fields Medalist: Vaughan F.R. Jones (1952–2020).

10.1.2.2 The Emmy Noether Lectures and the IMU

It is one thing to bemoan the poor statistics, i.e., the glaring underrepresentation of women in mathematics, and another to take action. The Association for Women in Mathematics (AWM) in the US seems to have been the first organization to take action specifically in favor of women in mathematics.[29] It came into being after a group of women formed a caucus at the Joint Mathematics Meetings in Atlantic City in 1971.[30]

> In those years the AMS was governed by what could only be called an "old boys network," closed to all but those in the inner circle. Mary [W. Gray (b. 1939)] challenged that by sitting in on the Council meeting in Atlantic City. When she was told she had to leave, she refused saying she would wait until the police came. (Mary relates the story somewhat differently: When she was told she had to leave, she responded she could find no rules in the by-laws restricting attendance at Council meetings. She was then told it was by "gentlemen's agreement." Naturally Mary replied "Well, obviously I'm no gentleman.") After that time, Council meetings were open to observers and the process of democratization of the Society had begun.

In March 1982, the AWM organized a Conference for the centennial of Emmy Noether's birth, at Bryn Mawr College, i.e., at the place where she had last worked.[31] Already two years earlier, at the San Antonio meeting in January 1980, the AWM Emmy Noether Lectures (chaired first by Karen Uhlenbeck) were inaugurated by Jessie MacWilliams (1917–1980). This series of lectures continued at the January meetings of the AMS. The first twelve lectures were given by Jessie MacWilliams, Olga Taussky-Todd, Julia Robinson (1919–1985), Cathleen S. Morawetz (1923–2017), Mary Ellen Rudin (1924–2013), Jane Cronin Scanlon (1922–2018), the

[28] Cf. the case study for France [Menger et al. 2020], which highlights, pp. 207–211, the extreme underrepresentation of women in mathematics in comparison with other sciences.

[29] See for instance [Barrow-Green 1994], p. 129, and the literature cited there.

[30] Cf. the particularly rich and varied September 1991 "Special Issue on Women in Mathematics" of the *Notices of the American Mathematical Society* (Vol. 38, No. 7) on the occasion of the twentieth anniversary of the AWM. See specifically Lenore Blum's account of the founding of the AWM, p. 740, from which the following quote is taken.

[31] See the special issue of the *Notices AMS* already quoted, pp. 744–748.

French mathematician Yvonne Choquet-Bruhat (b. 1923), Joan S. Birman, Karen K. Uhlenbeck, Mary F. Wheeler (b. 1938), Bhama Srinivasan (b. 1935), and Alexandra Bellow (b. 1935).[32] In April 2013 the lecture was renamed the "AWM-AMS Noether Lecture", and in 2015 it was jointly sponsored by the AWM and the AMS.

Extending this recurring American event to the international scene, the AWM would hold special Emmy Noether Lectures at ICMs starting in 1994. Thus the Russian authority on partial differential equations Olga Ladyzhenskaya would give an Emmy Noether Lecture at the 1994 ICM in Zürich. She is indeed listed as one of the participants of that Congress, but I have been unable to find any mention of this lecture of hers in the Proceedings. In particular, this Emmy Noether Lecture is not mentioned by the President of the Congress Henri Carnal (b. 1939) in his opening speech, even though he did mention Emmy Noether's plenary ICM lecture of 1932, and continued, apparently trying to be funny:

> I am therefore happy to observe not only that the number of plenary lectures by women will this time be greater than 0, and even greater than 1, but also that the highest federal and cantonal authorities are both represented here by women. This shows that we can always hope for positive changes![33]

Four years later in Berlin, the Emmy Noether Lecture did make it into the ICM Proceedings, in the weak sense that it was explicitly mentioned in Martin Grötschel's opening address, if only after comments on the social program of the Congress, and among events that "would not fit elsewhere":

> In accordance with the Program Committee and the IMU, the Organizing Committee opened a Section of Special Activities to cover topics of mathematical relevance that would not fit elsewhere in the official scientific program. These special activities included an afternoon session on electronic publishing with three talks and a panel discussion on "The Future of Electronic Communication, Information, and Publishing"; presentations of mathematical software on three afternoons; several special activities related to women in mathematics including the Emmy Noether Lecture given by Cathleen Synge Morawetz, and a panel discussion "Events and Policies: Effects on Women in Mathematics"; an afternoon on "Berlin as Centre of Mathematical Activity" (this workshop was suggested by the International Commission on the History of Mathematics); a roundtable discussion on "International Comparison of Mathematical Studies, University Degrees, and Professional Perspectives".[34]

As for 2002, the Proceedings of the Beijing ICM fail to mention Hu Hesheng's (b. 1928) Noether Lecture. However, that year marks the beginning of the integration of these lectures into IMU policy:

> The IMU General Assembly in Shanghai 2002 had adopted the following Resolution 5: "The General Assembly recommends continuing the tradition of the 1994, 1998, 2002 ICMs, by holding an Emmy Noether lecture at the next two ICMs (2006 and 2010), with selection of the speakers to be made by an IMU appointed committee."[35]

---

[32] Following the same special issue of the *Notices AMS*, p. 746. For the complete list, see [URL 37].

[33] See Proceedings ICM 1994, vol. 1, p. xxi.

[34] See Proceedings ICM 1998, vol. 1, p. 17.

[35] From the historical notes in [URL 38].

Even though such a committee was duly created for the Madrid ICM[36], the fact that Yvonne Choquet-Bruhat was chosen and gave the Emmy Noether Lecture was only mentioned in passing in the opening address of the President of the Congress, a bit like in Berlin back in 1998:

> Many other special activities were organized, a list of which would be too long to include in this introduction, although we may mention the scientific part of the Emmy Noether Talk, given by Ivonne [*sic*] Choquet-Bruhat, the special talk on Poincaré's Conjecture by John Morgan [b. 1946], and the talk given by Benoît Mandelbrot [1924–2010]. A joint scientific activity organized by the London Mathematical Society and the Real Sociedad Matemática Española was also held.[37]

The Hyderabad ICM of 2010 is the first ICM to include the Emmy Noether Lecture in the Proceedings, among the "Special Lectures." It was given by Idun Reiten (b. 1942) from Norway, on Cluster Categories.[38] The year also marked the decision to retroactively integrate earlier Emmy Noether Lectures given at—or rather: in the margin of—ICMs into the history of the IMU:

> At the General Assembly in Bangalore [in 2010] the Emmy Noether Lectures were adopted as a permanent ICM tradition via Resolution 8: "The General Assembly of the IMU recommends continuing the tradition of holding an Emmy Noether lecture at each ICM, with selection of the speaker to be made by a committee appointed by the IMU Executive Committee." To distinguish between the two series of Noether lectures it was decided to use the name ICM Emmy Noether Lecture for a lecture given at an ICM.[39]

The Emmy Noether Lecture in 2014 in Seoul was given by Georgia Benkart (b. 1949); in 2018 at Rio it was Sun-Yung Alice Chang's (b. 1948) turn. Today the ICM Emmy Noether Lecture is listed among the awards given by the IMU. It is a lifetime achievement award for women mathematicians. Yet, in view of its slightly complicated history as far as the IMU is concerned, we have decided not to include the lecturers, nor the corresponding committees, in our IMU database for the past seventy years. But in the future it will certainly have to be taken into account.

10.1.2.3 The Committee for Women in Mathematics (CWM) and its worldwide activities

The data we have quoted about the representation of women in Mathematics International show a difficult and irregular history, especially since 1967, when Van der Waerden wrote his letter to Auguste Dick. Looking at the last few years, though, in particular the short time span since the establishment of CWM in 2015, there is clear evidence that we are witnessing a new era. The numerous activities listed on the CWM website[40] conveys the kind of cultural internationalism that this Committee is working for and that is also reflected in the impressive list of the 150 CWM

[36] See Proceedings ICM 2006, vol. 1, p. 21.

[37] See Proceedings ICM 2006, vol. 1, pp. 10-11.

[38] See Proceedings ICM 2010, Vol. 1, pp. 558–594.

[39] From the historical notes in [URL 39].

[40] See [URL 40], as well as the activities reports posted on the site.

ambassadors.[41] The first *World Meeting for Women in Mathematics*, or $(WM)^2$ for short, took place in Rio as a satellite event of the 2018 ICM.[42] The second one is planned on the occasion of the 2022 ICM at St. Petersburg.[43]

This worldwide grassroots movement to the advantage of women mathematicians has at long last been built into the politics of Mathematics International backed by the IMU. But the women mathematicians' cause as we understand it today naturally undercuts divisions, headings and agencies which traditionally could be conceived of as separate concerns. Fighting the *glass ceiling* cannot limit itself to specific levels of education or career; the reflection has to address the whole spectrum, all the way from school education to the eligibility for the Fields Medal. And since the question of the career options for women mathematicians is intrinsically linked to cultural, national, and local constellations, the fight invites truly global networking. In this way, the CWM activities naturally meet with concerns pursued by ICMI and by the CDC.[44]

Furthermore, transversal connections about the cause of women in other directions extend well beyond the IMU. Under the roof of the International Science Council (ISC) an initiative has developed, which calls itself: *A Global Approach to the Gender Gap in Mathematical, Computing, and Natural Sciences. How to Measure It, How to Reduce It*. It involved mathematics, computing and several natural sciences.

> The mathematical and natural sciences have long benefited from the participation of excellent women scientists. However, at the end of the first decade of the twenty-first century, the percentage of women scientists remains shockingly low, and barriers to women's participation persist, leading to a gender gap at all levels and across all continents. It is against this backdrop that in 2016, the International Mathematical Union (IMU), through its Committee for Women in Mathematics, and the International Union of Pure and Applied Chemistry (IUPAC), supported by nine other ISC member unions and other partners, launched a project on the gender gap in science.
>
> The project comprised three main areas of research: a global survey of scientists, a databacked study on publications, and development of a database of good practice. The global survey asked scientists, both male and female, to reflect on their career experiences and any challenges they had encountered. It received responses from over 30,000 people in more than 150 countries, finding clear evidence for a gender gap in science.
>
> The project's second task was to develop an online tool to investigate the gender imbalance of scientific publications by women and men, across countries and fields of research. Shockingly, the study found that despite an increase in the proportion of women authors over time, women scientists were not publishing in top journals any more frequently than in the past, indicating that a gender barrier persists.

[41] See [URL 41].

[42] See [URL 42].

[43] See [URL 43].

[44] For an insightful presentation of many aspects of the problem we refer to the panel held at the 2018 ICM in Rio entitled: *The Gender Gap in Mathematical and Natural Sciences from a Historical Perspective*. See [URL 44].

> Finally, the project developed a 'database of good practices for girls and young women, parents, and organizations', to curate initiatives from all around the world that encourage the involvement of women in science. The database was made available on the IMU website in 2019, and is expected to expand in coming years.[45]

The Executive Committee of the initiative shows 23 members representing 11 bodies: The IMU, The International Union of Pure and Applied Chemistry (IUPAC), The International Union of Pure and Applied Physics (IUPAP), The International Astronomical Union (IAU), The International Union of Biological Sciences (IUBS), The International Council for Industrial and Applied Mathematics (ICIAM), The International Union of History and Philosophy of Science and Technology (IUHPST), UNESCO, the international initiative *Gender in science, innovation, technology and engineering* (GenderInSite), The Organization for Women in Science for the Developing World (OWSD), and The Association for Computing Machinery (ACM).

At a meeting at ICTP, Trieste, Marie-Françoise Roy commented: "We are happy with what we were able to do until now, but the long-term plan is to produce useful tools capable of living after the end of the project."[46] And in July 2020, it was decided to press ahead and set up the *Standing Committee for Gender Equality in Science*,

> a permanent organization formed by nine unions and partners that will start working in September 2020. Its goal will be to follow up the recommendations of the Gender Gap in Science project as well as maintaining and developing the tools created during the first years of the project.[47]

It has grown since its foundation and currently counts 16 unions as members.

Meanwhile, specifically for mathematics, the initiative launched by the 2018 meeting of $(WM)^2$ at Rio, to commemorate the twelfth of May, Maryam Mirzakhani's birthday, is now being followed by events in various parts of the world:

> For centuries women were disregarded as mathematicians, and the gender gap in mathematics remains very real. Celebratory events such as the ones supported by the May 12 Initiative bring about a crucial sense of belonging amongst women mathematicians and raise awareness throughout the entire mathematics community. The authors of this note belong to the coordinating group of the May 12 Initiative and tell the story of this international cooperation. We hope that next year you will join![48]

The Committee for Women in Mathematics is still rather young, so it will fall upon others to comment on its ongoing and future work. It may also be that the unquestioned bipolarity underlying our whole Section 10.1.2, which assumes a god-given dichotomy between two distinct genders and hence the possibility of sorting all mathematicians neatly into two disjoint drawers, will give way to an appreciation

---

[45] See International Science Council, *Annual Report* 2019, p. 18.

[46] See International Science Council, *Annual Report* 2019, p. 18.

[47] See [URL 45].

[48] See [Agarwal et al. 2019], p. 1879.

of the social nature of the notion of gender, and its consequences for the professional world in and beyond mathematics. If this comes to be, mathematicians will have come a very long way since Van der Waerden's letter.

10.1.3 The Commission for Developing Countries (CDC)

In Section 8.2.1 we have already mentioned the early beginnings of today's IMU Commission for Developing Countries (CDC). Its first predecessor, the *Commission on Exchange*, existed from 1952 until the mid 1970s, when it was reshaped under the name *Commission on Development and Exchange*, or CDE for short. We have pointed out the paradigm shift that occurred within this quarter century, from a general sponsoring of individual mobility, by which the IMU could walk in the footsteps of the philanthropic activities of the interwar period, to an increasing attention to structural problems of developing countries. This shift is a reflection, within the worldview practiced by the IMU, of the era of decolonization.

The following quote is from a 1976 letter of A. John Coleman (1918–2010) to the President of the IMU Deane Montgomery (1909–1992). Coleman had been chosen as chair of the Commission on Exchange during the 1974 meeting of the IMU Executive Committee at Harrison Hot Springs, BC, Canada. He would subsequently be a member of CDE, from 1979 to 1982.

> I shall begin by apologizing for my lack of activity as Chairman of the Exchange Commission during 1975 which was due to unusual pressure of work, consequent upon a variety of commitments which I had undertaken before my appointment to the Exchange Commission. As you are aware, at Harrison Hot Springs the nature of the Commission was radically changed, and its mandate was transformed from that of arranging a modest number of high level mathematical lectures to that of mobilizing the mathematics departments in developed countries to give meaningful help to our colleagues in underdeveloped countries. At the ICM, I did call a meeting attended by about 40 mathematicians from underdeveloped countries to initiate the discussion about what could or should be done. That meeting generated considerable enthusiasm. Even before the ICM, Professor [Henri] Hogbe-Nlend [b. 1939] of Bordeaux had conceived the idea of a Pan-African Mathematical Conference. I am sure you are aware that plans are well advanced for it to be held in Rabat, Morocco at the end of July. Professor [Yukiyosi] Kawada [1916–1993] has explored the possibility of a similar conference for Asia. Professor [Bernhard H.] Neumann [1909–2002] has been assiduous in circulating the IMU Canberra Bullet which provides very useful information to the mathematical community.[49]

In Parts I and II of this book, we have pointed out characteristic evolutionary steps taken by the professionalization of science in general, and mathematics in particular. The mathematical researcher of the nineteenth century did not have an office; singular exceptions apart, he was male and he worked from home. He was connected with the civilised world by a postal service, which in European cities guaranteed more than one home delivery per day; letters within Europe were only marginally slower than

[49] See Coleman to Montgomery, 11 February 1976, [Arch. IMU], SF 7, F 5, IMU_004.pdf. Cf. [Lehto 1998], pp. 179–183, as well as pp. 263–273 for certain further developments in the 1980s.

email is today. Institute buildings were the next step, which—as far as mathematics was concerned—was taken in the twentieth century. Between the World Wars, when the exclusively European domination of world mathematics had ended, the traveling (young) scientist was invented on a large scale, and the concept of Locally-grounded Transnational Research Site (LGTRS) emerged from the emblematic example of the IAS at Princeton—see 5.2. This LGTRS concept would spawn an international, would-be global network—see Section 8.3 above. But only countries well advanced in national, regional, and local scientific infrastructures could afford a node of their own in that network. The first reaction of this expanding network with respect to countries that were not sufficiently developed to have their own LGTRS, was to invite the most promising young talents to one of the sites of the network, thus reiterating an older pattern of which we have seen a few examples in Section 6.4. The early activities of the IMU Commission on Exchange fit this larger pattern.

The *African Mathematical Union* was founded in July 1976 during the first Pan-African Congress of Mathematics at Rabat, Morocco, which is mentioned in the above quote from Coleman's letter. The first volume of the journal *Afrika Mathematica* of the African Mathematical Union was published in 1978.[50]

Also in 1978—after the strongest wave of decolonization in the twentieth century, at a time when the IMU was revising its policy towards developing countries—France founded CIMPA, the *Centre international de mathématiques pures et appliquées* (or ICPAM in English) in Nice, a new kind of institute that would cooperate with UNESCO and with the IMU.

> According to its statutes, its mission is the training of mathematicians coming in priority from developing countries, by means of study visits during the university academic year and of summer schools, and with the help of the development of means of documentations.[51]

A comparable institute had already been created in Trieste, Italy, for Theoretical Physics back in 1964 on the initiative of the Nobel Laureate Abdus Salam (1926–1996); the *International Centre for Theoretical Physics* (ICTP), which today is named after its founder. Its mathematical branch started to play its important part in bringing together mathematical talent from developing countries in 1986. In this way, two mathematical centers with an explicit concern for developing countries were appended to the web of LGTRSs.

In January 1985, Vol. 1 of the Joint Bulletin of the IMU's *Commission on Development and Exchange* (CDE) and CIMPA was published, in the form of bound mimeographed typescripts, with the financial assistance of UNESCO, under the name *Mathematics and Development*. The main objective of this publication was "to serve as a liaison bulletin between mathematical institutions in the developing countries." The first issues, which appeared twice a year, were exclusively devoted to the following two projects:

---

[50] The journal would be relaunched with a new editorial board in 2010.

[51] See [URL 46].

- Selective Bibliography of Mathematics.
- Mathematical Directory of the Developing Countries; starting with Africa and the Arab Middle East, which was structured according to English speaking African countries, French and Portuguese speaking African countries, Arab speaking African countries, and The Arab Middle East.

The Selective Bibliography project, "launched by ICPAM and adopted by the General Assembly of IMU in Warsaw (August 1982) and the General Conference of UNESCO (October 1983), is a program to help developing countries to start constitution of their libraries in Pure and Applied Mathematics and Computer Sciences, taking in mind their financial difficulties and the lack of specialists of these various disciplines in these countries."[52] Jean Dieudonné—since 1964 professor in Nice, where CIMPA was based—started this project by proposing a first draft of a *Bibliographie sélective* in two parts. The first part listed what Dieudonné thought were the most urgent items: a little less than 100 titles from all branches of mathematics chosen in such a way as to make it possible to prepare a one year course on each subject. The second part contained about 300 titles, the fitting ones of which could be acquired depending on which more advanced courses were planned. Dieudonné's list was subsequently circulated, and discussed further among a number of colleagues. Dieudonné starts out with the section "Periodicals and Series", in which the first item is all the *Lecture Notes in Mathematics* of Springer Verlag, followed by the *Proceedings of Symposia in Pure Mathematics* of the AMS, and the French *Astérisque*. This hefty onset, and other proposed items, would be modified by colleagues involved in the later discussions, for instance by Jean-Pierre Serre. It is not clear to me to what extent these projects matured or were realized. A current analog of this kind of project is the Global Digital Mathematical Library, which we have mentioned in Section 10.1.1 above.[53]

Donating libraries, offprint collections or books to institutions of learning in developing countries was and is a frequent practice, but mostly in the form of individual initiatives. Among the many activities of today's CDC, there is a *Library Assistance Scheme* that offers to coordinate donations.[54] At the same time, online resources have modified the situation quite a bit. In 2010–2011, the IMU joined the European Mathematical Society EMS in organizing a series of workshops about "Finding Online Information in Mathematics" held in Ethiopia, Mali, Mozambique, and Cambodia.[55]

In archival documents related to the prehistory of CDC, one frequently encounters remarks about the enormous challenge, and about the importance of not slipping into a patronizing attitude.[56] This is strongly echoed in various reports elaborated over the last fifteen years, dealing with the situation of mathematics in Africa, in Latin

[52] Quoted *verbatim* from Hogbe-Nlend's editorial to *Mathematics and Development*, Vol. 1 (1985).

[53] See also *Bulletin of the IMU* 64 (July 2014), Appendix II, pp. 47–50.

[54] See [URL 47].

[55] See [URL 48].

[56] Coleman makes this point on several occasions in his letters.

America and the Caribbean, and in South-East Asia.[57] Let us quote for instance from the 2009 report on *Mathematics in Africa: Challenges and Opportunities* to the John Templeton Foundation. This report was coordinated by the IMU's *Developing Countries Strategy Group* (DCSG), which had itself been set up in 2003/2004 as a corollary of renewed interest in these questions on the part of the IMU Executive Committee since 2002. The DCSG merged with the CDE in 2010, thus creating today's CDC.[58]

> Given the enormity of this challenge, one might ask whether there are individual steps that offer exceptional leverage to jump-start the enterprise as a whole. For example, one step might be a program to support students of exceptional talent, identified perhaps by their participation in the mathematics Olympiads. While sending such students to top international universities, for example, is likely to produce great benefits for individuals, it was not suggested by our advisors. They felt unanimously that no "magic bullet" or quick fix could solve a problem that is systemic and institutional. Such a program might raise the visibility of mathematics among secondary school students, but this benefit could be reduced should privileged students decide to remain abroad rather than return home to unrewarding positions.
>
> The second suggestion is to *strengthen and expand successful training and research activities, especially regional networks of people and institutions.* There are several reasons our advisors highlight this option. First, successful networks by definition involve leaders of demonstrated talent and institutions capable of supporting creditable mathematics programs. Second, supporting a network helps build a critical mass of students and faculty who are otherwise likely to be professionally isolated. Third, by building on institutions and people already in place, networks use tools that are relatively inexpensive in relation to their power, such as partnerships, mentoring, distance learning, and internet-based collaboration.[59]

We invite the reader to browse the rich website of the CDC[60] and discover its current programs and activities, such as for instance the Volunteer Lecturer Program (VLP), which was established in 2008. It offers financial assistance to universities in developing countries to host a volunteer lecturer for an intensive course of several weeks.

Not less informative, often richer in detail, but a rather different type of text altogether, is the 2014 White Paper: "The International Mathematical Union in the Developing World: Past, Present, and Future," which was produced "for policy makers, funding agencies, constituencies of the IMU and ICMI, and for others who would like to learn more about the activities and objectives of the IMU."[61] The vantage point of this White Paper is the observation that it is—or it ought to be—in the best interest of every national government to improve the state of its mathematical education and profession at all levels.

---

[57] See [URL 49].

[58] For an overview of the achievements of the DCSG, see [URL 50].

[59] See [URL 51].

[60] See [URL 52].

[61] See the 64th *Bulletin of the International Mathematical Union*, July 2014, Appendix II, pp. 23–54. It is accessible at [URL 53].

10.1.4 The International Commission on the History of Mathematics (ICHM)

International scientific unions that have joined ICSU (in the past), resp. the ISC (since 2018), come in various forms. An interesting example is given by the *International Union of the History and Philosophy of Science* (IUHPS), which itself has expanded in 2010 into the *International Union of the History and Philosophy of Science and Technology* (IUHPST).

> There had been a long succession of international conferences of history of science and of philosophy of science since the beginning of the [twentieth] century. The historians [of science] founded their Union in 1947 and adhered to ICSU. The International Union of the Philosophy of Science IUPS was founded in 1949, but had not been admitted to ICSU by the time of the 1949 Fifth General Assembly. The two joined forces in 1956 to act as two divisions of one Union. IUHS had already developed several scientific sections and was to multiply them as time went on. Since many Unions have members with an interest in the history of their own science, Joint Commissions with some other Unions were also created. Enough to say that the way IUHS later merged into IUHPS is characteristic of the evolutionary character of many Unions.[62]

The Joint Commission to be discussed in the present section is attached both to the IUHPST and the IMU. It is the only Commission of the IMU that depends jointly on two international scientific unions. Since the IUHPST is the disjoint union of its two divisions: the *Division of History of Science and Technology* DHST and the *Division of Philosophy of Science and Technology* DPST, the *International Commission on the History of Mathematics* (ICHM) is in fact a joint commission of the IMU and the DHST.

ICHM was originally founded in 1971 by the DHST, which at the time—before the adjunction of the history of technology—was still simply the *Division of History of Science*, DHS. The DHST continues to be ICHM's primary affiliation; the commission continues to receive its basic annual grant from the DHST, and the official meetings every four years of the ICHM take place as part of the DHST international congresses.[63] The history of the ICHM is wrapped up in the congresses of the DHST: in Moscow (1971), Tokyo (1974), Edinburgh (1977), Bucharest (1981), Berkeley (1985), Hamburg & Munich (1989), Zaragoza (1993), Liège (1997), Mexico City (2001), Beijing (2005), Budapest (2009), Manchester (2013), Rio de Janeiro (2017), and the Congress at Prague, which had to be held online in 2021.

The ICHM achieves its greatest visibility through its official journal, *Historia Mathematica*. This journal was founded in 1974 by Kenneth O. May (1915–1977) in Toronto, who had earlier been one of the instigators of ICHM, and who published, jointly with Constance M. Gardner, the first edition of the *World Directory of Historians of Mathematics* in 1972. *Historia Mathematica* (or HM for short) publishes

---

[62] See [Greenaway 1996], pp. 79–80.

[63] Here and in the remainder of this Section, I rely on freely accessible information from the ICHM website [URL 54], which is embedded in the IMU website, as well as on personal communication from Craig Fraser and June Barrow-Green, the former and the current chair of the ICHM. Hearty thanks to both of them.

original research on the history of the mathematics of all periods and in all cultural settings. The journal is published by Elsevier and provides some funding for the ICHM.

At the International Congress of History of Science at Berkeley in 1985, the ICHM voted to approach the IMU regarding re-establishing itself as an inter-union commission between the IHPS(T)/DHS(T) and the IMU. This Joint Commission was established in 1987 following a ballot of members of the IMU, and began its work at the beginning of 1988. The DHST does not appoint representatives to the ICHM, because everyone on the Executive Committee of the ICHM, with the exception of the IMU representatives and the HM editors, is in some sense already part of the DHST. At present the ICHM counts 44 member countries.

In the last four years the ICHM has integrated itself somewhat more closely with the IMU from an operational point of view. Since 2018 the IMU arranged to include the ICHM accounts within their financial umbrella. Similarly the ICHM website is now managed by the IMU.

By comparison, the history of chemistry and of physics are sole commissions of the DHST, and are not part of the international unions for these sciences. The history of ancient astronomy left the International Astronomical Union and became a commission solely of the DHST, primarily because it felt that the IAU was not professional enough in its understanding of the older history. Occasionally there may have been concerns at the ICHM about the view of history held by working mathematicians and expressed to some extent in the IMU.

André Weil, for example, in his plenary lecture: "History of Mathematics: Why and How?" at the Helsinki ICM in 1978, stressed the history of ideas as the focal approach to the history of mathematics and concluded that "the craft of mathematical history can best be practiced by those of us who are or have been active mathematicians or at least who are in close contact with active mathematicians."[64]

This point of view almost seems to echo the indignation of the historian of mathematics Moritz Cantor addressing the history section of the Heidelberg ICM in 1904, who had serious doubts about a general history of the exact sciences. For him, the peculiarities of the various scientific disciplines made it inconceivable that a chemist by training and a mathematician by training could reasonably compete for the same history chair.[65]

By 1978, however, the community of historians of science, and of mathematics, had come a long way since the days of Moritz Cantor, and Weil's lecture, in spite of its erudition and in spite of the author's obvious sense of history, would provoke mixed feelings in the community of historians of mathematics, especially also among those who had themselves at some point left mathematical research and resolutely reoriented themselves as historians of mathematics.

[64] See Proceedings ICM 1978, Vol. 1, p. 234.

[65] See Proceedings ICM 1904, pp. 500–501.

Yet, ever since it became a joint commission, the ICHM has continued to maintain the link with the mathematicians. Thus the history of mathematics differs from the history of physics, chemistry and biology in being part of the associated disciplinary international union.

The ICHM co-sponsors events of high intellectual caliber with a view to encouraging quality research in the history of mathematics. ICHM co-sponsorship does not necessarily involve financial support, but applications for limited funding may be made. Special consideration is given to events organized by and/or for early career scholars. A recurring annual event based in Europe, which has been co-sponsored by the ICHM in recent years, is the so-called *Novembertagung* on the History of Mathematics. It is aimed at PhD and postdoctoral students in the history of mathematics and neighboring fields. The last five meetings were organized in Torino, Italy, in 2015, Sandbjerg, Denmark, in 2016, Brussels, Belgium, in 2017, Seville, Spain, in 2018, and Strasbourg, France, in 2019. The 2020 *Novembertagung*, organized by young researchers based in Berlin, had to be moved online.

At the International Congress of History of Science and Technology (ICHST) in 2017 at Rio de Janeiro, three Symposia on the History of Mathematics were co-sponsored by the ICHM: on The Resurgence of Applied Mathematics 1850–1950; on Mathematical Methods at Work in Ancient China; and on Global Mathematics.

The distinguished prize awarded by the ICHM since 1989 is the *Kenneth O. May Prize and Medal* in the History of Mathematics. Two of these Medals are usually awarded every four years at the ICHST, to colleagues whose work best exemplifies the high scholarly standards and intellectual contributions to the field that K.O. May worked so hard to achieve. The bronze Medal was designed by the Canadian sculptor Saulius Jaskus. The first woman to receive the Kenneth O. May Prize was Lam Lay Yong (b. 1936) from Singapore, in 2001; the Medal was actually given to her during a ceremony at the Beijing ICM in 2002.

Explicitly directed at young career researchers is the *Montucla Prize*. Since 2009 it has been awarded by the ICHM at each ICHST, to the author of the best article published by a young researcher in *Historia Mathematica* in the four years preceding the Congress. The prize money is generated by revenue of the journal. The first woman to receive the Montucla award was Jemma Lorenat (b. 1987) in 2017.

10.2 Framing ICMs

In Section 8.2.1.1 we have divided the sequence of the 28 ICMs held between 1897 and 2018 into two intervals of approximately 60 years each: from 1897 to 1958, and from 1962 to 2018, the distinction between the two periods being the participation of the IMU in organizing ICMs. There was no IMU to claim a share in the organization of the first five ICMs, between 1897 and 1912. The old IMU existed between 1920 and 1932, under the roof of the IRC. This Council managed to uphold its exclusion politics at the Strasbourg ICM in 1920 and in Toronto in 1924. But the three following ICMs—at Bologna in 1928, Zürich in 1932, and Oslo in 1936—were again organized independently, in defiance of the IMU, the IRC and ICSU. This autonomy of the local Organizing Committees continued at the first three ICMs after World War II. During the 1950 ICM in Cambridge, Mass., the new IMU was still in the making. In 1954 in Amsterdam and in 1958 in Edinburgh, the new IMU was established, but had no say in the organization of those ICMs.

> The mathematical program was determined before the 1962 Congress by the local Organizing Committee, for the ICM-62 and thereafter by a Consultative Committee (CC), which in 1982 was renamed Program Committee (PC). The members of the CC and PC are appointed partly by the IMU Executive Committee, partly by the local Organizing Committee. For the ICM-62, the CC was still advisory to the OC; thereafter, it had the sole authority for the scientific program. Since the 1962 Congress, the President of the IMU appoints its Chairman. For the ICMs 1966, 1970, and 1974, the IMU Executive Committee and the local OC each appointed four of the eight members. For the ICMs 1978, 1983, 1986, and 1990, the local OC could appoint two, three, or four members according to the decision of the IMU Executive Committee, which appointed the rest. Since 1990, the IMU Executive Committee has appointed seven members, the local OC, two.[66]

Since the 2002 ICM in Beijing, the Program Committee always counted 11 or 12 members. In 2002, two of them were from China. Likewise at the Madrid ICM in 2006, there were two Spanish colleagues among the members of the PC. For the 2010 ICM in Hyderabad, and in 2014 in Seoul, there was only one local member on the PC, and none in 2018 in Rio de Janeiro.

The progression of the IMU towards increasing control of the ICMs was not always smooth and uncontested. Major challenges took the form of political allegations against the IMU. Recall for instance from Section 8.2.3 Pontryagin's and Vinogradov's criticism of the IMU during the preparation of the ICM in Warsaw. Claiming that the IMU was favoring Western mathematicians with "Zionist ideology," they asked for "the procedure in force before the Stockholm ICM" to be restored, when the national Organizing Committee could control things.

Olli Lehto has included in his book on the history of the IMU an interesting section on the mounting of the 1978 ICM in Helsinki, based on his personal involvement.[67] Instead of focussing here in a similar way on another ICM of the recent past, based on archival material, we pass immediately to a new Committee that was set up recently upon the initiative of the current Secretary of the IMU: the *Structure Committee* (SC). Its creation highlights both the importance of invitations for the speakers' careers and the difficulty of balancing branches of mathematics, sections, and personal preferences of members on the progam committee. Such issues will also be reflected in our data analysis, which will occupy the remainder of this final Chapter 10.

---

[66] See [Lehto 1998], p. 320.

[67] See [Lehto 1998], Section 9.4.

Let us begin by quoting from the Guidelines for the ICM Structure Committee that were endorsed by the IMU General Assembly in 2018.[68]

> The International Congresses of Mathematicians (ICMs) are the most important IMU activity and need correspondingly careful preparation. Every ICM should reflect the current activity of mathematics in the world, present the best work being carried out in all mathematical subfields and different regions of the world, and thus, point to the future of mathematics. The invited speakers at an ICM should be mathematicians of the highest quality who are able to present current research to a broad mathematical audience.
>
> The ICMs have traditionally been organized in the form of a number of invited one-hour plenary lectures, to be held without other parallel activities. In addition, there is a number of sections defined in terms of different subfields of mathematics. In each section there is a number of 45-minute sectional lectures. The sections take place in parallel throughout the ICM. In addition, there are a small number of one-hour prize lectures associated with various prizes (Fields, Nevanlinna, Gauss, Chern, and Leelavati) and named lectures (Noether and Abel). The possible overlap of speaker for a prize lecture and plenary or sectional talk may result in changes in the program, as no person gives more than one talk at an ICM.
>
> Traditional target numbers are
>
> - 20 plenary lectures
> - 180 sectional lectures distributed over 18–20 sections
> - 10 prize and named lectures (in addition, there will be shorter laudatory talks in connection with the prizes)
>
> It is difficult to increase these numbers substantially without extending the duration of an ICM.
>
> The Structure Committee (SC) is responsible for the preparation of the Scientific Program of the ICM. It decides the structure of the Scientific Program, in particular,
>
> - the number of plenary lectures,
> - the sections and their precise definition,
> - the target number of talks in each section,
> - other kind of lectures, and
> - the arrangement of sections.
>
> The size and content of the sections should reflect the development of contemporary mathematics and should both reflect the importance and the volume of activity in the various subfields of mathematics.
>
> The prize lectures and named lectures cannot be altered by the SC. It is understood that the SC will employ the programs of previous ICMs as guidelines for its decisions. The SC may also propose other activities like discussion panels, non-mathematical talks, and talks aimed at the general public.
>
> If the SC wants to propose more radical changes in the structure of an ICM, it should make a proposal to the Executive Committee (EC), which then will decide in the matter.
>
> The responsibility to decide the speakers resides with the Program Committee.

Following work of an informal committee chaired by László Lovász, the inaugural Structure Committee was formed along the above guidelines in January 2019, with a view to preparing the 2022 ICM in St. Petersburg. Chaired by Terence Tao it counts

[68] See [URL 55].

14 members and has delivered a very substantial report about their work achieved in 2019, largely on the basis of comments from the mathematical community at large.[69] Here is a short extract.

> Many of the lectures at [an ICM] play dual roles, serving both as a prestigious recognition for the lecturer, and as a scientific talk disseminating the most important advances in a given field. For instance, the various prizes given out by the IMU at the Congress, such as the Fields Medals, are perceived as amongst the highest recognitions available in mathematics, and receive extensive attention outside of the mathematical community as well; but each of the prize laureates also gives an hour-long lecture on their work that is attended by a large fraction of the entire Congress. Similarly, the 20 or so plenary lectures are also regarded as highly prestigious, and each such lecture commands the undivided attention of the Congress. ... [These] plenary lectures are expected to be somewhat broader in order to appeal to the less specialist audience, but are still mostly given by eminent mathematicians who have been closely involved in recent advances in the field.
>
> While many aspects of the Congress appear to have been generally well received ..., several issues with the Congress were repeatedly raised by a number of participants and organizers. One frequent complaint was that expository quality of sectional and plenary talks was highly variable ... This was a particular concern for the plenary lectures, given that no other activities for Congress participants were scheduled during these extremely high-profile talks. ...
>
> Another recurring concern was that the subdivision of all of mathematics into a section structure that has evolved only very slowly over time affected the breadth of topics covered, with talks in well established traditional areas being favored over emerging, experimental or interdisciplinary areas. Related to this was a widespread perception that the Congress caters more to the "pure" disciplines of mathematics than the "applied" ones, with many in the applied mathematics community feeling that the prestige of an invitation to speak at the Congress, or the value of attending such a Congress, is less than what it would be for a member of the pure mathematics community.

Constituted at the end of the period studied in this book, the Structure Committee epitomizes the fact that ICMs are today unquestionably controlled by the IMU. We have thus come a long way. ICM routines have emerged for more than a century; the IMU has intervened since 1962. Now the IMU is monitoring the organizational success with respect to a blueprint of ICMs which has crystallized over the past 120 years.

In the following sections, we shall probe the functioning of the ICMs of the past seventy years.

---

[69] See [URL 56].

10.3 The Database

We shall now present the data-analysis that was realized for this book by Birgit Petri, Darmstadt. The aim of this quantitative investigation is to explore the image of mathematical excellency that the IMU has framed via its influence on the ICMs of the past seventy years. We are particularly interested in the way in which this image has changed over time.

As we saw in Section 8.2.1.1 and in the preceding section, the IMU took control of the organization of the ICMs only gradually. Since our analysis is based on ICM-related data, not everything we find necessarily reflects actions on the part of the IMU. This should be kept in mind when interpreting our results concerning the 1950s and early 1960s.

The Population Studied. We realized early on that the kind of quantitative analysis we were after would become too unwieldy if we included *all* speakers and committee members of all the ICMs since 1950.[70] Since our intention was to investigate the image of mathematical excellence projected by the IMU through the Congresses, we decided to isolate the top layer of all the ICMs between 1950 and 2018. Going through these 18 Congresses one by one, we thus restricted attention to those mathematicians who, at a given ICM, had one of the functions listed below. Note the underlying idea of symmetry of our criteria between those that are chosen and those that choose.

- Invited to deliver a Plenary Lecture.[71]
- Member of the Program Committee of the ICM.[72]
- Winner of the Fields Medal; Laudatory speaker on one (or more) Fields Medalist(s); Member of the Fields Medal Committee.
- Winner of the Nevanlinna Prize; Laudatory Speaker for the Nevanlinna Prize; Member of the Nevanlinna Prize Committee.[73]
- Winner of the Gauss Prize; Laudatory speaker for the Gauss Prize; Member of the Gauss Prize Committee.[74]

---

[70] The study [Mihaljević & Roy 2019] looks at all the speakers of all the ICMs, since 1897. As a result, the amount of available information about the individual persons in that database varies considerably. For instance, the authors had to partially resort to automated guessing of the gender of the speakers listed.

[71] We enter these people according to the IMU website [URL 57], cross-checked against the Proceedings of the corresponding ICM. The list includes speakers who could not, or would not attend the congress to which they were invited. In the earlier ICMs of our time span, Plenary Lectures were called "One Hour Lectures."

[72] This information is usually based on the Proceedings of the corresponding ICM; occasionally on [Lehto 1998], App. 8.

[73] The Nevanlinna Prize was awarded once at every ICM from 1983 until 2018. Following protests concerning Rolf Nevanlinna's (1895–1980) affinities with Nazi politics, the prize is called the IMU Abacus Medal as of 2022.

[74] For the Fields Medal, the Nevanlinna Prize, and the Gauss Prize, the winners and members of the Prize Committees are listed on the IMU website; the Laudatory Speakers in the ICM Proceedings.

The Nevanlinna Prize / IMU Abacus Award is given for outstanding contributions in Mathematical Aspects of Information Sciences. Compared to the Fields Medal it thus represents an explicit widening of scope of the achievements honored by the IMU. The same is true of the Gauss Prize, which was created to honor scientists whose mathematical research has had an impact outside mathematics.[75]

There are three more recent awards attributed by the IMU, which we have not taken into account for our analysis: The Chern Medal Award and the Leelavati Prize, which have only been awarded since 2010, and the ICM Emmy Noether Lecture, which was discussed above in Section 10.1.2.2.

The great majority of the persons in our database: 313 out of the 540, have presented a total of 334 plenary lectures. Going through the list of ICMs, there are of course many individuals who served in different functions as time went on. For instance, 83 among the 182 members of the ICM Program Committees have also given at least one Plenary Lecture at some point in their career. There are only 9 of the 58 Fields Medalists since 1950[76] who have never given a Plenary Lecture, nor served on a Program Committee or a Fields Medal Committee. In 25 cases, the Laudatory lecturers for Fields Medalists came from the corresponding Fields Medal Committee.

The information stored for each person. For each of the 540 persons in our population, the database records a certain amount of information gathered from openly accessible sources.[77]

The *biographical information* about each person in our population includes: gender, last name, first name(s); ICM(s) and function(s) which made the person part of our population; sectional talks given at other ICMs, if applicable; year of birth[78], year of death (when applicable), place of birth[79], country of birth, citizenship(s) held. There is also a column for free comments on biographical features, such as migration history, involvement in war-related projects, functions held inside the IMU.

For each person in our population, another goal was to record the following information concerning their *PhD*[80]: year, title, institution, thesis-advisor[81], (classification of the thesis, if available), explanatory comments as needed.

---

[75] The Gauss Prize, which is jointly awarded by the IMU and the German Mathematical Society DMV, was founded in 2006.

[76] Only Grigori Perelman declined the Fields Medal offered to him in 2006. He is nonetheless kept in our database, which records decisions of the IMU.

[77] Information from Wikipedia entries (in various languages) was checked and fine tuned with the help of institutional or personal websites open to the public, as well as published material such as necrologies.

[78] In a few exceptional cases, this could not be determined precisely.

[79] Not always given precisely.

[80] In countries like USSR, France, Germany, etc., where there are two theses of different levels, "PhD" refers to the first thesis. Information from the *Mathematics Genealogy Project* [URL 58] was cross-checked as best we could by other openly accessible sources.

[81] Given with name, first name, years of birth and death.

For all lectures at an ICM presented by a member of our population—be they plenary or sectional lectures, lectures given in honor of Fields Medalists or winners of a Nevanlinna or a Gauss Prize—we record: the lecturer, year of the ICM, subject classification[82], and the type of lecture delivered. The title of the lecture is given as well, if it is available.[83]

For each person in our population, we tried to determine their professional *affiliations* as follows. To each ICM we associate a symmetric time interval of 9 years; for instance, the interval associated to the 1990 ICM in Kyoto contains the years 1986 through 1994. For all those who entered our data because of a function at a given ICM, we list their affiliations (institution and corresponding duration) that have a non-empty intersection with the associated interval around the ICM. Here "affiliation" is understood as employment on a potentially permanent basis. Furthermore, we also note temporary research sojourns—insofar as they can be determined from easily accessible information—and participation at conferences in Oberwolfach.[84]

Periodization. In order to structure the material as well as the evidence detected, we shall routinely group the past 70 years into 5 consecutive periods each of which contains either 4 or 3 ICMs. This periodization was already used in Section 8.2 above.

Period 1. 1950 Cambridge (Mass.); 1954 Amsterdam; 1958 Edinburgh; 1962 Stockholm.

Period 2. 1966 Moscow; 1970 Nice; 1974 Vancouver; 1978 Helsinki.

Period 3. 1983 (moved from 1982) Warsaw; 1986 Berkeley; 1990 Kyoto.

Period 4. 1994 Zürich; 1998 Berlin; 2002 Beijing.

Period 5. 2006 Madrid; 2010 Hyderabad; 2014 Seoul; 2018 Rio de Janeiro.

Some of the questions studied in the sequel require consideration of slightly shifted or regrouped time intervals. This will be explained as we go along. However, the periods above will remain the principal frame of reference as we look at the past seventy years.

---

[82] According to the Mathematics Subject Classification (MSC) scheme. The relevant websites are [URL 59] jointly with [URL 60], and [URL 61], which indicates the classification at the time of publication. For laudatory lectures that are only classified under 'history, biography', we have determined an amended classification which reflects the mathematical subject.

[83] All this presupposes that we have sufficient information about the lecture. This is not always the case, for instance if a speaker actually did give a talk, but never submitted a manuscript. On the other hand, there were invited speakers who could not attend, but their talk was given by a proxy, or simply sent in as a manuscript.

[84] In this respect, Birgit Petri's survey actually went beyond the data that were finally used for this book. It also included activities as journal editors.

10.4 The Cupola of the ICMs

The population described in the previous section, about which we have collected all that information in our database, restricts attention to the most distinguished layer of mathematical achievements showcased at the ICMs. We call it the cupola of the ICMs.

10.4.1 Parts of the mathematical world

To get started it is instructive to look at the geographic distribution of our distinguished population, and how this changed over the five periods we have introduced above. In order to obtain such a global overview, diagrams showing all individual countries involved in the biographical data of our 540 mathematicians would be unreadable. On the other hand, certain individual countries are interesting to look at because they have been, or emerged as, leading countries on the mathematical scene during the twentieth century. Compromising as best we could between readability and respect for individual nations, the diagrams we present here employ the following 16 slots:

Individual countries shown are the USA, Russia[85], France, UK, Germany[86], Japan, and Israel. Hesitating to group Canada together with Central and South America, we also included it among the individual countries listed.

Regional groups of countries are:
the remaining countries of Western Europe (abbreviated as WE)[87];
the remaining countries of Eastern Europe (abbreviated as EE)[88];
the Middle East with the exception of Israel (abbreviated as ME)[89];
Central and South America (abbreviated as CSAm)[90];

---

[85] The meaning of this *ad hoc* term follows history. Until World War I it signifies the Russian Empire; during the existence of the Soviet Union it refers to the USSR; as of 1992 it stands for the Russian Federation. As a consequence, many Polish mathematicians for instance, such as Witold Hurewicz and Alfred Tarski, were born 'in Russia' in this technical sense.

[86] In the period 1949–1990, we use this label to refer only to the Federal Republic of Germany, grouping the German Democratic Republic with the rest of Eastern Europe.

[87] The countries of this regional group which we encounter in the biographical data of the 540 members of our population are: the four Nordic countries Norway, Sweden, Finland, Denmark; Belgium and the Netherlands; Switzerland, Austria-Hungary/Austria; Portugal, Spain, Italy, and Greece.

[88] In our biographical data we encounter the GDR, Poland, Ukraine, Czechoslovakia / The Czech Republic, Romania, and Hungary. Because of the historic variations of 'Russia' as well as Austria-Hungary/Austria, our regional group of 'Eastern Europe' makes no appearance, for instance, in the Period 1 diagrams of Figure 10.2 below.

[89] In our biographical data Iran, Turkey, Lebanon, and Saudi Arabia are mentioned.

[90] In our biographical data we find Mexico, Ecuador, Brazil, Argentina, and Chile.

India[91];
East Asia with the exception of Japan (abbreviated as EA)[92];
Australia and New Zealand (abbreviated as ANZ);
Africa.[93]

A person may have a variety of links with a country, or a part of the world. In this section we look at the worldwide distribution of the members of our population according to three different breakdowns: their *nationality at birth* (Fig. 10.2), the place where they obtained their *PhD* (Fig. 10.3), and the place of *professional affiliation* at the moment of the ICM, or ICMs, which brought them into our list (Fig. 10.4).[94]

As mentioned before, our total population is selected in a symmetric way between those that are chosen and those that choose. For instance, being chosen as a plenary speaker at an ICM qualifies a person to be in our population, just as serving on the program committee that chooses plenary speakers does. The same is true for the various other distinctions we are keeping track of. Both types of functions have to come together to craft an ICM, and its most distinguished stratum.

The plenary speakers form the biggest subgroup of our population. It turns out to be instructive—if only to avoid hasty conclusions about the relative strength of national groups in a given period—to always accompany a breakdown of our total population, given in a big annotated pie chart, by the corresponding colored chart, smaller and without annotations, for the subpopulation of the plenary speakers.

In order to see the evolution in time, the breakdowns are done separately for each of the five periods introduced above. The total numbers, relative to our whole population, of which the big charts show percentages, varies between 136 and 175 for the first four periods. It jumps up to 260 for the most recent interval 2006–2018, which comprises four ICMs like the first two periods.[95] The respective total numbers

[91] From the whole Indian subcontinent only the area of today's Republic of India occurs in biographical data of our population.

[92] In our biographical data we find China (as of 1949 the People's Republic), Taiwan, Hong Kong, Vietnam, South Korea, and Singapore.

[93] Of all African countries, South Africa is the only one that occurs in the biographical data of our population.

[94] Each person counts with multiplicity 1 for each ICM that brought him or her into our population. Given such an ICM, if a person either holds n simultaneous professional affiliations in different countries or parts of the world during that ICM, or if a person switches jobs n times between different nations or parts of the world during the year of the ICM considered, then each of the corresponding nations or parts of the world are counted with multiplicity $1/n$.

[95] More precisely, the total numbers per period underlying the various pie charts of the whole population are 175, 164, 136, 155, 260 for the charts in Fig. 10.2 and 10.4. The arithmetic mean of the first four of these numbers is 157.5. Thus the factor $\frac{260}{157.5} = 1.65$ allows us to approximately translate percentages given for the first four periods, resp. for the fifth period, into comparative sizes of the underlying groups of mathematicians. Since not every member of our population had a PhD, the underlying total numbers per period for the big charts in Fig. 10.3 are slightly different: 160, 161, 136, 154, respectively 260.

per period concerning only the plenary speakers, underlying the small charts, are as follows: 78, 68, 47, 58, 83 for the small charts in Figure 10.2 and 10.4; respectively 74, 65, 47, 57, 83 for the small charts about PhDs in Figure 10.3.

Here is an example: In Fig. 10.2 we represent our mathematicians according to the regions of the world they come from. For each one of our five periods, the corresponding part of our whole population is broken up according to regions of birth in a big chart with annotations for all parts whose size exceeds a reasonable lower bound of about 4%. Each one of these five big charts is accompanied by a small chart that uses the same colors as the corresponding big chart, but for reasons of space does not specify any percentages. For periods 1 and 2, the small chart is placed below the big chart of the periods. The small chart for period 3 figures on the right of the big annotated chart covering the interval 1983–1990. And the small charts for periods 4 and 5 are placed above their corresponding big annotated pies. These small pie charts show the breakup of all the mathematicians that gave plenary lectures at at least one of the ICMs of its period. Looking for instance at period 2, 1966–1978, we see from the small chart that the plenary speakers born either in the US or in 'Russia' account for more than half of all plenary lecturers; whereas only 20.1 + 20.1 = 40.2% of our total population for this period were born in the US or 'Russia.'

Looking at the origins of our mathematicians (Fig. 10.2), we can see the world of mathematics gradually opening up. For instance, the total share of Europe (including Russia) in our whole population shrank during the seventy years we are looking at from 77.2% in the first period to 49.2% of the population in the most recent timespan. In the charts accounting for professional affiliations (Fig. 10.4), the European share dwindles even more, from almost 59% to only about one third of the population.

East Asia on the other hand is a part of the mathematical world (in our national grouping) which advances slowly but steadily in Fig. 10.2, i.e., as a region of origin. India does as well, except for a lower presence in Period 3. A nation which makes its first appearance as a country of birth in the third period, and then establishes itself firmly, is Israel. The cupola of the ICMs is obviously much more diverse today than seventy years ago. Nonetheless, the distribution can hardly be called global, considering for instance the share of mathematicians from Africa. Indeed, Fig. 10.4 shows that not a single mathematician of our population had his or her professional affiliation in Africa, not even during the last period 2006–2018.

The motley diversity of national origins indicated in the pie charts for Period 5 of Fig. 10.2 has to be confronted with the dominating place held by the USA in the career-oriented charts. For the places where PhDs were obtained (Fig. 10.3) this US dominance starts in Period 2. In professional affiliations (Fig. 10.4) the USA clearly dominates all of the five periods. The US dominance tends to be even more pronounced within the subgroup of plenary speakers—see the small pie charts. For instance, mathematicians with a professional affiliation in the USA account for more than half of all plenary speakers in the two most recent periods.

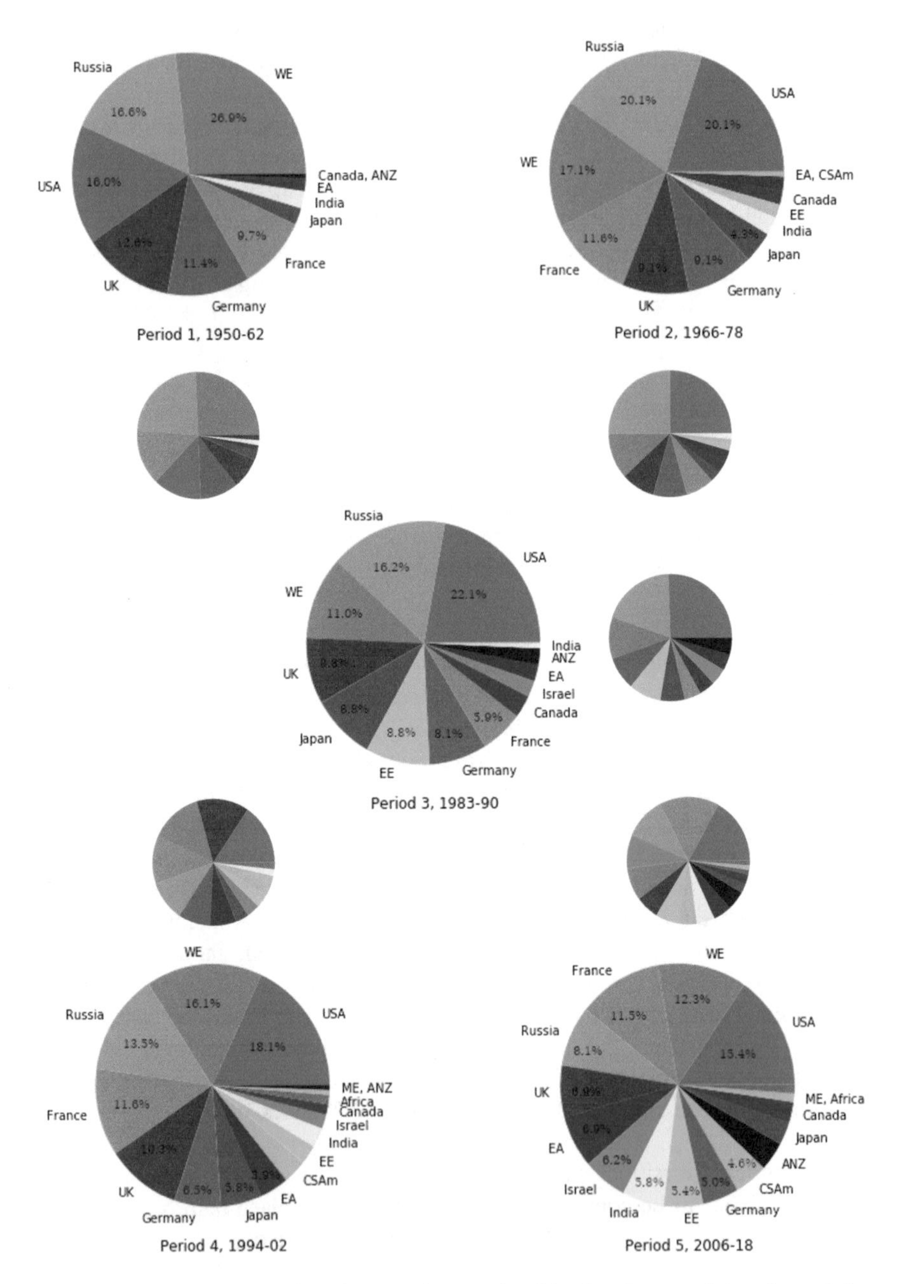

Fig. 10.2 Geographic distribution of our total population (big charts), resp. of the subpopulation of plenary speakers (small charts), according to countries of origin.

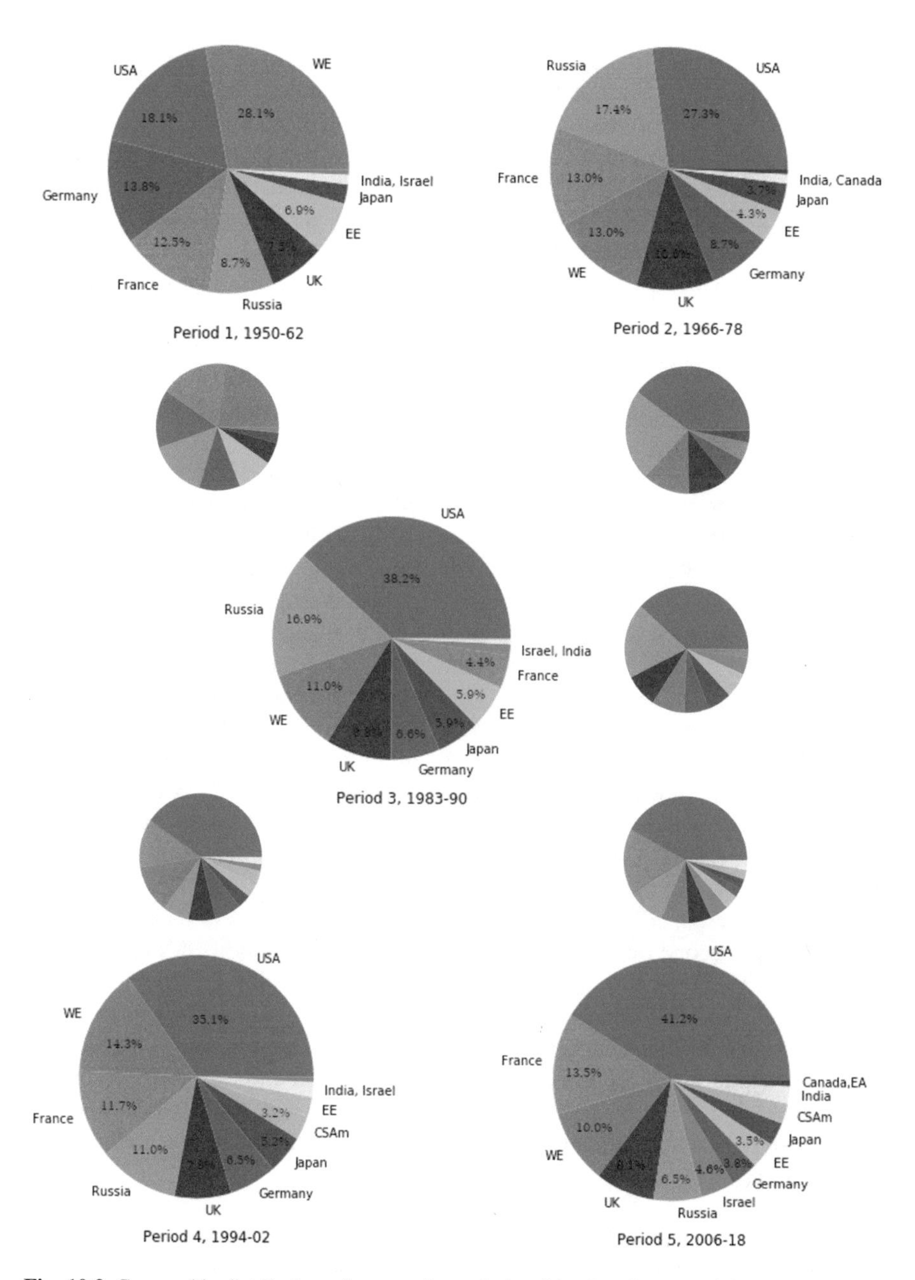

Fig. 10.3 Geographic distribution of our total population (big charts), resp. of the subpopulation of plenary speakers (small charts), according to places where their PhD was obtained.

The share of the USA as a country of origin (Fig. 10.2) grows over the first three periods, but then recedes roughly to the level of the first period. Thus the true nature of the United States for the cupola of the ICMs only comes to the fore when looking at the careers of the members of our population. Excellent careers in mathematics today have more than a 45% chance of at least passing through the US. However, the data we have are not sufficient to analyze the flow of researchers passing through the US, for instance as graduate students or young researchers, and installing themselves in permanent positions. At any rate, such a finer analysis, in order to be useful, would have to be performed on a much bigger population of mathematicians.

10.4.1.1 The Soviet Union, seen through the lens of plenary ICM lectures

As noted above, our definition of "Russia" leads to an unusually strong representation as a country of origin—including for instance mathematicians of the Polish school—in the first periods of Figure 10.2. In spite of this hitch, our data do reflect the tremendous weight of mathematics practiced in Russia, not only in Period 2, with the ICMs at Moscow and Helsinki, but also in Period 3 of Fig. 10.4. Let us look at this more closely, focussing on the subpopulation of plenary speakers.

For political reasons, the USSR was not present at the 1950 ICM at Harvard. The Soviet delegation at the 1954 ICM in Amsterdam consisted of four mathematicians three of whom gave one hour lectures: Pavel Alexandrov, Andrey Kolmogorov, and Sergey Nikolsky (1905–2012). Going through all the plenary speakers of the first period 1950–1962, and looking at their countries of birth—without assembling them into parts of the world—Russia and France lead with an equal score. Considering the plenary speakers of the first period 1950–1962 according to their countries of affiliation, the Soviet Union holds the second place, with 14.1% of all plenary speakers, and is topped only by the supremely dominating USA (44.2%).

During the second timespan, 1966–1978, the position of the USSR asserts itself strongly. Both as a country of origin and as a country of affiliation, Russia accounts for 23.5% of all plenary lectures of the second period. The number of plenary lectures given by mathematicians born in the USA outnumbered those given by Russians by just 1. And in terms of professional affiliations, the gap between the USSR and the USA observable in the first period was substantially reduced, while that between the USSR and the next smaller country France widened considerably—see the small pie charts for Periods 1 and 2 in Fig. 10.4.

The subjects addressed by Soviet plenary speakers at the ICMs of the first two periods are substantially more inclined towards applied mathematics than those of the speakers from other nations. Indeed, the invitation of colleagues from the Soviet Unions represented 15.2% of all the plenary talks of the first period; and 18% of them belonged to Mathematical Physics, a domain which appears on the whole in only 6% of all the plenary lectures.[96] In other words, 45.6% of all the plenary lectures of the first period touching on Mathematical Physics were given by Soviet mathematicians.

Similarly, in the second period, when the Soviets account for 21.5% of all the plenary lectures, 44% of all the plenary lectures touching on Optimization/Numerical Analysis/Computer Science or Algorithms were given by Soviet mathematicians.

During the third period, 1983–1990, the new official meanings of *Glasnost* and *Perestroika* entered Soviet politics. The share of the Soviet Union at the ICMs began to recede. Among the plenary lecturers of this period we find for the first time a mathematician who was born and had obtained his PhD in the USSR, but was no longer working there by the time he gave his ICM talk: Mikhail L. Gromov had left the USSR already in 1974; in the early 1980s he settled in Paris; he gave a plenary talk at the 1986 ICM in Berkeley. Nonetheless the 17% of plenary lectures given by Soviet mathematicians in the third period is still the second highest score of all countries, and it is bigger than those of Israel and France taken together—see the small chart for Period 3 in Fig. 10.4.

The decline of plenary talks given by Russian mathematicians during the last two periods reflects the brain drain of Russian mathematics during and after the end of the Soviet Union, and contributes to the process mentioned above that would offer the USA more than a 50% share of all plenary lecturers given between 1994 and 2018, according to their affiliation—see the last two small charts in Fig. 10.4. If one looks at the origins of the plenary speakers during Period 4, 1994–2002, we see 6 Russians, which puts the Russian Federation fourth among all nations, between France and Germany. But only 2 of those 6 mathematicians were still working in Russia. Two others had already obtained their PhDs abroad: the Fields Medalists Maxim L. Kontsevich (b. 1964), whose thesis director was Don Zagier (b. 1951) in Bonn, and Vladimir A. Voevodsky, who obtained his PhD at Harvard University under the direction of David Kazhdan (b. 1946).

During the most recent period, 2006–2018, Russian-born mathematicians were invited to give 9 plenary talks (approximately 11%), thus placing the country third, behind the USA and France, among the countries of origins of plenary lecturers. But only two of them were still actually based in Russia: Alexei N. Parshin, and Grigori Y. Perelman who in 2006 refused the Fields Medal and the invitation to speak. The green Russian wedge in the Period 5 charts of Fig. 10.4 has become rather slim.

If we look, not just at the plenary speakers, but at all the members of our population which had some function at one of the ICMs of the last two periods, almost precisely 40% of those colleagues who were born in the Soviet Union were still working there, whereas almost exactly 40% of them now had a job in the USA.

[96] We anticipate here the rough classification scheme into 10 major subfields of mathematics, which will be introduced in detail in Section 10.5.1 below.

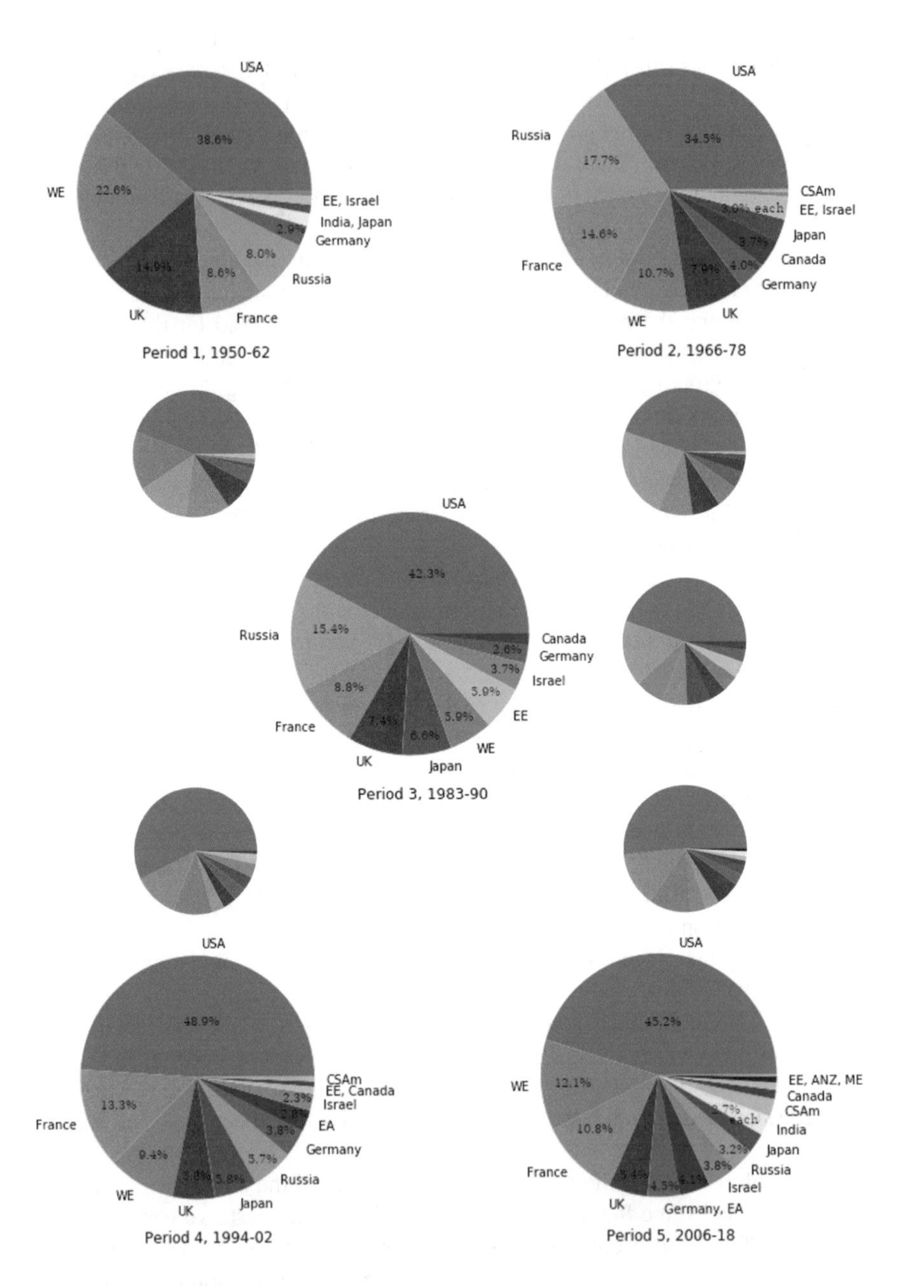

Fig. 10.4 Geographic distribution of our total population (big charts), resp. of the subpopulation of plenary speakers (small charts), according to professional affiliation.

So much about the Soviet Union. As to other individual countries, France managed to impose itself strongly after a weak spell during which fewer French mathematicians rose to the cupola. This low tide is particularly visible in the third period in Fig. 10.4.[97]

As to the other West European nations, including the UK and Germany, they did manage to reassert themselves, even though not to the point of regathering the historic weight which is still visible in the first period.

Comparing in general the big, annotated charts, which give the geographic distribution of our population as a whole—including all committee members[98]—, with the corresponding small pie charts obtained by restricting to the subpopulation of plenary speakers, we see that this crucial subpopulation tends to confirm the observations made for the whole population. However, many of the effects come out more blatantly, as we already pointed out for the US dominance in Figure 10.4. The sagging and subsequent comeback of France starts earlier and is more marked in the small charts of Figure 10.3.

10.4.1.2 The Program Committees

The plenary speakers at an ICM were selected by the Program Committees (PC).[99] Their altogether 182 members—of whom 13 have served twice, in Program Committees at two different ICMs—are part of our population. We can thus look at their geographic breakdown according to origin or affiliation.

Since we have already analyzed the plenary speakers, and since 93 of the 195 seats on the Program Committees were held by colleagues who were also plenary speakers at some ICM,[100] it makes sense to restrict to the 102 committee members who were never invited as plenary lecturers, between 1950 and 2018. Among these colleagues we tend to find in particular the PC members who were appointed by the local Organizing Committee of the corresponding ICM.

The geographical distribution according to professional affiliations of those PC members who were not plenary speakers is noticeably different from what we have seen in Fig. 10.4. Not surprisingly, it tends to be strongly influenced by the places where the ICMs were held. Thus in the first period, with its ICMs at Cambridge (Mass.), Amsterdam, Edinburgh, and Stockholm, the four countries USA, the Netherlands, UK, and Sweden, account for more than 94% of the members of PCs who were

[97] For an overview of the situation in France, see [Menger et al. 2020]. Cf. the *Rapport national de conjoncture scientifique 1969* commissioned by the CNRS, which was prepared by several leading French mathematicians with a view to improving the situation of mathematical research in France.

[98] . . . and also the prize winners; but their statistical effect is relatively small.

[99] Cf. Section 10.2 above for the gradual evolution over time, its composition and its name, of what is known today as the ICM Program Committee.

[100] There are seven cases where a member of the PC also gave a plenary lecture *at the same ICM*. The last time this happened was in 2002 at Beijing. The typical case is of course a plenary speaker who will be recruited for the PC at a later ICM. Note that our data do not record any information about plenary lectures given at ICMs before World War II.

never plenary speakers. Similarly, in the second period, the Soviet Union, France, Canada, and Finland represent 76.4%. In Period 3, Poland, the USA, and Japan account for 70%; adding to this the Soviet Union covers 90% of all PC members of the third period. The effect is much less pronounced in the last two periods, with a share of 40.2%, respectively 52%, of members employed in one of the countries where an ICM of the corresponding period was held. But here too, between 1994 and 2018, the geographical breakdown of the professional affiliations of PC members who never were plenary speakers looks very different from what we see in Fig. 10.4. For instance, the USA only has a share of 13.6% in Period 4, and 10% in Period 5.

10.4.2 Institutions of the Cupola

Having looked at the way in which countries, respectively parts of the world, contributed to the Cupola of the ICMs, let us now try and scale down our attention to individual academic institutions. To be chosen as a plenary speaker for an ICM conveys considerable prestige to a mathematician. It is the sort of distinction that affords a bright mention in a CV. In the world of the ICMs it is topped only by the public celebrity that comes with a Fields Medal, or maybe a Nevanlinna Prize.[101] Relative to a given ICM, we introduce the initialism *FNP* to refer to all those persons who were awarded a Fields Medal, received a Nevanlinna Prize, or were invited to be a Plenary Speaker at that ICM. Having one of its members chosen as FNP at an ICM consolidates an academic institution's standing in the world of mathematics, and inviting or recruiting such a mathematician is in their best interests.

The set of all the mathematical institutes that were home to an FNP is too big and variegated to be conveniently surveyed and followed through the five periods. Therefore we try to single out mathematical institutions that had a notable share of FNP colleagues. For each period we will thus determine a set of research-oriented mathematical institutes whose recruitment politics and research agenda accommodated particularly well the mathematical excellence framed by the ICMs, and by the IMU, for the corresponding timespan.

After testing several variants, we finally settled on the *ad hoc* approach described below, which proved to be adapted to the purpose. It turns out to account for more than half of the FNPs in all our five time periods. The nitty-gritty details of the analysis do not make for a particularly pleasant read. The reader may skip them and go straight away to the subsequent discussion of the five periods, where we also mention a few of the names behind the institutes that occur.

[101] We do not consider the Gauss Prize in this section, because this would have a potential bearing only on the last of our five periods. The Nevanlinna Prize was awarded during the last three periods.

Technicalities. Let us fix one of our five periods.[102] We then pose the following

Definition. An academic institution employing mathematicians is called an *Institution of the Cupola* relative to this period—or a *CI* for short—if at least one of the following two conditions is satisfied:

- There exists at least one ICM of this period at which at least two mathematicians affiliated with the institution were chosen as FNPs;
- at most one ICM of the period has not seen any FNP from among the mathematicians affiliated with the institution.[103]

In order for this definition to be effective, we have to make precise how we count (i) institutions, and (ii) mathematicians from the FNP group.

(i) When checking the conditions of the definition, in most cases it is obvious which institutions need to be considered. Clearly mathematical institutes of universities should qualify just as well as LGTRSs.

There are two subtle cases, which we solve by inclusion: The first problem we have encountered in mustering the affiliations of all the FNP colleagues concerns the University, or the Universities of Paris. Indeed, until the 1960s there was a unique *Université de Paris*, which also included the *Ecole Normale Supérieure* (ENS) as one of its components. It was then split up into thirteen individual universities in 1970 as part of the political reaction to the student revolt of May 1968. In recent years, regrouping universities has again gained ground in France. For the sake of coherence over the seventy years studied here, in the present section we have decided to lump together all Paris universities, including the Paris branch of the ENS, into one 'institution' for all of our five periods. We shall call this synthesis *Paris, University*.

The second place that calls for explanation is Moscow, more precisely the relationship between Lomonosov State University and the Academy of Sciences at Moscow. Simultaneous affiliations with both institutions were very common; about half of the FNPs at Moscow State University were also linked to the Academy. Furthermore, both Moscow State University and the Academy at Moscow qualify individually as CIs during our first two periods. Moscow State University also qualifies as a CI by itself for the third period.[104] Considering their large intersection, we treat the union of both institutions as a single CI, which we call *Moscow State University & Academy of Sciences*, or *Moscow U&A* for short. This does not include other institutions in

[102] Our analysis in this section proceeds period by period. Looking at other time intervals would conceivably yield slightly different lists of distinguished institutions. Indeed, our goal is by no means to establish a ranking of institutional excellence. We simply continue to spell out the image of mathematics projected by the sequence of ICMs, and how it highlights certain institutes.

[103] To render this second condition coherent over time, we artificially adjust our five periods so that all of them contain four ICMs. Therefore, whenever we check the second condition, we keep the 1978 ICM in Helsinki in Period 2, but we also count it for Period 3; and likewise we include the 1990 ICM in Kyoto both in Period 3 and in Period 4 when checking this second condition.

[104] To verify this, one has to apply the second condition of CIs, adding the Helsinki ICM as explained in the previous footnote.

Moscow—see for instance the occurrence of the Moscow Institute for Information Transmission (IPPI) in Period 2.

(ii) Here is how we count the FNPs of a given period. When looking at one ICM of the period we just count heads. That is to say, for instance, that a person who receives a Fields Medal and also gives a plenary lecture at the same ICM is counted as one FNP. However, going through the 3 or 4 ICMs of a given period, a mathematician may be selected as an FNP at several of them, say at m of the ICMs ($1 \leq m \leq 4$). It turns out that $m \leq 2$ for all members of our population and for every period selected.[105] For the whole period, each FNP person will be counted with its multiplicity m.

It would theoretically be possible that our definition establishes an institution as a CI, for a given period, due to a single FNP mathematician, who received such an honour at several ICMs of the period. This might be considered awkward. As a matter of fact, such a case never presents itself in any of our five periods.

Finally, a mathematician who is chosen as an FNP at an ICM may be affiliated during the year of that ICM with more than one institution. Indeed, there are cases where an FNP is affiliated, during the year of the ICM in question, with two different CIs of the corresponding period. (Such a circumstance does not affect the verification of either of the criteria of our definition from the point of view of any of the institutions to which the person is bound.) These cases, however, are not frequent enough to warrant a detailed investigation.

Let us now walk through the five periods and see in each of them which institutions distinguish themselves with respect to the ICMs of that period. A hard core of CIs based in the USA will emerge throughout the seventy year span. We shall briefly comment on them at the end of this section. Other interesting CIs will be discussed as we go along.

Period 1, 1950–1962. For the first period, we find ten Institutions of the Cupola, i.e., ten CIs. We list them here in descending order according to the number of FNPs of Period 1 affiliated with the establishment in question.

- Moscow State University & Academy of Sciences
- The Institute for Advanced Study, Princeton
- Princeton University
- Paris, University
- University of Chicago
- Columbia University
- The Swiss Federal Institute of Technology, ETH, Zürich
- Harvard University
- Massachusetts Institute of Technology, MIT
- University of California at Berkeley, UCB.

[105] Several mathematicians have been chosen as FNPs at three different ICMs in the course of their career, but never within one of our periods.

Altogether we count[106] for the first period 49 FNPs affiliated to these ten CIs.[107] This amounts to almost 60% of altogether 82 FNPs[108] in Period 1.

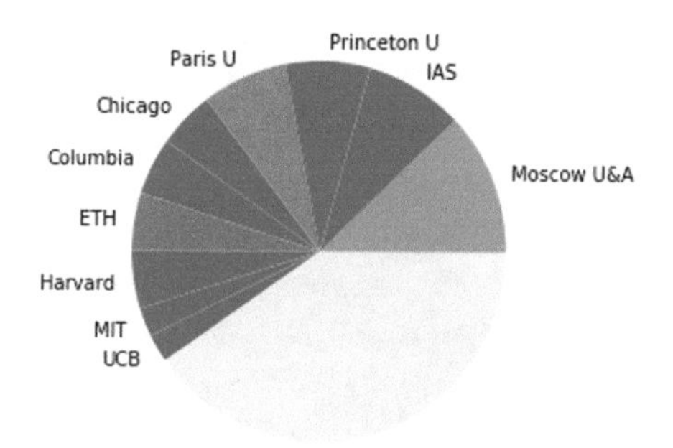

Fig. 10.5 The FNP share of CIs during Period 1; colors correspond to the countries of the institutions.

Note the strong presence of Moscow U&A in spite of the fact that the first ICM of the period, in 1950, took place without any Russian participation.

The ETH in Zürich qualifies as a CI in this period on account of both conditions of our definition, thanks to plenary lectures given by Heinz Hopf in 1950, Eduard Stiefel (1909–1978) at the 1954 ICM, and by Beno Eckmann (1917–2008) and Peter Henrici (1923–1987) at the 1962 ICM. In fact, 1962 was also the year when Henrici moved from UCLA to Zürich; but UCLA does not qualify as a CI for the first period.

Period 2, 1966–1978. Thirteen CIs are borne out by the four ICMs between 1966 and 1978. Here they are, again in descending order of the number of FNPs affiliated with these institutes.

- Moscow State University & Academy of Sciences
- Princeton University
- Harvard University
- Massachusetts Institute of Technology, MIT
- *Institut des Hautes Études Scientifiques*, IHES, Bures-sur-Yvette (near Paris).
- University of California at Berkeley, UCB
- The Institute for Advanced Study, Princeton
- University of Chicago
- Stanford University
- Paris, University

[106] In the sense explained above: a person chosen as FNP at m distinct ICMs of the period is counted with multiplicity m.

[107] The number of individual persons giving rise to these 49 FNPs is 44.

[108] Corresponding to 76 physical persons.

- *Institut problem peredachi informatsii* (Institute for Problems in Information Transmission), IPPI, Moscow
- *Collège de France*, Paris
- University of Cambridge, UK

We count, with multiplicities as before, 53 FNPs linked to these thirteen CIs of the second period. This represents almost 68% of the 78 FNP total of Period 2.[109]

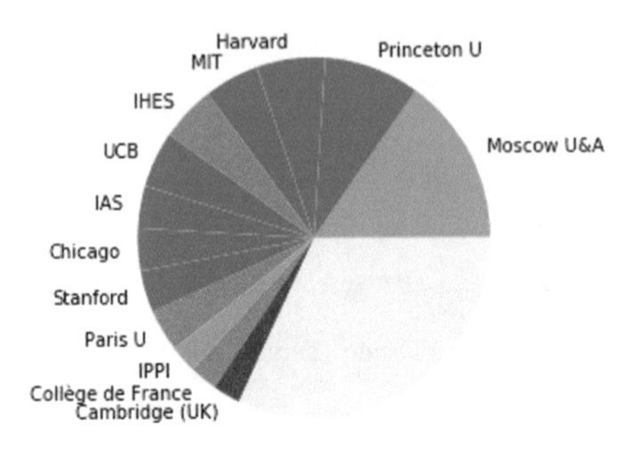

Fig. 10.6 The FNP share of CIs during Period 2.

IHES was founded in 1958—cf. Section 8.3 above. It makes its first appearance as a CI in this list for the second period, thanks to Alexander Grothendieck's 1966 Fields Medal (*in absentia*), Pierre Deligne's plenary lecture in 1974 as well as his 1978 Fields Medal, and Dennis Sullivan's (b. 1941) plenary lecture in 1974 at Vancouver.

The *Collège de France* happens to meet the first condition of a CI in this second period because two of the plenary speakers at the 1974 ICM in Vancouver held chairs at this venerable French institution (founded in 1530, open to the French public and independent of the French system of higher education): Jacques-Louis Lions and Jacques Tits (1930–2021).

The IPPI at Moscow owes its presence in our list, next to the Moscow U&A, to Grigory A. Margulis, who was awarded the Fields Medal in 1978, and to Roland Lvovich Dobrushin (1929–1995), who gave a plenary lecture at the 1978 ICM in Helsinki.

At the 1970 ICM in Nice, France, it happened for the first time that two out of four Fields Medals were awarded to mathematicians from the University of Cambridge, UK. The same constellation would repeat itself at the Berlin ICM in 1998, in Period 4. Periods 2 and 4 are the only ones where Cambridge University rose to CI status. In 1970 the two winners of the Fields Medal were Alan Baker (1939–2018) and John G. Thompson (b. 1932). In fact, Thompson only moved to Cambridge in 1970 where

[109] The count involves 48 physical persons affiliated with a CI, i.e., more than three quarters of the altogether 63 mathematicians chosen for FNPs during the second period.

he was offered the Rouse Ball Professorship. He had already given a plenary lecture on the classification of finite simple groups in Moscow in 1966, when he was still at the University of Chicago, another CI of the second period.

Period 3, 1983–1990. We find twelve CIs for this period of only three ICMs. They are listed hereafter in descending order of the number of FNPs affiliated with them. For the first time, institutions from outside of the USA and Europe make their way into the list, in this period that saw the very first ICM held neither in Europe, nor in North America: in Kyoto, Japan, in 1990.

- Princeton University
- Harvard University
- *Institut des Hautes Études Scientifiques*, IHES, Bures-sur-Yvette (near Paris).
- Massachusetts Institute of Technology, MIT
- University of California at Berkeley, UCB
- Research Institute for Mathematical Sciences, RIMS, at Kyoto
- The Hebrew University, Jerusalem
- Moscow State University & Academy of Sciences
- Courant Institute, New York University, NYU
- Brown University,
- University of California at San Diego, UCSD
- The Institute for Advanced Study, Princeton.

We count, with multiplicities as before, 36 FNPs affiliated with one of these twelve CIs of the third period. This represents about 64% of the 56 FNP total of Period 3.[110] Note that the third period is the first one that comprises only three ICMs.

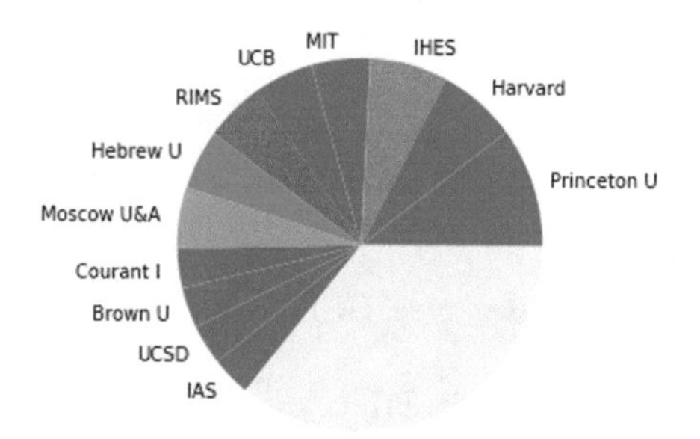

Fig. 10.7 The FNP share of CIs during Period 3.

[110] The count involves 34 physical persons affiliated with the CIs, out of the altogether 54 mathematicians chosen for FNPs during the third period.

The Hebrew University at Jerusalem makes its first appearance as a CI in period 3, in view of the plenary lectures given by Michael O. Rabin (b. 1931) in Warsaw in 1983, and the fact that Saharon Shelah was invited for a plenary talk at the same ICM. Shelah did not attend the Warsaw Congress, though, but was invited again and gave a plenary talk in 1986 at Berkeley.

The RIMS in Kyoto enters the above list of CIs for the third period on several accounts. To start with the 1990 ICM, which brought the mathematical world to Kyoto and which closes this period, Shigefumi Mori was awarded the Fields Medal, and Yasutaka Ihara (b. 1938) gave a plenary lecture. This already checks the first condition of our definition of a CI for the third period.

Furthermore, Mikio Sato (b. 1928) presented a plenary lecture at the Warsaw ICM in 1983, and Masaki Kashiwara (b. 1947) had given a plenary talk at Helsinki in 1978 (which enters into the verification of the second condition of our definition of a CI for the third period, even though it belongs to period 2). Thus RIMS also satisfies the second condition of a CI for Period 3.

Michael H. Freedman's (b. 1951) Fields Medal in 1986 was one of the two events that brought the University of California at San Diego into our list for Period 3. The other one is Richard M. Schoen's (b. 1950) plenary lecture at the Berkeley ICM in 1986.

New York University owes its place in the period to Robert Tarjan's (b. 1948) Nevanlinna Prize and Peter Lax's (b. 1926) plenary address in 1983.

Brown University is listed in view of the 1983 plenary talks presented by Wendell H. Fleming (b. 1928) and Robert MacPherson (b. 1944).

Period 4, 1994–2002. In this second period consisting of only 3 ICMs, we find the following ten CIs, ordered as before.

- Paris, University
- Massachusetts Institute of Technology, MIT
- Princeton University
- Harvard University
- University of California at Berkeley, UCB
- Stanford University
- The Institute for Advanced Study, Princeton
- *Institut des Hautes Études Scientifiques*, IHES, Bures-sur-Yvette (near Paris).
- University of Chicago
- University of Cambridge, UK

For this fourth period, we count—with multiplicities, as before—41 FNPs affiliated with (at least) one of the ten CIs. This represents about 62% of the 66 FNP total of the penultimate period.[111]

[111] The count involves 39 physical persons affiliated with the CIs, out of the altogether 64 mathematicians chosen for FNPs during the fourth period.

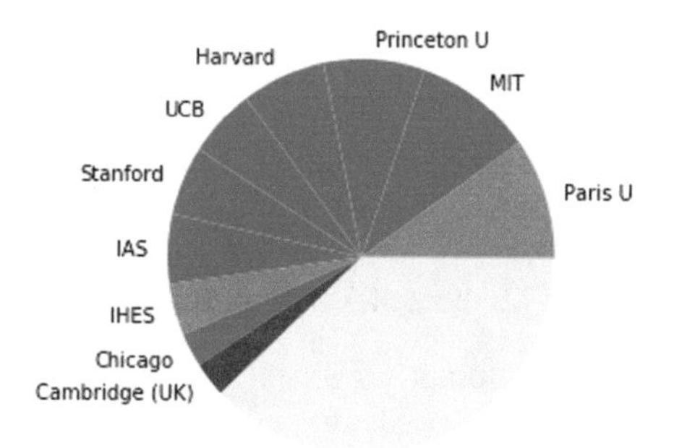

Fig. 10.8 The FNP share of CIs during Period 4.

We have already mentioned above the double score of Cambridge University for the Fields Medals awarded at Berlin in 1998: Timothy Gowers and Richard Borcherds (b. 1959).

Note that Moscow State University & Academy has disappeared from the list of CIs in Period 4, after the end of the Soviet Union. Already in the third period, Moscow U&A barely passed the criterion of a CI. This fact is related to a gradual redistribution among institutions. In the first period, 86.4% of the Soviet FNPs were affiliated with Moscow U&A. During the second period, 14 out of 18 FNPs from the USSR were employed in Moscow; 12 of them were affiliated with Moscow U&A. Our short third period sees 7 out of 11 Soviet FNPs employed in Moscow, but only 3 affiliated with Moscow U&A.

Seven of the ten CIs of Period 4 are based in the USA, and the fourth period is the only one in which all CIs are based in only three different countries: the USA, France, and the UK. This may have to do with the fact that the period only comprises three ICMs. At any rate, it does not indicate a trend towards national concentration, as is shown by the subsequent period:

Period 5, 2006–2018. For this most recent period, we find fifteen CIs, listed as before in descending order of their FNP count.

- Paris, University
- Princeton University
- University of California at Berkeley, UCB
- Stanford University
- Yale University
- Courant Institute, New York University, NYU
- Massachusetts Institute of Technology, MIT
- The Hebrew University, Jerusalem

- The Institute for Advanced Study, Princeton
- University of Chicago
- *Instituto Nacional de Matemática Pura e Aplicada*, IMPA, Rio de Janeiro
- The Swiss Federal Institute of Technology, ETH, Zürich
- University of California at Los Angeles, UCLA
- University of Oxford, UK
- The Hausdorff Center of Mathematics, HCM, Bonn, Germany.

For this last period, which comprises four ICMs, we count—with multiplicities, as before—58 FNPs affiliated with (at least) one of the 15 CIs. This amounts to about 60% of the 96 FNP total of the period.[112]

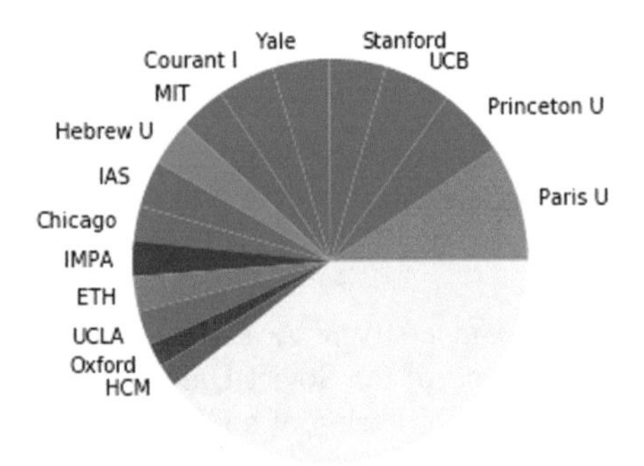

Fig. 10.9 The FNP share of CIs during Period 5.

This most recent period displays the biggest number and the greatest diversity of CIs. The presence of the Courant Institute, New York, marks a kind of opening towards more applied mathematics. Its strong presence among the CIs of Period 5 is afforded by various events, which took place at three different ICMs: Subhash Khot's (b. 1978) Nevanlinna Prize in 2014; the 2006 plenary lectures by Percy Deift (b. 1945) and Robert V. Kohn (b. 1953); finally the 2018 plenary talks by Sylvia Serfaty (b. 1975) and Lai-Sang Young (b. 1952).

The IMPA in Rio de Janeiro makes its appearance among the ICs of the last period, not because of the Rio ICM in 2018, but because of Artur Avila (b. 1979), who was awarded the Fields Medal in Seoul in 2014—he had already presented a plenary lecture in Hyderabad in 2010—, and Fernando Marques's (b. 1979) plenary lecture in Seoul.

The example of the Swiss ETH is interesting because none of its three FNP members during the fifth period is of Swiss origin. Two of them, the Greek mathematical physicist Demetrios Christodoulou (b. 1951) and Rahul Pandharibande (b. 1969), of

[112] The count involves 57 physical persons affiliated with the CIs, out of the altogether 94 mathematicians chosen for FNPs during the fifth period.

Indian origin, started their careers in the US and had been professors at Princeton University—one of our perennial CIs—before coming to Zürich. They gave plenary lectures in 2014, resp. 2018. The 2018 Fields Medalist Alessio Figalli (b. 1984) grew up in Italy and received most of his advanced research training in France. In 2016, he moved to the ETH.

Peter Scholze's Fields Medal in 2018 would not have sufficed to make the Hausdorff Center at Bonn a CI for the fifth period. The other person needed for that was Geordie Williamson (b. 1981) from the University of Sydney. After his earlier stay at the Max-Planck-Institute for Mathematics at Bonn, Williamson was still a Bonn Research Fellow in 2018.

These two CIs of the fifth period, ETH and HCM, remind us of the intrinsically international mathematical culture that is generally implemented today at all major research centers, not only at the institutions we are looking at here. This is usually taken for granted, even though it is the result of a fairly recent historical process—see Sections 5.2 and 8.3 above. All these research-oriented institutes are of course locally based in their respective countries. For our small selection this is shown by the colors of their pieces in the pie charts. Not all countries can pride themselves of such institutes, let alone of CIs that make it into our selection. However, all major existing research centers cultivate their international dimension, often with an almost global reach. Focussing again on the CIs, this twofold reflection of today's global academic world: in the colors of the charts, and in the origins of individual researchers, was particularly pronounced in the latest period. It holds the promise of a continuing worldwide mathematical network for the future.

Surveying all five periods, we find exactly four institutions that turned out to be CIs in every single period: the Institute for Advanced Study as well as Princeton University, the University of California at Berkeley, and the Massachusetts Institute of Technology MIT. All of them are located in the USA, and each of them is a visible competitor for outstanding mathematical talent on the global market. Altogether, over all the five periods, these four institutions have been the home base of 90 FNPs. This is not far from a quarter of the 378 FNPs of all periods. (About two thirds of this total count of 378 FNPs—249 of them, to be precise—were affiliated with some CI of their respective period.)

Three other institutions are CIs in all but one time period: the Universities of Chicago, Harvard, and Paris, University.

Two further institutions managed to rise to the cupola in three consecutive periods: Moscow U&A in the first three periods, and IHES (which was founded only in 1958) in periods 2 through 4.

Stanford University also appears three times, if not in consecutive periods.

Whereas French institutions are certainly visible in the first three periods, the conglomerate of the Paris Universities turns out to be the biggest CI worldwide in the two most recent periods 4 and 5. Even more is true: In Period 4, IHES is also a CI; its FNPs for that period were all Fields Medalists: Jean Bourgain (1954–2018) in 1994, Kontsevich in 1998, and Laurent Lafforgue in 2002. Let us for a moment amalgamate Paris University and IHES.[113] Then it turns out that, both in Period 4 and in Period 5, this combined institution counts precisely as many—namely, ten—FNPs as were located at Princeton (taking IAS and Princeton University together), during each of the last two periods.

France is a particularly centralized country, and what we call "Paris University" is a synthesis of all Paris universities. This may partly explain the extraordinary performance of this CI since 1994. Nevertheless, already the sheer list of mathematicians professionally affiliated with Paris when they received their Fields Medal in those years—apart from the three mentioned above, there were Pierre-Louis Lions (b. 1956) and Jean-Christophe Yoccoz (1957–2016) in 1994, Wendelin Werner (b. 1968) in 2006, Cédric Villani (b. 1973) in 2010, and Artur Avila in 2014—establishes Paris as the mathematical hotspot that has resonated most intensely with the Cupola of the ICMs during the last quarter century.

Traces of Mathematical Genealogies in the Cupola. We have investigated all the members of our population whose thesis advisors are also in the database. The resulting *PhD graph* of advisorships inside of our population has 68 connected components, 38 of which are just couples. The idea of a PhD thesis has changed according to historical and national contexts, and the type of relationship between thesis student and advisor depends on local cultures as well as personal idiosyncrasies. Yet, even if there are also other influences in mathematical careers than those exerted by advising a thesis, the PhD graph does illustrate a basic transmission of academic mathematical excellence inside our population. This justifies showing a few remarkable connected components.

By far the biggest connected component of our PhD graph is that formed by Kolmogorov's thesis students who would themselves enter our population at a certain point in time—see Fig. 10.10. In fact, only about 1/8 of all of Kolmogorov's thesis students belong to our database. Here and in the following graphs each dot in our diagrams represents a person from our population. The persons that show up in this component, besides Andrey Kolmogorov, are Anatoli Vitushkin, Israel Gelfand, Sergey Nikolsky, Anatoly Maltsev (1909–1967), Roland Dobrushin, Albert N. Shiryaev (b. 1934), Yuri Prokhorov, Yuri Rosanov (b. 1934), Yakov Sinai (b. 1935), Eugene Dynkin, and Vladimir Arnold. The following 'generations' include Alexander Varchenko (b. 1949), Victor Vassiliev (b. 1956), Anatoliy Skrokhod (1930–2011), Grigory Margulis, and Marina Ratner (1938–2017). The only person in this component who did not get her degree at Moscow University is the South Korean mathematician Hee Oh (b. 1969), who obtained her PhD at Yale University in 1997 under the direction of Margulis.

[113] For Period 5, it is enough to only consider Paris University, since IHES did not have a single FNP during those years.

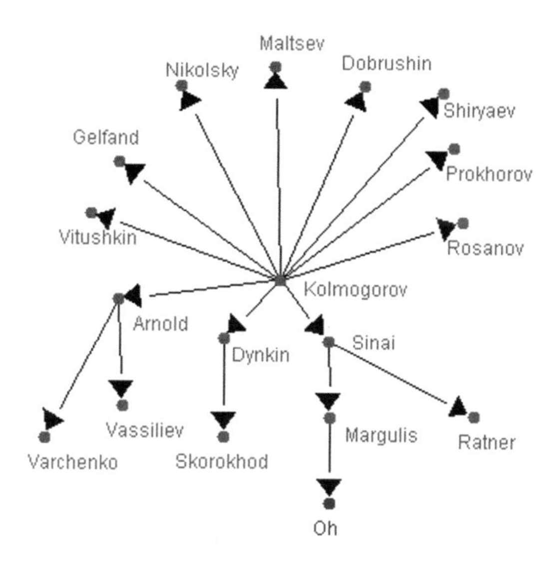

Fig. 10.10 Kolmogorov's component of the PhD graph in our population.

In Fig. 10.11 we present all the other connected components of the PhD graph within our population that have at least six vertices. The most frequent nationality in each component is shown in red; the blue vertices correspond to nation origins different from the dominant one.

The page starts with Shokichi Iyanaga's component, which is completely situated in Japan. It comprises Kenkichi Iwasawa, Kiyoshi Ito (1915–2008), Mikio Sato, Yasutaka Ihara; and from there to Michio Jimbo (b. 1951), Masaki Kashiwara, Tetsuji Miwa (b. 1949), and to Ihara's student Kazuya Kato (b. 1952).

To the right, we start from Laurent Schwartz, who leads us to Alexander Grothendieck, Gilles Pisier (b. 1950), Bernard Malgrange (b. 1928), Jacques-Louis Lions, Michel Raynaud (1938–2018), Pierre Deligne, Jean-Michel Bismut (b. 1948); and via Deligne to Michael Rapoport (b. 1948) and Peter Scholze.

William Hodge's progeny includes Michael Atiyah, Simon Donaldson (b. 1957), George Lusztig (b. 1946), Frances Kirwan (b. 1959), Peter Kronheimer (b. 19639), and Corrado de Concini (b. 1949).

Another genealogy starts in the UK with Harold Davenport. It includes Alan Baker, Hugh L. Montgomery (b. 1944), John Conway (1937–2020), and Richard Borcherds. From Baker we get to John Coates (b. 1945), whence Catherine Goldstein (b. 1958), as well as the branch of Andrew Wiles (b. 1953), with Richard Taylor, and Manjul Bhargava (b. 1974).

Heinz Hopf leads us to Hans Freudenthal—who got his PhD when Hopf was still in Berlin. Later in Zürich he was one of the thesis advisors of Friedrich Hirzebruch, even though the latter obtained his degree in Münster, Germany with Heinrich

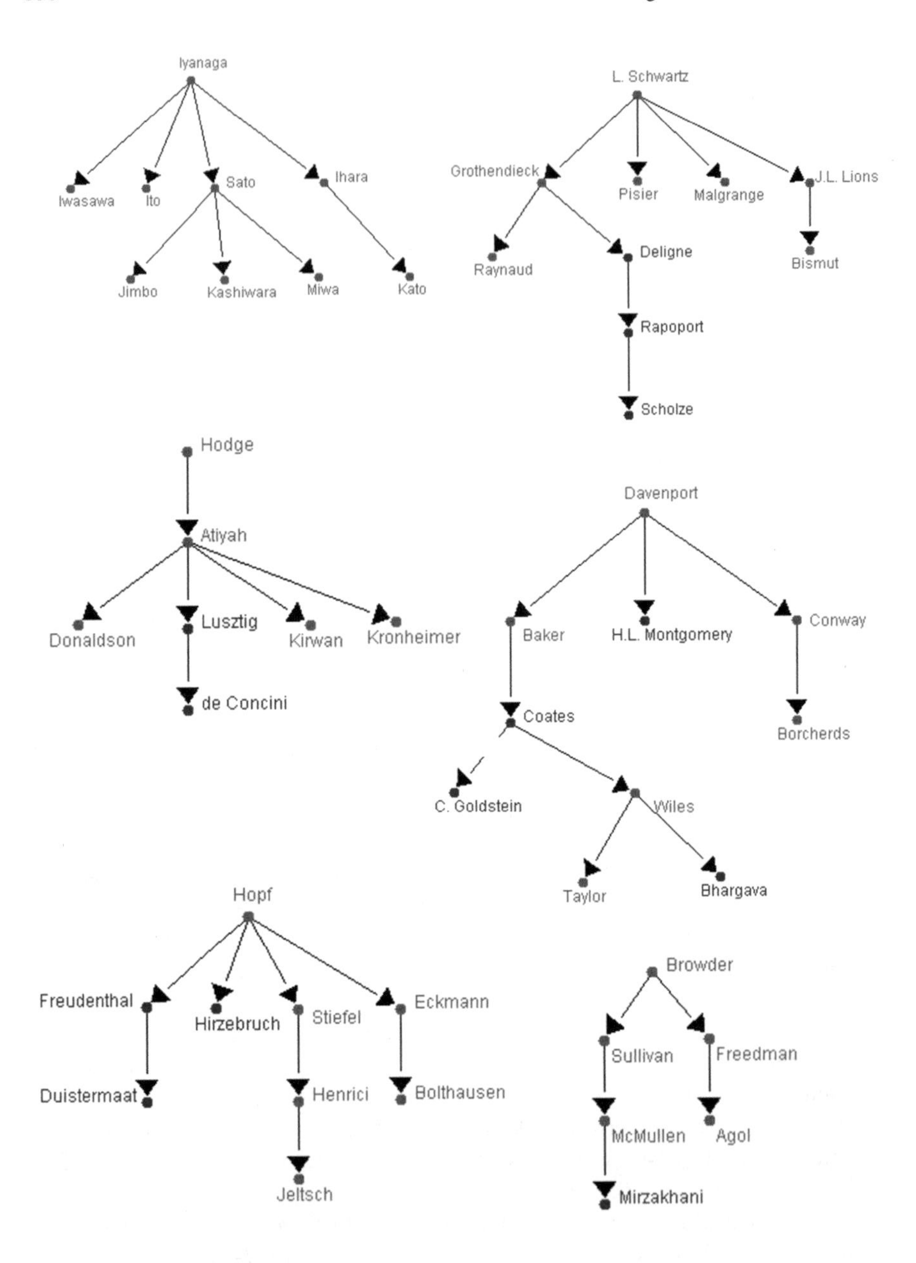

Fig. 10.11 Various components of the PhD graph in our population.

Behnke. Further there are Eduard Stiefel and Beno Eckmann. From there we pass to Johannes Duistermaat (1942–2010), Peter Henrici, Erwin Bolthausen (b. 1945), and finally to Rolf Jeltsch (b. 1945).

We conclude the samples in Fig. 10.11 with a graph situated in the USA, starting from William Browder (b. 1934): Dennis Sullivan, Michael Freedman (b. 1951), Curtis McMullen (b. 1958), Ian Agol (b. 1970); and Maryam Mirzakhani.

10.5 Framing Domains of Mathematics

Mathematicians enter our database because their mathematical creativity or expertise is recognized as outstanding or particularly useful for a successful ICM. In the preceding section we have tried to portray this group of people geographically according to their origins and professional affiliations. The present section addresses their mathematical specialties. Since we are dealing with the cupola of the ICMs, the breakdown of the domains of expertise upheld by our population, and its evolution over time, reflects the domains of mathematical research that received particular attention on the part of the framers of the ICMs.

10.5.1 Mathematical Subdomains

In order to screen for mathematical specialties we shall use the following rough breakup of mathematics into major subdomains.[114] Note the corresponding abbreviations that will be used for quick reference in the sequel.

- **Gen**: General Mathematics; History; Foundations. This corresponds to sections 00, 01, 03, 06, 08, and 18 of the Mathematics Subject Classification MSC.[115]
- **Discr**: Discrete Mathematics & Convex Geometry; MSC sections 05, 52.
- **NTAG**: Number Theory. Algebra. Algebraic Geometry. Group theory; MSC sections 11, 12, 13, 14, 15, 16, 17, 19, 20.
- **Ana**: Real and Complex Analysis; MSC sections 26, 28, 30, 31, 32, 33, 40, 41.
- **OpTh**: Harmonic and Functional Analysis; Operator Theory; MSC sections 42, 43, 44, 46, 47.
- **DIEq**: Differential and Integral equations; MSC sections 34, 35, 37, 39, 45.
- **OptCS**: Optimization. Numerical Analysis. Computer Science. Algorithms; MSC sections 49, 65, 68, 90, 93, 94.
- **ProbStat**: Probability Theory and Statistics. Applications to Economics, Biology and Medicine; MSC sections 60, 62, 91, 92.
- **TopGeo**: Topology and Geometry; MSC sections 22, 51, 53, 54, 57, 58.
- **MaPh**: Mathematical Physics; MSC sections 70, 74, 76, 78, 80, 81, and 82.

---

[114] It has been used in a similar manner before, for instance in [Mihaljević & Teschke 2014].

[115] As mentioned before, he relevant websites are [URL 59] jointly with [URL 60], and [URL 61]. This classification has been systematically used by zbMATH Open since 1980.

In order to be able to appreciate how our cupola differs from the overall mainstream of mathematical production, let us start with the total distribution of the almost 3.6 million publications refereed in the *Zentralblatt*—or rather zbMATH Open, as it is now called—between 1949 and 2020; their breakup is shown in Fig. 10.12.[116] The four leading domains—each of which represents more than 10% of the total—are Optimization (including Computer Science), Mathematical Physics, Probability and Statistics, and Differential & Integral Equations. Even though it gives a first impression of the mathematical production per domain, this pie chart can also be misleading, if only because publication strategies vary from one mathematical speciality to another. Some may, for example, tend to prefer a greater number of shorter pieces in specialized journals to a smaller number of major papers in highly visible periodicals.[117] We nonetheless use this chart as a signpost of the mathematical production at large.

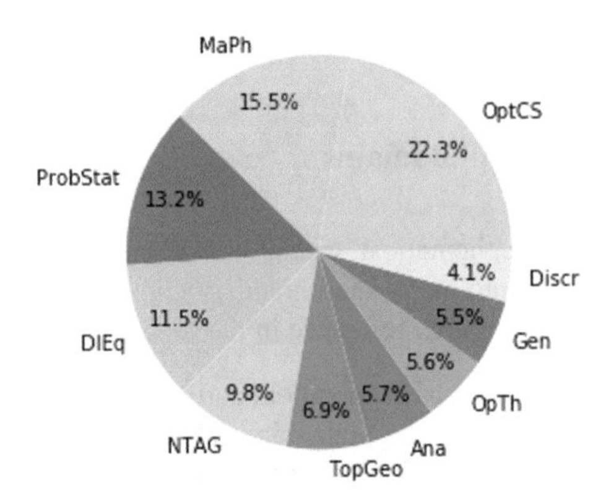

Fig. 10.12 The distribution of all publications refereed in zbMATH Open between 1949 and 2020.

[116] Here and in the sequel of this section we profit from the generous massive access to zbMATH Open data as well as additional information, which was granted us for the preparation of this book. Personal thanks go to Olaf Teschke for being such a reliable partner in this collaboration. As of 2021, the major part of these data are available via the API [URL 62] under CC-BY-SA 4.0 license. This should make it possible to reproduce the analyses presented here, or to perform similar ones.

[117] The classification used by *Zentralblatt*, resp. zbMATH Open, has also evolved over time. However, all older papers classified in a way that is no longer used today can be sorted unambiguously with respect to our ten subdomains. As for multiple classifications, which do occur frequently, if a paper is classified to belong to n different subdomains of our list, each of these subdomains is counted for that paper with weight $1/n$.

10.5.2 Fields Medalists

One of the first ideas one may have, if one wants to compare Fig. 10.12 to the Cupola of the ICMs, is to look at the sequence of Fields Medal, and how the work for which they have been awarded is distributed among our ten subdomains of mathematics. Indeed, no other distinction in the domain of mathematics catches the public eye as much as the Fields Medal.[118] An oft-heard comment points out that the choice of the Fields Medals expresses a strongly biased image of the broad advance of the mathematical sciences, highlighting certain areas of pure mathematics disproportionally.

> All 56 winners [of the Fields Medal] so far have been phenomenal mathematicians, but such biases have contributed to 55 of them being male, most being from the United States and Europe and most working on a collection of research topics that are arguably unrepresentative of the discipline as a whole.[119]

Some such information about the Fields Medals can indeed be seen immediately in our data. For instance, about two thirds of all Fields Medalists have worked in the domains *NTAG* or *TopGeo*.[120] Still, we abandoned this sort of inquiry after a few initial attempts. The principal reason is that the sample is too small to allow for an enlightening study of distributions. This continues to hold true when one tries to enlarge the group studied by adding the Fields Medal Committee members. Independently of the method one would like to apply, it should also be remembered that in most cases the actual Fields Medalist had to be chosen from among a small group of comparable contenders.[121]

Given all these difficulties, we think that the serious study of the attribution of Fields Medals over the years has to wait until the archival evidence concerning the work of the Fields Medal Committees is accessible for historical scrutiny. Unfortunately, in view of the extravagant 70-year embargo imposed by the IMU on the files of all of its Prize Committees, this means that we still have to wait quite a long time. This renders occasional insights gleaned from other, accessible sources, as in Barany's work, all the more exciting.

---

[118] The comparison of the Fields Medal with the Nobel Prize sounds obvious today, but probably only dates back to 1966—see [Barany 2015].

[119] See [Barany 2018], p. 271.

[120] Incidentally, the 16 Fields Medalists of the most recent period, 2006–2018, came from 12 different parts of the world (in the sense introduced in Section 10.4.1), and were still employed in 8 different parts of the world at the moment of their award. This is by far the most geographically diverse group of all the time periods.

[121] Cf. the corresponding loose discussion in [Bannister & Teschke 2018].

10.5.3 Plenary Speakers

Instead of the medals, we turn to the plenary lecturers at the ICMs during the past seventy years. The following Fig. 10.13 shows two possible classification breakups of their production. On the left, we simply count the plenary lectures themselves according to the subdomains they can be associated with.[122] The second way of counting the production of our plenary speakers is by looking at all the papers they published at about the same time as their plenary talk.[123] The breakdown of these publications is shown on the right in Fig. 10.13.

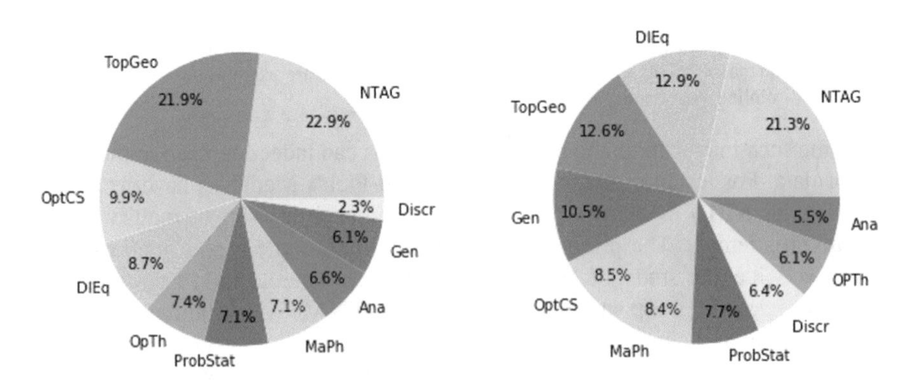

Fig. 10.13 Plenary Lectures, 1950–2020 (left); all publications of plenary speakers around their plenary talk, 1949–2020 (right).

The difference between Fig. 10.12 and 10.13 is blatantly obvious. Of the four domains that take the lead in Fig. 10.12, only Differential & Integral Equations reaches again a score above 10% in one of the charts of Fig. 10.13. Optimization (including Computer Science) still comes in third in the count of the Plenary Lectures (the left-hand chart). Mathematical Physics, and Probability & Statistics definitely lose their prominent positions. The new leader is Number Theory/Algebra/Algebraic Geometry/Group theory, which we call *NTAG*. This and Topology/Geometry are the

[122] Only for the lectures that were published in the ICM Proceedings do we have an MSC classification. This, by the way, is independent of whether the talk was actually delivered at the ICM or not. (In one exceptional case, the classification is that of an independent publication with the same title as the lecture.) In this way, no classification data are available for 20 invited lectures. The total number of talks available with their classifications is 315.

[123] Specifically, for a speaker who delivered a plenary lecture at the ICM in the year N, we look at the classifications of all the papers (co-)authored by that speaker that appeared in the four year interval $[N-1, N+2]$. This includes the plenary lecture itself, if it was published. We again acknowledge the generous access to the zbMATH Open data without which this analysis could not have been realized.

two newcomers from pure mathematics in the upper tier of the survey of publications by plenary speakers.

Both charts shown in Fig. 10.13 can be broken up according to our five time periods, instead of considering all the seventy years at once. When one does this for the plenary lectures themselves (i.e., the chart on the left), the dominance of *NTAG* and *TopGeo* shows in every single period, and *OptCS* comes in third in all periods but the first one (1950–1962), which is the only period where both *OpTh* (9.7%) and *Ana* (9.0%) compete with *DIEq* (9.0%) to break the 10% threshold, whereas *OptCS* does not even attain 5%.

However, going through our five periods with a view to the publications of our plenary speakers at about the time of their ICM lecture, like in the breakdown on the right of Fig. 10.13, yields pie charts that vary a great deal.

For example, in Period 3: 1983–1990, the biggest share goes to the category General Mathematics/History/Foundations, which we call *Gen*. This is largely due to one particularly prolific person among the plenary lecturers, Saharon Shelah. He published 156 papers classified in this category around the same time as the 1983 and 1986 ICMs, 154 of them concern set theory. This personal contribution represents 16% of all papers published by plenary speakers of the third time period at about the time of their lectures, and boosts the *Gen* category to 19.8% among those publications.[124]

In Period 4: 1994–2002, the three strongest specialities are Mathematical Physics, Differential & Integral Equations, and Optimization/Numerical Analysis/Computer Science; only then follow *NTAG* and *TopGeo*.

We have looked a bit more into the most prolific authors of our population, and into publication patterns according to the different categories, in particular the frequency of co-authored papers. The proportion of co-authored papers in our population increases gradually over time, from altogether less than 30% in Period 1 to more than 70% in Period 5. But the variation between the different specialities is considerable. The domains *Discr* and *OptCS* show the highest proportions of co-authored papers.

---

[124] Incidentally, considering the total publication record, Shelah is the second most prolific author of our whole population, topped only by Erdős. Both of them were awarded the Wolf Prize; Erdős in 1983, and Shelah in 2001, when he was the first mathematician born in Israel to win this award.

10.5.4 Filtering the Mathematical Production

We could end here. Instead, let us pry into the matter from a different point of view. We would like to capture the subject distribution of the plenary lectures in terms of a selection procedure which is not immediately linked with the IMU or the ICMs. The hope is to gain a new perspective on the choices made for the ICMs. To do this, we have turned to the internal work flow of zbMATH Open.[125]

No working mathematician can keep abreast of all mathematical publications; everyone has to prioritize her or his attention, according to her or his special interests, within the large field of mathematics, and through a personal ranking of the mathematical journals she or he will try to follow. In other words, we all apply our personal filters in monitoring the incessant production of the mathematical literature. What happens to the Cupola of the ICMs when we look at it through such a lens? To be sure, biases in favor of certain branches of mathematics have to be avoided in the analysis; the effect of being keenly interested by plenary lectures on topics near one's own research domain—however well presented other talks may be, to the large crowds gathering at the ICMs—is as natural as it is uninteresting for the kind of filtering we are looking for. Is it possible to trace the production of all the plenary speakers by carefully selecting journals without any prejudice with respect to subdomains of mathematics?

The first idea could be to look only at *Generalist Journals*, in the sense explained in [Mihaljević & Teschke 2014].[126] These journals try to publish good mathematics in an unbiased way with respect to mathematical subdomains. The problem with this approach is that the percentage of the papers of our plenary speakers published in generalist journals turns out to be too small to be a fair reflection of their productivity. Therefore we had to look for other filters adapted to our problem.

All mathematical journals whose articles are treated by zbMATH Open, with a view to being reviewed, are categorized by the zbMATH Open editorial board according to their expected scientific quality, and with a view to keeping a reasonable balance between specialized journals and those that try to cover many branches of mathematics. The most prestigious category, for which every editor was allowed to make a limited number of proposals, is internally called *Fast Track*; the papers in these journals receive the most speedy treatment. Once all the Fast Track slots are filled, the board decides on the next best journals, called *Category 1*. And so forth. When journals change their profile as time goes by, the zbMATH Open editors try to react swiftly and re-categorize them if necessary. Our access to the zbMATH Open data included this categorization of all the journals.

The zbMATH Open procedures just described go back to the first years of the twenty-first century. In spite of individual journals that may change categories, the hierarchy is generally quite stable. It essentially still reflects a configuration that

[125] Once more we thank Olaf Teschke for providing the necessary background information.

[126] Cf. [Grcar 2010].

was recognized by the editorial board in the early years of this century. The years before the turn of the century had clearly contributed to shaping this configuration. Indeed, we have checked that no major discontinuity occurred around 2000 for the breakdowns we have been studying. All this encouraged us to look at the period 1993–2020, and filter papers of the Plenary Speakers according to the internal zbMATH Open categories of journals they were published in.

For the time interval 1993–2020, the Fast Track journals published about 13.6% of all mathematical papers. If one adds to this the Category 1 journals, we attain 38.6% of the total mass of publications. Altogether 989,332 papers have been treated in FT & Cat. 1 journals between 1993 and 2020, of which 1,404 were (co-)authored by plenary speakers in chronological vicinity to their ICM talks, as explained above. The corresponding classification breakdowns are shown in Fig. 10.14.

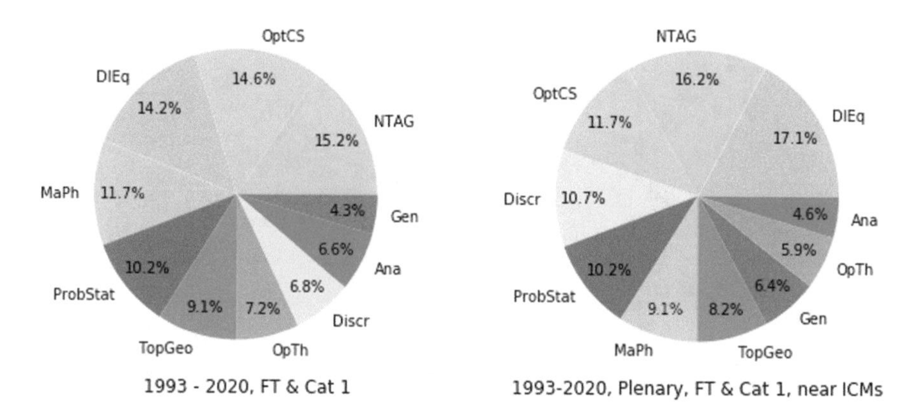

Fig. 10.14 Comparing all publications in Fast Track and Cat. 1 journals between 1993 and 2020 (left), to those (co-)authored by Plenary Speakers around their plenary talk (right).

In spite of a slight reshuffling of several shares, the overall resemblance of the two sets of papers is remarkable. This suggests that the class of journals chosen for this comparison is rather well adapted to the production of the plenary speakers of the last seven ICMs. In other words, the selection procedures for the Cupola of those ICMs appear to be by and large compatible with the internal hierarchization of mathematical journals practiced by zbMATH Open.

There are two special phenomena visible in Fig. 10.14 that should be mentioned. The first one is the unexpectedly strong share of Discrete Mathematics & Convex Geometry among the publications of the Plenary Speakers. This is due to the fact that Plenary Speakers in this domain, between 1993 and 2020, tended to be markedly more prolific than those of the other specialities. Indeed, for each mathematical domain we have computed the rate of publications of Plenary Speakers, in the four year interval about their ICM lecture as was explained above.[127] The domain *Discr* distances itself from all other specialities, with a median value of 259 papers per speaker, more than twice the median of all domains taken together.

The second peculiarity becomes apparent when one compares the charts in Fig. 10.14 and 10.13. Even though the latter apply to the total period since the 1950s, one immediately wonders why Topology/Geometry, which gets a conspicuous share of the cake, does so poorly in Fig. 10.14. As a matter of fact, the discrepancy is just as dramatic if one replaces the diagrams of Fig. 10.13 by the corresponding ones for the last two time periods. The explanation of this effect lies again in the publication pattern of the domain. In fact, two factors contribute: The 'papers per speaker' rate for *TopGeo* is only 11, the lowest rate of all domains. Furthermore, it turns out that the speakers in this domain, and in those years, tend to publish an unusually high share of their papers not in journals, but rather in conference proceedings and other multi-author volumes.

[127] More precisely, all speakers are fractionally counted with equal weight for each one of the domains that appear in the classification of their plenary lectures, and the same is done for every publication in the $[N - 1, N + 2]$ time interval.

References

These references list: 1. Illustration sources, 2. Archives, 3. Publications, and 4. Websites referred to in the text.

1. ILLUSTRATION SOURCES

Front cover picture. Opening session of the ICM 2014 in Seoul. Unknown photographer. Thanks to Hyungju Park for providing this photograph for free use by the International Mathematical Union, Hausvogteiplatz 11a, 10117 Berlin, Germany.

Back cover picture. General group photograph of the 1932 ICM in Zürich. Photographer Johannes Meiner (1867–1941). Credit: ETH Library, Zürich. Public domain: `https://de.wikipedia.org/wiki/Internationaler_Mathematikerkongress#/media/Datei:Internationaler_Mathematikerkongress_Z%C3%BCrich_1932_-_ETH_BIB_Portr_10680-FL_(Johannes_Meiner).jpg` (consulted 9 March 2022). Courtesy of the International Mathematical Union, Hausvogteiplatz 11a, 10117 Berlin, Germany.

Fig. 1.1. The two sides of the Fields Medal. Photographer Stefan Zachow. Courtesy of the International Mathematical Union, Hausvogteiplatz 11a, 10117 Berlin, Germany.

Fig. 1.2. Above. Alexander von Humboldt, daguerreotype by Hermann Biow (1804–1850), 1847. Public domain; {{PD-US-expired}}: `https://commons.wikimedia.org/wiki/File:Humboldt,_Alexander_von_1847.jpg` (consulted 7 March 2022).
Bottom left. August Leopold Crelle ca. 1825, portrait of unknown origin. Public domain; {{PD-US-expired}}: `https://commons.wikimedia.org/wiki/File:Crelle_August_Leopold.jpg` (consulted 7 March 2022).
Bottom right. Adolphe Quetelet, portrait by J.-B. Madou, 1839. Credit: Felix Archiev, Antwerp, Belgium, Accession No.12 # 12861. License: CC0 1.0 Universal Public Domain Dedication; `https://commons.wikimedia.org/wiki/File:Quetelet,_Adolphe_(1796--1874);_astronoom,_wiskundige,_socioloog,_Madou,_Jean_Baptiste,_Felixarchief,_12_12861_recto.jpg` (consulted 7 March 2022).

Fig. 1.3. The trip of Betti, Brioschi, and Casorati. Source: [Durand 2018], p. 220. Thanks to Antonin Durand for providing his original collage; only the title line was changed to English.

N. Schappacher, *Framing Global Mathematics*,
https://doi.org/10.1007/978-3-030-95683-7

Fig. 1.4. Marie Curie in conversation with Henri Poincaré. Standing on the right Paul Langevin and Albert Einstein. Cropped from a group photograph, taken on 2 November 1911 by Benjamin Couprie, of the 1911 Solvay Conference in Brussels. Public Domain; {{PD-US-expired}}: `https://commons.wikimedia.org/wiki/File:1911_Solvay_conference.jpg` (consulted 6 March 2022).

Fig. 1.5. First General Assembly of the IAA. Cover page of the invitation to the *Soirée su 20 Avril 1901*. Credit: [Arch. AWG], Scient 9382, Fasz. 2, Nr. 34. Thanks to the Archive of the Göttingen Academy of Sciences for the permission and the scan. (Cf. [Gierl 2004], Tableau Festkultur, Abb. 4 and 5.)

Fig. 1.6. The delegates of the IAA at their meeting in Rome in May 1910. Source: Frontispiece of Ernesto Mancini, *L'Associazione internazionale delle Accademie e le sue adunanze generali in Roma.* Rome 1910. (Cf. [Gierl 2004], no. 7 of the *Gruppensitzungen* illustrations.)

Fig. 1.7. Felix Klein and some of his correspondents. Drawing of unknown origin. Source: [Rowe 1985], p. 75. Thanks to David Rowe and Springer Verlag for permission to reproduce this picture. License ordered through CCC.

Fig. 1.8. The Göttingen Mathematical Society in 1902. In handwriting on the back of the picture, presumably written by Martin Schwarzschild: *Photo by Gebr. Noelle, Königl. Württemb. Hofphotographen. Inhaber: William Noelle, Göttingen, Kurzestrasse 5a.* Source: [Arch. SUBG], Cod. Ms. K. Schwarzschild 23 : 1, 16. Thanks to the Handschriftenabteilung of the Göttingen Staats- und Universitätsbibliothek for the permission and the scan.

Fig. 1.9. Rikitaro Fujisawa. Photograph cropped from a group photograph of the "Congress of Thinkers" at Williams College published in *The Evening Star* (Washington, D.C.), 15 August 1922. Source: George Grantham Bain Collection, Library of Congress. LC-DIG-ggbain-34783 DLC (digital file from original negative). Flickr Commons project, 2020.

Fig. 2.1. Hermann Weyl on a teeter-totter, uncertain date and photographer. I conjecture that this picture was taken on the same occasion as the well-known "Gasthof Vollbrecht Photograph": `https://opc.mfo.de/detail?photo_id=9265` (consulted 8 March 2022). If this is indeed the case, then, according to [Eckes & Schappacher 2016], it was taken at Nikolausberg near Göttingen about 15 July 1933. In this case the photographer may have been Natascha Artin (1909–2003). Source: [Pólya 1987], p. 109.

Fig. 3.1. Mauro Picone in 1903. Unknown photographer. Public domain; {{PD-US-expired}}: `https://commons.wikimedia.org/wiki/File:Picone1903.jpg` (consulted 8 March 2022).

Fig. 4.1. Vito Volterra and Émile Picard at the Strasbourg ICM in 1920, drawn by the Strasbourg artist Eugène-Michel Maeckler (1872–1925); see Proceedings ICM 1920, p. XL, footnote 1. All these drawings were printed separately and delivered as loose sheets with the 1920 ICM Proceedings. Reproduction of prints kept in the Bibliothèque mathématique de l'IRMA, Strasbourg.

Fig. 4.2. The new buildings of Strasbourg University from the 1870s / 1880s. Engravings by unknown author(s). Source: *L'Univers illustré*, journal hebdomadaire, 29 November 1884, p. 765. Scan of my personal copy of this page, which I bought in a Strasbourg art store many years ago. Thanks to Catherine Goldstein for having found out its source.

Fig. 4.3. Gabriel Kœnigs in 1920(?). This photograph of unknown date and origin was printed separately and delivered as a loose sheet with the 1920 ICM Proceedings, along with Eugène-Michel Maeckler's drawings. Reproduction of a print kept in the Bibliothèque mathématique de l'IRMA, Strasbourg.

Fig. 4.4. Leonard E. Dickson during his general lecture at the 1920 ICM in Strasbourg, drawn by Eugène-Michel Maeckler (1872–1925); see Proceedings ICM 1920, p. XL, footnote 1. All these drawings were printed separately and delivered as loose sheets with the 1920 ICM Proceedings. Reproduction of a print kept in the Bibliothèque mathématique de l'IRMA, Strasbourg.

Fig. 4.5. John Charles Fields. Photograph, before 1912, unknown author. Source: Acta Mathematica 1882–1912: Table générale des tomes 1–35; Konrad Jacobs, Erlangen. Public domain: `https://commons.wikimedia.org/wiki/File:John_charles_fields.jpg` (consulted 9 March 2022).

Fig. 4.6. General group photograph of the Toronto ICM, 1924. Photographer P.E. McDonald. Credit: University of Toronto Archives. Public domain: `https://utarms-online.library.utoronto.ca/islandora/object/utarmsIB%3A2008-5-1aMS` and `https://utarms-online.library.utoronto.ca/islandora/object/utarmsIB%3A2008-5-1bMS` (consulted 9 March 2022). Courtesy of the International Mathematical Union, Hausvogteiplatz 11a, 10117 Berlin, Germany.

Fig. 4.7. Salvatore Pincherle ca. 1900. Unknown photographer. Public domain; {{PD-US-expired}}: `https://commons.wikimedia.org/wiki/File:Salvatore_Pincherle.jpg` (consulted 10 March 2022).

Fig. 5.1. Rockefeller map of mathematical centers in Europe, 1927. Credit: [Arch. RAC], Rockefeller Archive Center, International Education Board, Series 1.1 (Appropriations – Numerical) Box 10, Folder 143, Scientific Center Maps, 1926-1927. Cf. [Siegmund-Schultze 2001], p. 44. Thanks to the Rockefeller Archive Center for the permission to reproduce this map.

Fig. 5.2. Rockefeller map of mathematical centers in the USA, 1927. Credit: [Arch. RAC], Rockefeller Archive Center, International Education Board, Series 1.1 (Appropriations – Numerical) Box 10, Folder 143, Scientific Center Maps, 1926-1927. Cf. [Siegmund-Schultze 2001], p. 54. Thanks to the Rockefeller Archive Center for the permission to reproduce this map.

Fig. 6.1. Emmy Noether's personal questionnaire of 19 April 1933, filled in by herself, with two female gender symbols added by the administration. Credit: [Arch. GStA], GStA PK, I. HA Rep. 76 Va Nr. 10081; folio 10, reverse side. Creative commons license CC-BY-SA 3.0.

Fig. 6.2. *Leitfaden* from [Van der Waerden 1930–31], Vol. 1, p. VIII. Courtesy of Springer Verlag.

Fig. 6.3. Group picture of the Bourbaki Congress at Besse la Chandesse in July 1935. Standing from left to right: Henri Cartan, René de Possel, Jean Dieudonné, André Weil; Luc Olivier (biologist). Sitting: A. Mirlès, Claude Chevalley, Szolem Mandelbrojt. Photographer: Marie-Thérèse Bastien (secretary Faculté des sciences, Nancy). Credit: Fonds Pierre Dugac. Creative commons license (BY-NC-ND): `http://archives-bourbaki.ahp-numerique.fr/exhibits/show/80-ans/congres-besse/photos` (consulted 10 March 2022).

Fig. 6.4. Group picture of the First International Congress on Topology, Moscow 4–10 September 1935. Unknown Photographer. Credit: [Arch. ETH], Portrait 07561. Public domain: https://www.e-pics.ethz.ch/index/ETHBIB.Bildarchiv/ETHBIB.Bildarchiv_579714.html (consulted 10 March 2022).

Fig. 6.5. Shiing-Shen Chern (left) and the Wei-Liang Chow couple in Hamburg, 1936. Unknown photographer. Source: [Yau 2012], picture album. Thanks to Shing-Tung Yau and to International Press Boston, Inc., for permission to reproduce the photo.

Fig. 7.1. Beppo Levi about 1930. Unknown photographer. Courtesy of Laura Levi. Cf. [Levi 2000], cover and p. 33.

Fig. 7.2. Electronic Computer Project, The Institute for Advanced Study. Julian Bigelow, Herman Goldstine, J. Robert Oppenheimer, and John von Neumann, 1952. Photographer Alan Richards. Credit: [Arch. IAS], Photograph Collection; v. Neumann & computer photos folder. See: https://albert.ias.edu/handle/20.500.12111/3494 Thanks to the Shelby White and Leon Levy Archives Center, IAS, for the permission and the scan.

Fig. 8.1. On the train to Nikko for the 1955 conference on algebraic number theory. Shown are (left to right) T. Tamagawa, J.-P. Serre, Y. Taniyama, and A. Weil. Unknown photographer. Source: [Shimura 1989], photo immediately preceding p. 187. Credit: John Wiley & Sons, Inc. License ordered through CCC.

Fig. 8.2. Two of the four Fields Medalists of 1966: (1) Alexander Grothendieck on the left, at the Montreal ICM in 1970. Photographer Konrad Jacobs. Creative commons, public domain, https://commons.wikimedia.org/wiki/File:Alexander_Grothendieck.jpg (consulted 6 March 2022).
(2) Stephen Smale in 1966. Photographer Caroline Abraham. Source: [Smale 1984], p. 21. Credit: Springer Verlag New York. License ordered through CCC.

Fig. 8.3. José Luis Massera and Marta Valentini in March 1984, after his liberation. Unknown photographer. Source: [Markarian & Mordecki 2010]; see http://www.cmat.edu.uy/$\sim$mordecki/massera/libro/fotos/09.jpg (consulted 12 March 2022). Thanks to Ernesto Mordecki for granting permission, also on behalf of Marta Valentini, to reproduce this picture.

Fig. 8.4. East front of Hausvogteiplatz, Berlin, with the new IMU headquarters indicated. Photographer Stefan Zachow. Courtesy of the International Mathematical Union, Hausvogteiplatz 11a, 10117 Berlin, Germany.

Fig. 8.5. The Borromean link in the mosaic floor of Room 18 of the Roman Villa del Casale near Piazza Armerina, Sicily. On the left, overview of Room 18. Photographer Tyler Bell. Attribution 2.0 Generic (CC BY 2.0). See https://www.flickr.com/photos/tylerbell/43053916921 (consulted 20 November 2021). On the right, detail. Photographer Thomas Delzant, Strasbourg. Thanks to him for authorizing the reproduction. Below, the current IMU logo, Courtesy of the IMU.

Fig. 9.1. Entry in the Oberwolfach Guestbook of a meeting about the renewal of (the German subcommittee of) ICMI, held in Oberwolfach, 23–24 August 1952. Source: [Arch. MFO], Gästebuch No. 1 (1946 – 1954), p. 32. See [URL 68]. Courtesy of Mathematisches Forschungsinstitut Oberwolfach.

Fig. 10.1. The first page of Van der Waerden's letter to Auguste Dick dated 8 March 1967. Distribution of 'talent for mathematics and physics' among all students, resp. female students. Credit: [Arch. ÖAW], Folder: Nachlass Auguste Dick, 11/35. Thanks to the Austrian Academy of Sciences, Vienna, for the permission and the scan.

Figs. 10.2 – 10.14. Diagrams produced for this book by Birgit Petri, Darmstadt, on the basis of her quantitative study of the ICM population since 1950.

2. ARCHIVES

Archive AWG. Archiv der Akademie der Wissenschaften zu Göttingen, Germany. Staats- und Universitätsbibliothek Göttingen, Abteilung Spezialsammlungen und Bestandserhaltung. Folder: Scient 9382, Fasz. 2, Nr. 34.

Archive GStA. Geheimes Staatsarchiv Preußischer Kulturbesitz, Berlin-Dahlem, Germany. Folder: GStA PK, I. HA Rep. 76 Va Nr. 10081.

Archive ETH. Archiv der Eidgenössischen Technischen Hochschule Zürich / Archive of the Swiss Federal Institute of Technology, Zürich, Switzerland. Folders (quoted with document numbers): Weyl papers. Van der Waerden papers.

Archive IAS. Shelby White and Leon Levy Archives Center, Institute for Advanced Study, Princeton, NJ, USA. Folders: Records of the School of Mathematics / Members, Visitors, Assistants / Box 35 / Hua Luogeng. Nr. 56883. Records of the Office of the Director: Member Files: Box 20: Chern Shiing Shen. Nr. 56484; Photograph Collection—v. Neumann & computer photos folder.

Archive IMU. Archive of the International Mathematical Union (IMU), Berlin, Germany. Folders: Subfond IMU - Helsinki, Ser 12 IMU Conferences, 12.1 Colloquia 1952-1955, File 7 Symposium Tokyo 1955, File 7.2 On the preparation of the symposium, Discussion on invitation of mathematicians (participation of German Mathematicians). Subfond IMU – Helsinki, Ser 11 Union Lectures. Subfond IMU – Helsinki, Ser 21 Human rights. Subfond Jacques-Louis Lions, F 5 Commission on Development and Exchange, IMU_004.pdf. The Jessen papers [Arch. Jessen].

Archive JESSEN. Børge Jessen Papers at the Archive of the Department of Mathematical Sciences, University of Copenhagen, Denmark. See [URL 64]. Copies of the parts of this archive pertaining to the history of the IMU have been given to the IMU Archive. There they constitute the Subfond Børge Jessen – Copenhagen Archive, Ser 1 digital, Box 62, Folder 3 IMU 1953-1955 correspondences, 3.pdf, from which we cite.

Archive JHU. Special Collections, Sheridan Libraries, Johns Hopkins University. BLC-2045, 3400 N. Charles Street, Baltimore MD 21218. Folder: Wei-Liang Chow papers, MS.0762, box 1.

Archive ÖAW. Archiv der Österreichischen Akademie der Wissenschaften, Vienna, Austria. Folder: Nachlass Auguste Dick, 11/35.

Archive MFO. Mathematisches Forschungsinstitut Oberwolfach. Digital Archive. Folder: Gästebuch No. 1 (1946 – 1954). Signature: E 20 / 00148. See https://oda.mfo.de/handle/mfo/55 (consulted 9 March 2022).

Archive RAC. Rockefeller Archive Center, 15 Dayton Avenue, Sleepy Hollow, NY 10591. Folder: International Education Board, Series 1.1 (Appropriations – Numerical) Box 10, Folder 143, Scientific Center Maps, 1926-1927.

Archive SUBG. Staats- und Universitätsbibliothek Göttingen, Abteilung für Handschriften und seltene Drucke. Papendieck 14, D-37073 Göttingen, Germany. Folder: Cod. Ms. K. Schwarzschild.

Bibliothèque MIR. La bibliothèque de Mathématiques-Informatique recherche de Sorbonne Université, site Pierre et Marie Curie, patio 15–26, 4 place Jussieu, 75005 Paris, France. Folder: Lettres de Juliusz Schauder à Jean Leray.

3. PUBLICATIONS

ICM PROCEEDINGS. Throughout the book, a quote of the form: "Proceedings ICM xxxx" refers to the Proceedings of the International Congress of Mathematicians that took place in the year xxxx. All these Proceedings can be freely downloaded at `https://www.mathunion.org/icm/proceedings` (consulted 20 October 2021).

Nikita AGARWAL, Carolina ARAUJO, Petra BONFERT-TAYLOR, Mojgan MAHMOUDI, Marie Françoise OUEDRAOGO, Olga PARIS-ROMASKEVICH, Marie-Françoise ROY, Elisabetta STRICKLAND, and Andrea Vera GAJARDO. May 12: Celebrating Women in Mathematics From One Idea to One Hundred Events. *Notices of the AMS* Dec. 2019, 1879–1886.

Andrea ALBRECHT. Mathematische und ästhetische Moderne. Zu Robert Musils Essay 'Der mathematische Mensch'. *Scientia Poetica* 12 (2008), 218–250.
—. "Die Kunst ist nur der Affe dieser Gedankenkämpfe". Erkenntnisprozesse in literarischer Darstellung bei Hermann Broch und Robert Musil. In *Literarische Denkformen* (Claus Zittel, Marcus Born, eds.). Wilhelm Fink München 2018; pp. 273–298.
— & Franziska BOMSKI. Mathematik, Logik, Geometrie, Wahrscheinlichkeitstheorie. In *Robert Musil-Handbuch* (Birgit Nübel, Norbert Christian Wolf, eds.). De Gruyter Berlin 2016; pp. 510–516.

Daria E. APUSHKINSKAYA, Alexander I. NAZAROV, Galina I. SINKEVICH. In Search of Shadows: the First Topological Conference, Moscow 1935. arXiv:1903.02065v3 25 May 2019 (consulted 2 February 2021.)

Carolina ARAUJO, Georgia BENKART, Cheryl E. PRAEGER, Betül TANBAY (eds.). *World Women in Mathematics 2018. Proceedings of the First World Meeting for Women in Mathematics* $(WM)^2$. Springer 2019.

Raymond Clare ARCHIBALD. *A Semicentennial History of the American Mathematical Society 1888–1938*. AMS 1938.

Tom ARCHIBALD. Charles Hermite and German Mathematics in France. In [Parshall & Rice 2002]; pp. 123–137.

Mitchell G. ASH. Forced migration and scientific change in the Nazi era. In *Oberwolfach Reports* No. 51/2011, 2897–2901.

David AUBIN, Catherine GOLDSTEIN (eds.). *The War of Guns and Mathematics: Mathematical Practices and Communities in France and Its Western Allies around World War I*. AMS 2014.

Michèle AUDIN. *Correspondance entre Henri Cartan et André Weil (1928–1991).* Documents Mathématiques vol. 6. SMF Paris 2011.

Stéphane AUDOIN-ROUZEAU & Jean-Jacques BECKER (eds.). *Encyclopédie de la Grande Guerre 1914–1918.* Bayard Paris 2004.

Charles BABBAGE. *Reflections on the Decline of Science in England, and on Some of its Causes.* B. Fellows London 1830.

F.W.G. BAKER. *ICSU – UNESCO. Forty Years of Cooperation.* Brochure. ICSU Secretariat Paris, November 1986. 30 pages.

Stefan BANACH. *Théorie des opérations linéaires.* Garasiński Warszawa 1932.

Adam BANNISTER & Olaf TESCHKE. Can Statistics Predict the Fields Medal Winners? *EMS Newsletter* June 2018, 40–43.

Michael BARANY. The Myth and the Medal. *Notices of the AMS* January 2015, 15–20.
—. *Distributions in Postwar Mathematics.* PhD. Thesis Princeton U 2016[a].
—. Fellow Travelers and Traveling Fellows: The Intercontinental Shaping of Modern Mathematics in Mid-Twentieth Century Latin America. *Historical Studies in the Natural Sciences*, 46–5 (2016[b]), 669–709.
—. The Fields Medal should return to its roots. *Nature* 553 (18 January 2018), 271–273.
—, Anne-Sandrine PAUMIER, Jesper LÜTZEN. From Nancy to Copenhagen to the World: The internationalization of Laurent Schwartz and his theory of distributions. *Historia Mathematica* 44 (2017), 367–394.

June BARROW-GREEN. International Congresses of Mathematicians from Zürich 1897 to Cambridge 1912. *The Mathematical Intelligencer* 16–2 (1994), 38–41.
—. Gösta Mittag-Leffler and the Foundation and Administration of *Acta Mathematica*. In [Parshall & Rice 2002], pp. 139–164.
—. The Historical Context of the Gender Gap in Mathematics. In [Araujo et al. 2019]; pp. 129–145.

George BASALLA. The spread of Western Science. *Science* 156 (1967), 611–622.

Steve BATTERSON. *Pursuit of Genius. Flexner, Einstein, and the Early Faculty at the Institute for Advanced Study.* AK Peters, Wellesley Mass 2006.

Charles BAUDELAIRE. *"The Painter of Modern Life" and Other Essays*, edited and translated by Jonathan Mayne. London Phaidon Press 1964.

Heinrich BECKER, Hans-Joachim DAHMS, Cornelia WEGELER (eds.). *Die Universität Göttingen unter dem Nationalsozialismus.* Second edition. Saur München 1998.

Robert A. BEELER & Rick NORWOOD. Polish Mathematics in the First Half of the Twentieth Century. *The Mathematical Scientist* 39 (2014), 1–14.

Heinrich BEHNKE & Peter THULLEN. *Theorie der Funktionen mehrerer komplexer Veränderlichen.* Springer 1934.

Bruno BELHOSTE. *La formation d'une technocratie. L'École Polytechnique et ses élèves de la Révolution au Second Empire.* Belin Paris 2003.

Jeremy BENTHAM. *An Introduction to the Principles of Morals and Legislation.* T. Payne & Son London 1780/1789.

Birgit BERGMANN, Moritz EPPLE, Ruti UNGAR (eds.). *Transcending Tradition. Jewish Mathematicians in German-Speaking Academic Culture.* Springer 2012.

Kurt-R. BIERMANN. *Die Mathematik und ihre Dozenten an der Berliner Universität 1810–1933. Stationen auf dem Wege eines mathematischen Zentrums von Weltgeltung.* Akademie Verlag Berlin 1988.

Adriaan BLAAUW. *History of the IAU. The Birth and First Half-Century of the International Astronomical Union.* Springer 1994.

Harald BOHR. *Fastperiodische Funktionen.* Springer 1932.

John BOLI & George M. THOMAS. *Constructing World Culture. International Nongovernmental Organizations Since 1875.* Stanford U Press 1999.

Andrey A. BOLIBRUCH, Yuri S. OSIPOV, Yakov G. SINAI (eds.). *Mathematical Events of the Twentieth Century.* Springer / Phasis 2006.

Christian BONAH. *Instruire, guérir, servir. Formation, recherche et pratique médicales en France et en Allemagne pendant la deuxième moitié du XIXe siècle.* Strasbourg 2000.

Benedetto BONGIORNO & Guillermon P. CURBERA. *Giovanni Battista Guccia. Pioneer of International Cooperation in Mathematics.* Springer 2018.

Tommy BONNESEN & Werner FENCHEL. *Theorie der konvexen Körper.* Springer 1934.

George BOOLE. *A treatise on the calculus of finite differences.* Macmillan London 1860.

Bernhelm BOOSS-BAVNBECK & Jens HØYRUP (eds.). *Mathematics and War.* Birkhäuser Basel 2003.

George J. BORJAS & Kirk B. DORAN. Prizes and Productivity: How Winning the Fields Medal Affects Scientific Output. *The Journal of Human Resources* 50–3 (2015), 728–758.

Umberto BOTTAZZINI. From Paris to Berlin: contrasted images of Nineteenth Century mathematics. In [Bottazzini & Dahan 2001]; pp. 31–47.
— & Amy DAHAN DALMEDICO. *Changing Images in Mathematics. From the French Revolution to the New Millenium.* Routledge London-New York 2001.
— & Jeremy GRAY. *Hidden Harmonies. The Rise of Complex Function Theory.* Springer 2013.

Frédéric BRECHENMACHER. Self-Portraits with Évariste Galois (and the shadow of Camille Jordan). *Revue d'histoire des mathématiques* 17 (2011), 273–371.

Éric BRIAN & Marie JAISSON. *The Descent of the Human Sex Ratio at Birth. A Dialogue between Mathematics, Biology and Sociology.* Springer 2007.

Alexander BRILL & Max NOETHER. Die Entwicklung der Theorie der algebraischen Functionen in älterer und neuerer Zeit. Bericht erstattet der Deutschen Mathematiker-Vereinigung. *Jahresbericht der Deutschen Mathematiker-Vereinigung* 3 (1892–93), I–XXIII and 107–566.

Bernard BRU. Souvenirs de Bologne. *Journal de la Société Française de Statistique* 144 (2003), 135–226.

Otto BRUNNER, Werner CONZE, Reinhart KOSELLECK. *Geschichtliche Grundbegriffe. Historisches Lexikon zur politisch-sozialen Sprache in Deutschland.* 8 volumes. Klett-Cotta Stuttgart 1982; Studienausgabe 2004.

Annalisa CAPRISTO. French mathematicians at the Bologna Congress (1928). Between participation and boycott. In *The Latin Sisters. Images of Italian mathematics in France from Risorgimento to Fascism.* (F. Brechenmacher, G. Jouve, L. Mazliak, R. Tazzioli, eds.) New York Springer 2016, pp. 289–309.

James CARLSON, Arthur JAFFE, Andrew WILES (eds.). *The Millenium Prize Problems.* Clay Mathematical Institute & American Mathematical Society 2006.

Ana CARNEIRO & Natalie PIGEARD. Chimistes alsaciens à Paris au 19ème siècle: un réseau, une école ? *Annals of Science* 54 (1997), 533–546.

Gianfranco CASNATI, Alberto CONTE, Letterio GATTO, Livia GIACARDI, Marina MARCHISIO, Alessandro VERRA (eds.). *From Classical to Modern Algebraic Geometry. Corrado Segre's Mastership and Legacy.* Springer-Birkhäuser 2016.

Rémi CATELLIER & Laurent MAZLIAK. The emergence of French probabilistic statistics. Borel and the Institut Henri Poincaré around the 1920s. *Revue d'histoire des mathématiques* 18 (2012), 271–335.

Cinzia CERRONI & Aldo BRIGAGLIA. The 'Circolo Matematico di Palermo' and the First World War: the crisis of scientific internationalism. A view through the unedited correspondence of De Franchis with Edmund Landau and other mathematicians. *Historia Mathematica* 55 (2021), 64–94.

Emmanuelle de CHAMPS. *Enlightenment and Utility. Bentham in French, Bentham in France.* Cambridge U Press 2015.

Christophe CHARLE. *La république des universitaires 1870–1940.* Edition du Seuil Paris 1994.
—. *Discordance des temps. Une brève histoire de la modernité.* A. Colin Paris 2011.

Claude CHEVALLEY. Valeur de la science. *Ordre Nouveau* 8, 15 février 1934, 22–24.
—. Variations du style mathématique. *Revue de métaphysique et de morale* 42 (1935), 375–384.
— & Arnaud DANDIEU. Logique hilbertienne et psychologie. *Revue philosophique* 113 (1932), 99–111.

Wei-Liang CHOW. *The Collected Papers of Wei-Liang Chow.* Edited by S.S. Chern & V.V. Shokurov. World Scientific 2002.
— & Bartel L. Van der WAERDEN. Zur Algebraischen Geometrie XII. Über zugeordnete Formen und algebraische Systeme von algebraischen Mannigfaltigkeiten. *Mathematische Annalen* 113 (1937), 692–704. — For a condensed translation of this paper into English by Wen-ling Huan, see [Chow 2002], pp. 1–13.

Paul A. COHEN. *China Unbound. Evolving Perspectives on the Chinese Past.* Routledge Curzon London & New York 2004.

Bruno COLBOIS, Christine RIESTMANN, Viktor SCHROEDER (eds.). *math.ch/100. Schweizerische Mathematische Gesellschaft / Société Mathématique Suisse / Swiss Mathematical Society 1910–2010.* EMS 2010.

La COMMISSION permanente du répertoire. *Index du Répertoire Bibliographique des Sciences Mathématiques.* Gauthier Villars Paris 1893.

Julian COOLIDGE. Corrado Segre. *Bulletin of the American Mathematical Society* 34 (1927), 352–357.

Leo CORRY. *Modern Algebra and the Rise of Mathematical Structures.* Springer Basel 2004.

— & Norbert SCHAPPACHER. Zionist Internationalism through Number Theory; Edmund Landau at the Opening of the Hebrew University in 1925. *Science in Context* 23 (4) (2010), 427–471.

John E. CRAIG. *Scholarship and Nation Building. The Universities of Strasbourg and Alsatian Society, 1870–1939.* The U of Chicago Press 1984.

Elisabeth CRAWFORD. *Nationalism and Internationalism in Science, 1880–1939. Four Studies of the Nobel population.* Cambridge U Press 1992.

—, Terry SHINN, Sverker SÖRLIN (eds.). *Denationalizing Science. The Context of International Scientific Practice.* Springer-Science+Business Media 1992.

— & Josiane OLFF-NATHAN (eds.). *La science sous influence. L'université de Strasbourg, enjeu des conflits franco-allemands 1872–1945.* La Nuée Bleue Strasbourg 2005.

Guillermon P. CURBERA. *Mathematicians of the World, Unite! The International Congress of Mathematicians. A Human Endeavor.* A K Peters Wellesley, Mass, 2009.

Pierre & Marie CURIE. Sur une substance nouvelle radioactive, contenue dans la pechblende. *Comptes rendus de l'Académie des sciences* vol. 127, no. 3 (1898), 175–178.

Amy DAHAN DALMEDICO. An image conflict in mathematics after 1945. In [Bottazzini & Dahan 2001]; pp. 223–253.

Amy DAHAN. Axiomatiser, modéliser, calculer: les mathématiques, instrument universel et polymorphe d'action. In [Dahan & Pestre 2004]; pp. 49–81.

— & Dominique PESTRE (eds.). *Les sciences pour la guerre 1940–1960.* Edition EHESS Paris 2004.

Hans-Joachim DAHMS. Otto Neuraths "International Encyclopedia of Unified Science" als Torso. Bemerkungen über die geplanten, aber nicht erschienenen, Monographien der Enzyklopädie. In Elisabeth Nemeth, Richard Heinrich (eds.), *Otto Neurath: Rationalität, Planung, Vielfalt.* Oldenburg Wien, Berlin 1999; pp. 184–227.

Lorraine DASTON. The Ideal and Reality of the Republic of Letters in the Enlightenment. *Science in Context* 4 (1991), 367–386.

Joseph W. DAUBEN. Internationalizing Mathematics East and West: Individuals and Institutions in the Emergence of a Modern Mathematical Community in China. In [Parshall & Rice 2002]; pp. 253–285.

Claude DEBRU (ed.). *Akademien im Kriege. Académies en Guerre. Academies in War.* Acta Historica Leopoldina, vol. 75, 2019.

Claude DEBUSSY. *Correspondance 1884–1918.* (François Lesure, ed.) Hermann Paris 1993.

Sergei Sergeevich DEMIDOV. 70 years of the journal Uspekhi Matematicheskikh Nauk. *Uspekhi Matematicheskikh Nauk* 61/4 (2006), 203–207 (Russian); *Russian Mathematical Surveys* 61/4 (2006), pp. 793–797 (English).

Marie-Noële DENIS. Les statues de l'Université impériale de Strasbourg et la pédagogie du pangermanisme. *Revue des Sciences Sociales* 34 (2005), 84–93.

Alain DESROSIÈRES. *The Politics of Large Numbers. The History of Statistical Reasoning.* (Translated by Camille Naish.) Princeton U Press 1998.

Max DEURING. *Algebren.* Springer 1935.

Auguste DICK. *Emmy Noether. 1882–1935.* Beiheft Nr. 13 zur Zeitschrift *Elemente der Mathematik*. Birkhäuser Basel 1970.

Hermann DIELS. Die Organisation der Wissenschaft. In *Die allgemeinen Grundlagen der Kultur der Gegenwart* (Paul Hinneberg, ed.). Teubner Berlin & Leipzig 1906.

Shisun DING, Ming-Chang KANG, Eng-Tjioe TAN. Chiungtze C. Tsen (1898–1940) and Tsen's Theorems. *Rocky Mountain Journal of Mathematics* 29 (1999), 1237–1269.

Diane DOSSO. Le plan de sauvetage des scientifiques français. New York, 1940–1942. *Revue de Synthèse* 127 (2006), 429–451.

Vladimir DRAGOVIĆ & Irina GORYUCHKINA. Polygons of Petrović and Fine, algebraic ODEs, and contemporary mathematics. *Archive for History of Exact Sciences* 74 (2020), 523–564.

Pierre DUHEM. *La théorie physique. Son objet, sa structure.* Second ed. Librairie Marcel Rivière Paris 1914. Présentation et édition par Sophie Roux. ENS Editions Paris 2016.
—. *La Science Allemande.* Hermann Paris 1915.

A. Hunter DUPREE. *Science in the Federal Government. A History of Politics and Activities to 1940.* Harvard U Press Cambridge, MA, 1957.

Antonin DURAND. *La quadrature du cercle. Les mathématiciens italiens et la vie parlementaire (1848–1913).* Editions rue d'Ulm Paris 2018.

Peter DUREN, Richard A. ASKEY, Uta C. MERZBACH (ed.s). *A Century of Mathematics in America. Part I.* History of Mathematics, Vol. 1, AMS Providence RI 1988.

Wolfgang ECCARIUS. August Leopold Crelle als Herausgeber des Crelleschen Journals. *Journal für die reine und angewandte Mathematik* 286/287 (1976), 5–25.

Christophe ECKES. Organiser le recrutement de recenseurs français pour le *Zentralblatt* à l'automne 1940. Les premiers liens entre Harald Geppert, Helmut Hasse et Gaston Julia sous l'occupation. *Revue d'histoire des mathématiques* 24 (2018), 259–329.
— & Norbert SCHAPPACHER. Dating the Gasthof Vollbrecht Photograph. Commentary in the Oberwolfach Photo Collection 2016. `https://opc.mfo.de/files/GVanglais_def.pdf` (consulted 8 March 2022).

Moritz EPPLE. Branch Points of Algebraic Functions and the Beginnings of Modern Knot Theory. *Historia Mathematica* 22 (1995), 371–401.
—. *Die Entstehung der Knotentheorie. Kontexte und Konstruktionen einer modernen mathematischen Theorie.* Vieweg Braunschweig 1999.

— & Volker REMMERT. 'Eine ungeahnte Synthese zwischen reiner und angewandter Mathematik': Kriegsrelevante mathematische Forschung in Deutschland während des Zweiten Weltkrieges. In Doris Kaufmann (ed.), *Geschichte der Kaiser-Wilhelm-Gesellschaft im Nationalsozialismus. Bestandsaufnahme und Perspektiven der Forschung*. Vol. 1. Wallstein Göttingen 2000; pp. 258–295.

—, Andreas KARACHALIOS, Volker REMMERT. Aerodynamics and Mathematics in National Socialist Germany and Fascist Italy: A Comparison of Research Institutes. *Osiris* 20 (2005), 131–158.

Karl Dietrich ERDMANN. *Toward a Global Community of Historians. The International Historical Congresses and the International Committee of Historical Sciences, 1898–2000.* J. Kocka & W. Mommsen (eds.). Berghahn Books, New York/Oxford 2005.

Paul ERICKSON, Judy L. KLEIN, Lorraine DASTON, Rebecca LEMOV, Thomas STURM, Michael D. GORDIN. *How Reason Almost Lost Its Mind. The Strange Career of Cold War Rationality.* U of Chicago Press 2013.

Marie FARGE. Scholarly Publishing and peer-reviewing in open access. In *Europe's Future: Open Science, Open Innovation, and Open to the World* (Carlos Moedas, ed.). The European Commission 2017. See [URL 63].

Danielle FAUQUE & Robert FOX. Binding the Wounds of War. Internationalism, National Interests, and the Order of World Science, 1919–1931. In *Blockades of the Mind. Science, Academies, and the Aftermath of the Great War.* (Wolfgang U. Eckert & Robert Fox, eds.) *Acta Historica Leopoldina* vol. 78 (2021); pp. 41–68.

Solomon FEFERMAN. Weyl Vindicated: *Das Kontinuum* Seventy Years Later. In S. Feferman, *In the Light of Logic*. Oxford U Press 1998; pp. 249–283.

Fausto FERRARI. Weyl and Marchaud Derivatives: A Forgotten History. *Mathematics* 6 (2018). 25 pp. `https://www.mdpi.com/2227-7390/6/1/6/htm` (consulted 6 February 2022).

Ludwig FINSCHER (ed.). *Die Musik in Geschichte und Gegenwart (MGG). Allgemeine Enzyklopädie der Musik begründet von Friedrich Blume.* Zweite, neubearbeitete Ausgabe herausgegeben von Ludwig Finscher. *Sachteil* in 9 Bänden. Bärenreiter Kassel etc. 1998.

Kurt FLASCH. *Die geistige Mobilmachung. Die deutschen Intellektuellen und der Erste Weltkrieg.* Alexander Fest Verlag Berlin 2000.

Craig FRASER. Mathematics in Library Subject Classification Systems. In *Research in History and Philosophy of Mathematics* (M. Zack & D. Schlimm, eds.). Proceedings of the Canadian Society for History and Philosophy of Mathematics, La Société Canadienne d'Histoire et de Philosophie des Mathématiques 2017; pp. 181–197.

Maurice FRÉCHET. Les mathématiques à l'Université de Strasbourg. *La Revue du mois* 10 avril 1920, 337–362.

— & Horace Bryon HEYWOOD. *L'équation de Fredholm et ses applications à la physique mathématique.* Avec une préface et une note de Jacques Hadamard. Hermann Paris 1912.

David FROMKIN. *A Peace to End All Peace. The Fall of the Ottoman Empire and the Creation of the Modern Middle East.* Henry Holt & Co. New York 1989.

Fulvia FURINGHETTI & Livia GIACARDI. People, Events, and Documents of ICMI's First Century. *Actes d'Història de la Ciència i de la Tècnica, nova època* 3–2 (2010), 11–50.

— & —. ICMI in the 1950s and 1960s: Reconstruction, Settlement, and "Revisiting Mathematics Education". In F. Furinghetti & L. Giacardi (eds.), *The International Commission on Mathematical Instruction, 1908–2008: People, Events, and Challenges in Mathematics Education.* Cham, Switzerland: Springer. *To appear* 2022.

Fulvia FURINGHETTI & Livia GIACARDI & Marta MENGHINI. Actors in the changes of ICMI: Heinrich Behnke and Hans Freudenthal. In E. Barbin, K. Bjarnadóttir, F. Furinghetti, A. Karp, G. Moussard, J. Prytz, G. Schubring (Eds.). *Dig where you stand, 6. Proceedings of the sixth International Conference on the History of Mathematics Education, September 19-22, 2019, at Luminy, France.* (pp. 247-260). Münster WTM-Verlag 2020.

Peter GALISON. The Ontology of the Enemy: Norbert Wiener and the Cybernetic Vision. *Critical Inquiry* 21 (Autumn 1994), 228–266.

Livia GIACARDI & Rossana TAZZIOLI. Dibattiti nella comunità dei matematici italiani: l'apporto dell'Archivio dell'Unione Matematica Italiana. *Accademia delle Scienze Torino. Atti Classe di Scienze Fisiche* 152 (2018), 73–98.

— & —. The UMI Archives. Debates in the Italian Mathematical Community, 1922–1938. *EMS Newsletter* September 2019, 37–44.

— & —. Le *Bollettino dell'Unione Matematica Italiana* (BUMI) et ses enjeux politiques et idéologiques (1922–1943). In *Circulations mathématiques dans et par les journaux : histoire, territoires et publics* (H. Gispert, P. Nabonnand, J. Peiffer, eds.). ISTE London 2021 (*to appear*).

Martin GIERL. *Geschichte und Organisation. Institutionalisierung als Kommunikationsprozess am Beispiel der Wissenschaftsakademien um 1900.* Abhandlungen der Akademie der Wissenschaften zu Göttingen, Philologisch-Historische Klasse, vol. 233. Vandenhoeck & Ruprecht Göttingen 2004.

—. "Little Big Science". Die Reform der Göttinger Akademie der Wissenschaften und die Wissenschaftsorganisation um 1900. In *Wissenschaftsakademien im Zeitalter der Ideologien. Politische Umbrüche, wissenschaftliche Herausforderungen, institutionelle Anpassungen. Arbeitstagung des Projektes zur Geschichte der Leopoldina.* (R. Vom Bruch, S. Gerstengarbe. J. Thiel, eds.). Halle (Saale), Stuttgart 2014; pp. 91–108.

Charles Coulston GILLISPIE (ed.). *Dictionary of Scientific Biography.* 16 volumes. American Council of Learned Societies 1970–1980.

Hélène GISPERT. Une comparaison des journaux français et italiens dans les années 1860–1875. In [Goldstein, Gray, Ritter 1996], pp. 390–406.

—. The German and French Editions of the Klein-Molk Encyclopedia: Contrasted Images. In [Bottazzini & Dahan 2001]; pp. 93–112.

—. The Effects of War on France's International Role in Mathematics, 1870–1914. In [Parshall & Rice 2002]; pp. 105–121.

—. *La France mathématique de la IIIe République avant la Grande guerre.* Société mathématique de France Paris 2015.

— & Renate TOBIES. A comparative study of the French and German Mathematical Societies before 1914. In [Goldstein, Gray, Ritter 1996], pp. 407–430.

Roger Godement. *Analyse Mathématique IV: Intégration et théorie spectrale, analyse harmonique, le jardin des délices modulaires.* Springer 2003.

Catherine GOLDSTEIN. Un arithméticien contre l'arithmétisation : les principes de Charles Hermite. In *Justifier en mathématiques* (D. Flament et P. Nabonnand, eds.). Maison des Sciences de l'Homme Paris 2011; pp. 129–165.

—. The Hermitian Form of Reading the *Disquisitiones*. In [Goldstein, Schappacher, Schwermer 2007], pp. 377–410.

—. Long-term History and Ephemeral Configurations. In *Proceedings of the International Congress of Mathematicians*, Rio de Janeiro 2018; vol. 1, pp. 487–522.
—, Jeremy GRAY, Jim RITTER (eds.). *L'Europe mathématique / Mathematical Europe*. Edition de la Maison des sciences de l'homme Paris 1996.
—, Norbert SCHAPPACHER, Joachim SCHWERMER (eds.). *The Shaping of Arithmetic after C.F. Gauss's* Disquisitiones Arithmeticae. Springer Verlag 2007.
— & Norbert SCHAPPACHER. A Book in Search of a Discipline. In [Goldstein, Schappacher, Schwermer 2007]; pp. 3–65.

Luis Español GONZÁLEZ (ed.). *Historia de la Real Sociedad Matemática Española (RSME)*. Real Sociedad Matemática Española, Sevilla 2011.

Francisco A. GONZÁLEZ REDONDO. La reorganización de la Matemática en España tras la Guerra Civil. La posibilitación del retorno de Esteban Terradas y Julio Rey Pastor. *La Gaceta de la Real Sociedad Española* 5–2 (2002), 463–490.

Sarvepalli GOPAL, Sergei L. TIKHVINSKY et al. (eds.). *History of Humanity. Volume VII: The Twentieth Century*. UNESCO Publishing 2008.

Constantin GOSCHLER. *Rudolf Virchow. Mediziner – Anthropologe – Politiker*. Böhlau Verlag Köln, Wien 2002.
—. Deutsche Naturwissenschaft und naturwissenschaftliche Deutsche. Rudolf Virchow und die "deutsche Wissenschaft". In *Wissenschaft und Nation. Universalistischer Anspruch und nationale Identitätsbildung im europäischen Vergleich* (Ralph Jessen & Jakob Vogel, eds.) Frankfurt a.M. 2002; pp. 97–114.

Loren GRAHAM & Jean-Michel KANTOR. A Comparison of Two Cultural Approaches to Mathematics. France and Russia, 1890–1930. *Isis* 97 (2006), 56–74.
— & —. *Naming Infinity. A true story of religious mysticism and mathematical creativity*. Harvard U Press 2009.

Ivor GRATTAN-GUINNESS. A mathematical union: William Henry and Grace Chisholm Young. *Annals of Science* 29–2 (1972), 105–185.

George W. GRAY. *Education on an international scale. A history of the International Education Board 1923–1938*. Harcourt Brace New York 1941.

Jeremy GRAY. Anxiety and Abstraction in Nineteenth-Century Mathematics. *Science in Context* 17 (2004), 23–47.
—. *Plato's Ghost. The Modernist Transformation of Mathematics*. Princeton U Press 2008.

Joseph F. GRCAR. Topical Bias in Generalist Mathematics Journals. *Notices of the American Mathematical Society*, vol. 57, no. 11, December 2010, 1421–1424.

Frank GREENAWAY. *Science International. A history of the International Council of Scientific Unions*. Cambridge U Press 1996.

Angelo GUERRAGGIO & Pietro NASTASI. *Italian Mathematicians Between the Two World Wars*. Birkhäuser 2005.
— & Giovanni PAOLONI. *Vito Volterra* (transl. Kim Williams). Springer 2013.

Heini HALBERSTAM. An Obituary of Loo-keng Hua. *Mathematical Intelligencer* 8 (1986), 63–65.

Maurice HALBWACHS & Alfred SAUVY. *Le point de vue du nombre, 1936*. Édition critique Marie Jaisson & Éric Brian. Institut national d'études démographiques Paris 2005.

Godfrey H. HARDY. *Bertrand Russell & Trinity: a College Controversy of the Last War*. Cambridge U Press 1942.
— & Marcel RIESZ. *The General Theory of Dirichlet's Series*. Cambridge U Press 1915.

Michael HARRIS. *Mathematics without apologies. Portrait of a Problematic Vocation*. Princeton U Press 2015.
—. *Virtues of Priority*. arXiv:2003.08242v1 (18 March 2020).

Bettina HEINTZ. *Die Innenwelt der Mathematik. Zur Kultur und Praxis einer beweisenden Disziplin*. Springer Verlag Wien 2000.

Ruth HENIG. *The Peace that Never Was. A History of the League of Nations*. Haus Publishing London 2019.

Arend HEYTING. *Mathematische Grundlagenforschung. Intuitionismus, Beweistheorie*. Springer 1934.

Horace Bryan HEYWOOD (H.B.H.). The International Congress of Mathematicians. *Nature* vol. 106, no. 2658 (7 October 1920), 196–197.

Ann HIBNER KOBLITZ. A Biographical Sketch. In *The Legacy of Sonya Kovalevskaya. Proceedings of a Symposium Sponsored by The Association for Women in Mathematics and The Mary Ingraham Bunting Institute, held October 25–28, 1985* (Linda Keen, ed.). AMS 1987; pp. 3–16.

David HILBERT. Grundlagen der Geometrie. In *Festschrift zur Feier der Enthüllung des Gauss-Weber-Denkmals in Göttingen, herausgegeben von dem Fest-Comitee*. Teubner Leipzig 1899.

Wilhelm HIS. Zur Vorgeschichte des deutschen Kartells und der internationalen Association der Akademien. *Berichte der mathematisch-physischen Klasse der Königlich Sächsischen Gesellschaft der Wissenschaften zu Leipzig*, Sonderheft 1902, 12–15.

Bernard R. HODGSON. ICMI in the post-Freudenthal era: moments in the history of mathematics education from an international perspective. In *"Dig where you stand". Proceedings of the Conference on On-going Research in the History of Mathematics Education* (K. Bjarnadóttir, F. Furinghetti, G. Schubring, eds.). Reykjavik, University of Iceland, School of Education 2009; pp. 79–96.
—. Whither the Mathematics/Didactics Interconnection? Evolution and Challenges of a Kaleidoscopic Relationship as Seen from an ICMI Perspective. In Sung Je Cho (ed.). *The Proceedings of the 12th International Congress on Mathematical Education. Intellectual and Attitudinal Challenges*. Springer 2015; pp. 41–61.

Christopher HOLLINGS. *Mathematics Across the Iron Curtain. A History of the Algebraic Theory of Semigroups*. AMS 2014.
—, Ursula MARTIN, Adrian RICE. The Lovelace – De Morgan mathematical correspondence: A critical re-appraisal. *Historia Mathematica* 44 (2017), 202–231.
—, Ursula MARTIN, Adrian RICE. The early mathematical education of Ada Lovelace. *BSHM Bulletin* 32 (2017), 221–234.

—, Reinhard SIEGMUND-SCHULTZE. *Meeting under the integral sign? The Oslo Congress of Mathematicians on the eve of the Second World War.* AMS 2020.

Annick HORIUCHI. Sur la recomposition du paysage mathématique japonais au début de l'époque Meiji. In [Goldstein, Gray, Ritter 1996], pp. 248–268,

Albert Geoffrey HOWSON. Seventy Five Years of the International Commission on Mathematical Instruction. *Educational Studies in Mathematics* 15 (1984), 75–93.

Kyra T. INACHIN. "Märtyrer mit einem kleinen Häuflein Getreuer". Der erste Gauleiter der NSDAP in Pommern, Karl Theodor Vahlen. *Vierteljahrshefte für Zeitgeschichte* 49 (2001), 31–51.

Patrick D.F. ION, Thierry BOUCHE, Gadadhar MISRA, Alf A. ONSHUUS, Stephen M. WATT, Liu ZHENG. International Mathematical Knowledge Trust IMKT: An Update on the Global Digital Mathematics Library. In *Proceedings ICM 2018*, Vol. 1, pp. 1157–1176.

Akira IRIYE, Petra GOEDDE, William I. HITCHCOCK. *The Human Rights Revolution. An International History.* Oxford U Press 2012.

Giorgio ISRAEL & Pietro NASTASI. *Scienza e razza nell'Italia fascista.* il Mulino Bologna 1998.

Shokichi IIYANAGA. *Collected Papers*. Iwanami Shoten Tokyo 1994.

Catherine JAMI. In Memoriam Jospeh Needham (December 9, 1900 – March 24, 1995). *Historia Mathematica* 24 (1996), 1–5.

Harold Spencer JONES. The Early History of ICSU 1919–1946. *ICSU Review* 2–4 (1960) , pp. 169–187. Quoted from the reprint: The Royal Society, London, 1961.

The JOURNEY to the West. Translated and Edited by Anthony C. Yu. U of Chicago Press 2012.

Robert KANIGEL. *The Man Who Knew Infinity. A Life of the Genius Ramanujan.* Washington Square Press New York 1991.

Jean-Michel KANTOR (ed.). *Jean Leray (1906–1998).* Gazette des mathématiciens Édition spéciale SMF Paris 2000.

Shaul KATZ. Berlin Roots—Zionist Incarnation. The Ethos of Pure Mathematics and the Beginnings of the Einstein Institute of Mathematics at the Hebrew University of Jerusalem. *Science in Context* 17 (2004), 199–234.

Victor J. KATZ. The History of Stokes's Theorem. *Mathematics Magazine* 52–3 (May, 1979), 146–156.

Eva KAUFHOLZ-SOLDAT, Nicola OSWALD (eds.). *Against All Odds. Women's Ways to Mathematical Research Since 1800.* Springer 2020.

Akitsugu KAWAGUCHI. Sokan no kotoba. *Tenzoru* 1–1 (1938), 1–2.

Pauline KERR & Geoffrey WISEMAN. *Diplomacy in a Globalizing World: Theories and Practice.* Oxford U Press 2013, 2018.

Daniel J. KEVLES. George Ellery Hale, the First World War, and the Advancement of Science in America. *Isis* 59–4 (1968), 427–437.

John Maynard KEYNES. *The Economic Consequences of the Peace.* London Macmillan 1920.

Tinne Hoff KJELDSEN. New Mathematical Disciplines and Research in the Wake of World War II. In [Booß-Bavnbeck & Høyrup 2003]; pp. 126–152.

—. The Development of Nonlinear Programming in Post War USA: Origin, Motivation, and Expansion. In *The Way Through Science and Philosophy. Essays in Honour of Stig Andur Pedersen* (H.B. Andersen, F.V. Christiansen, K.F. Jørgensen, V.F. Hendricks, ed.s). College Publications London 2006; pp. 31–50.

—. A Multiple Perspective Approach to History of Mathematics: Mathematical Programming and Rashevsky's Early Development of Mathematical Biology in the Twentieth Century. In *Interfaces between Mathematical Practices and Mathematical Education* (G. Schubring, ed.) Springer 2019; pp. 143–167.

Jurjen Ferdinand KOKSMA. *Diophantische Approximationen.* Springer 1936.

Gina Bari KOLATA. Anti-Semitism Alleged in Soviet Mathematics. *Science* 202–4373 (15 December 1978), 1167–1170.

Andrei Kolmogoroff [KOLMOGOROV]. *Grundbegriffe der Wahrscheinlichkeitsrechnung.* Springer Berlin 1933.

Wolfgang KRULL. *Idealtheorie.* Springer 1935.

Harald KÜMMERLE. Hayashi Tsuruichi and the success of the Tôhoku Mathematical Journal as a publication. *Advanced Studies in Pure Mathematics* 79 (2018), pp. 347–358.

—. *Die Institutionalisierung der Mathematik als Wissenschaft im Japan der Meiji- und Taishô-Zeit (1868–1926).* Acta Historica Leopoldina 77, Deutsche Akademie der Naturforscher Leopoldina. Nationale Akademie der Wissenschaften, Halle (Saale) *to appear* 2021.

—. Tannaka Tadao's 1938 paper on the duality of non-commutative topological groups and its historical background. In R. Krömer, E. Haffner, K. Volkert (eds.), *Duality in 19th and 20th century mathematical thinking.* Birkhäuser Basel *to appear.*

Ernst Eduard KUMMER. *Collected Papers*, ed. André Weil. Two volumes. Springer 1975.

Kazimierz KURATOWSKI. *A Half Century of Polish Mathematics: Remembrances and Reflections.* Pergamon Press 1980.

Jérôme LAMY (ed.). *La carte du ciel. Histoire et actualité d'un projet scientifique international.* Observatoire de Paris / EDP Science Les Ulis 2008.

Serge LANG. *Algebra.* First printing. Addison-Wesley Reading, Mass. 1965.

René LAVOLLÉE. Les unions internationales. *Revue d'histoire diplomatique* 1 (1887), 331–362.

Everett S. LEE. A Theory of Migration. *Demography* 3–1 (1966), 47–57.

Joep LEERSEN. Notes towards a Definition of Romantic Nationalism. *Romantik. Journal for the Study of Romanticisms* 2 (2013), 9–35.

Jean-Jacques LEFRÈRE & Patrick BERCHE. Un cas de délire scientifico-patriotique : le docteur Edgar Bérillon. *Annales Médico-Psychologiques* 168 (2010), 707–711. `https://hal.archives-ouvertes.fr/hal-00690282` (consulted 18 July 2021).

Solomon LEFSCHETZ. *Algebraic Topology.* AMS New York 1942.

Olli LEHTO. *Mathematics Without Borders. A History of the International Mathematical Union.* Springer Verlag 1998.

Franz LEMMERMEYER & Peter ROQUETTE. *Helmut Hasse und Emmy Noether. Die Korrespondenz 1925–1935.* With an introduction in English. U Verlag Göttingen 2006.

António José F. LEONARDO, Décio R. MARTINS, Carlos FIOLHAIS. Costa Lobo and the Study of the Sun in Coimbra in the First Half of the Twentieth Century. *Journal of Astronomical History and Heritage* 14 (1) (2011), 41–56.

Rebecka LETTEVAL, Geert SOMSEN, Sven WIDMALM. *Neutrality in Twentieth-Century Europe. Intersections of Science, Culture, and Politics after the First World War.* Routledge New York & London 2012.

Trevor H. LEVERE. *Transforming Matter. A History of Chemistry from Alchemy to the Buckyball.* The Johns Hopkins U Press Baltimore & London 2001.

Laura LEVI. *Beppo Levi. Italia y Argentina en la vida de un matemático.* Libros del Zorzal 2000.

Paul LÉVY. *Quelques aspects de la pensée d'un mathématicien.* Blanchard Paris 1970.

Francisco Miranda da COSTA LOBO (ed./author; name not printed on cover). *Congresso International de Matemáticas realisado em Strasburgo em 30 de Setembro de 1920.* Coimbra Imprensa da Universidade 1921. (38 pages.)

François Le LIONNAIS (ed.). *Les grands courants de la pensée mathématique.* Cahiers du Sud 1948.

Sir Henry LYONS (ed.). *Fifth Assembly of the International Research Council and the First Assembly of the International Council of Scientific Unions held at Brussels, July 11th, 1931. Reports of Proceedings.* London 1932.
—. *Second General Assembly of the International Council of Scientific Unions held at Brussels, July 9th to 13th, 1934. Reports of Proceedings.* London 1935.

Jean MAHWIN. In Memorian Jean Leray. *Topological Methods in Nonlinear Analysis, Journal of the Juliusz Schauder Center* 12 (1998), 199–206.

Karl-Heinz MANEGOLD. *Universität, Technische Hochschule und Industrie. Ein Beitag zur Emanzipation der Technik im 19. Jahrhundert unter besonderer Berücksichtigung der Bestrebungen Felix Keins.* Duncker & Humblot Berlin 1970.

Camille MARBO. *A travers deux siècles. Souvenirs et rencontres (1883–1967).* Editions Bernard Grasset Paris 1968.

Roberto MARKARIAN & Ernesto MORDECKI (eds.). *José Luis Massera. Ciencia y compromiso social.* Pedeciba Montevideo, Uruguay, 2010.

A.A. MARKOFF (=Andrey Andreyevich MARKOV). *Differenzenrechnung*. Authorized German translation by T. Friesendorff and E. Prümm. Teubner Leipzig 1896.

Laurent MAZLIAK. The beginnings of the Soviet encyclopedia. The utopia and misery of mathematics in the political turmoil of the 1920s. *Centaurus* 60 (2018), pp. 25–51.

—. A whole new vigor: About Montel's book 'Les mathématiques et la vie' (1947). *Teaching Mathematics and Computer Science* 18–3 (2020), 51–60.

— & Rossana TAZZIOLI. *Mathematicians at War. Volterra and His French Colleagues in World War I.* Springer 2009.

— & Rossana TAZZIOLI (eds.). *Mathematical Communities in the Reconstruction After the Great War 1918–1928. Trajectories and Institutions.* Birkhäuser 2021.

Barry MAZUR. Conjecture. *Synthese* 111 (1997), 197–210.

Herbert MEHRTENS. *Moderne – Sprache – Mathematik. Eine Geschichte des Streits um die Grundlagen der Disziplin und des Subjekts formaler Systeme.* Suhrkamp Frankfurt a.M. 1990.

Pierre-Michel MENGER, Colin MARCHIKA, Yann RENISIO, Pierre VERSCHUEREN. Formations et carrières mathématiques en France: un modèle typique d'excellence ? *Revue française d'économie* 35 (2020), 155–217.

Marta MENGHINI, Fulvia FURINGHETTI, Livia GIACARDI, Ferdinando ARZARELLO (eds.). *The First Century of the International Commission on Mathematical Instruction (1908–2008). Reflecting and Shaping the World of Mathematics Education.* Istituto della Enciclopedia Italiana Roma 2009.

Angelo MESSEDAGLIA. Le statistiche criminali dell'Impero austriaco nel quadriennio 1856–59, con particolare riguardo al Lombardo-Veneto e col confronto dei dati posteriori fino al 1861 inclusivamente. Esposizione critica. *Atti Reale Istituto Veneto di scienze, lettere ed arti*, 3rd ser., 11 (1865–1866). `https://babel.hathitrust.org/cgi/pt?id=uc1.a0000978973&view=1up&seq=15` (consulted 5 March 2020).

Helena MIHALJEVIĆ-BRANDT & Olaf TESCHKE. Journal Profiles and Beyond: What Makes a Mathematics Journal "General"? *European Mathematical Society (EMS) Newsletter*, March 2014, 55–56.

Helena MIHALJEVIĆ & Marie-Françoise ROY. A Data Analysis of Women's Trails Among ICM Speakers. In [Araujo et al. 2019]; pp. 111–128.

Hermann MINKOWSKI. *Briefe an David Hilbert* (L. Rüdenberg & H. Zassenhaus, eds.). Springer 1973.

Ivan MOERLEN & Lucien BECHELEN. *La lutte de la jeunesse estudiantine alsacienne et lorraine contre l'emprise allemande de 1871 à 1918.* Chez l'auteur L. Bechelen Strasbourg-Cronenbourg 1957.

Felix MÜLLER. *Vocabulaire mathématique Français-Allemand et Allemand-Français: Contenant les termes techniques employés dans les mathématiques pures et appliquées.* Teubner Leipzig 1900.

Robert MUSIL. *Precision and Soul: Essays and Addresses.* (Edited by B. Pike and D.S. Luft) U of Chicago Press 1990.

Philippe NABONNAND, Jeanne PEIFFER, Hélène GISPERT. Circulation et échanges mathématiques (18e – 20e siècles). *Philosophia scientiae* 19–2, 2015, 7–16.

Johann [John] von NEUMANN. *Mathematische Grundlagen der Quantenmechanik*. Springer 1932.

James William NIXON. *A History of the International Statistical Institute 1885–1960*. ISI The Hague 1960.

Emmy NOETHER. *Gesammelte Abhandlungen — Collected Papers*. Nathan Jacobson (ed.). Springer 1983.

Klaus NOHLEN. *Baupolitik im Reichsland Elsaß-Lothringen 1871–1918*. Berlin 1982.

Luboš NOVÝ. Les mathématiques et l'évolution de la nation tchèque (1860–1918). In [Goldstein, Gray, Ritter 1996], pp. 499–515.

Father F.L. ODENBACH SJ. The International Seismological Association. *Bulletin of the Seismological Society of America* 1–3 (Sept. 1911), 103–106.

Eduardo ORTIZ (ed.). *The works of Julio Rey Pastor*. London 1988.

Jürgen OSTERHAMMEL. *The Transformation of the World. A Global History of the Nineteenth Century*. Princeton U Press 2014.

Richard OVERY. *The Morbid Age. Britain Between the Wars*. Allan Lane London 2009.

Larry OWENS. Mathematicians at War. Warren Weaver and the Applied Mathematics Panel, 1942–1945. In *The History of Modern Mathematics* (David E. Rowe & John McCleary, eds.). Academic Press 1989. Vol. II; pp. 287–305.

Carol PARIKH. *The Unreal Life of Oscar Zariski*. Academic Press Boston etc. 1991.

Karen Hunger PARSHALL. *James Joseph Sylvester. Life and Work in Letters*. Oxford U Press 1998.
—. "A New Era in the Development of Our Science:" The American Mathematical Research Community, 1920–1950. In *A Delicate Balance: Global Perspectives on Innovation and Tradition in the History of Mathematics. A Festschrift in Honor of Joseph W. Dauben* (David E. Rowe & Wann-Sheng Horng, eds.). Birkhäuser 2015; pp. 275–308.
— & Adrian C. RICE (eds.). *Mathematics Unbound. The Evolution of an International Mathematical Research Community, 1800–1945*. History of Mathematics vol. 23. AMS / LMS 2002.

Aleksei PARSHIN. Numbers as Functions: The Development of an Idea in the Moscow School of Algebraic Geometry. In [Bolibruch et al. 2006]; pp. 297–329.
—. Mathematik in Moskau. Es war eine große Epoche. *Mitteilungen der Deutschen Mathematiker-Vereinigung* 18 (2010), 43–48.

Silvana PATRIARCA. *Numbers and Nationhood. Writing Statistics in Nineteenth-century Italy*. Cambridge U Press 1996.

Elizabeth Chambers PATTERSON. *Mary Somerville and the Cultivation of Science*. International Archives of the History of Ideas / Archives internationales d'histoire des idées, Vol. 102. Springer 1983.

Patrick PETITJEAN. Needham and UNESCO: perspectives and realizations. In P. Petitjean,V. Zharov, G. Glaser, J. Richardson, B. de Padirac, G. Archibald (eds), *Sixty Years of Sciences at Unesco, 1945–2005.* Unesco 2006; pp. 43–47. `https://halshs.archives-ouvertes.fr/halshs-00166502/document` (consulted 20 May 2021).

—, Catherine JAMI, Anne Marie MOULIN (eds.). *Science and Empires. Historical Studies about Scientific Development and European Expansion.* Kluwer Academic Publishers Dordrecht etc. 1992.

Birgit PETRI & Norbert SCHAPPACHER. On Arithmetization. In [Goldstein, Schappacher, Schwermer 2007], pp. 343–374.

Émile PICARD. *L'histoire des sciences et les prétentions de la science allemande.* Pour la verité 1914–1915 (Etudes publiées sous le patronage des Secrétaires perpétuels des cinq Académies). Perrin & Cie. Paris 1916.

Michele PICCIRILLO, Patricia Maynor BIKAI, Thomas A. DAILEY. *The Mosaics of Jordan.* American Center of Oriental Research, Amman 1993.

Herbert PIEPER. A Network of Scientific Philanthropy: Humboldt's Relations with Number Theorists. In [Goldstein, Schappacher, Schwermer 2007], Chapter III.1; pp. 201–233.

Henri POINCARÉ. Über transfinite Zahlen. In H. Poincaré, *Sechs Vorträge über ausgewählte Gegenstände aus der reinen Mathematik und mathematischen Physik; auf Einladung der Wolfskehl-Kommission der Königlichen Gesellschaft der Wissenschaften gehalten zu Göttingen vom 22.–28. April 1909.* B.G. Teubner Leipzig, Berlin 1910; pp. 43–48.

George PÓLYA. *The Pólya Picture Album. Encounters of a Mathematician*, edited by G.L. Alexanderson. Birkhäuser Boston & Basel 1987.

Lev S. PONTRYAGIN. Soviet Anti-Semitism: Reply by Pontryagin. *Science* 205–4411 (14 September 1979), 1083–1084.

Jean-Guy PRÉVOST. *A Total Science. Statistics in Liberal and Fascist Italy.* McGill-Queens U Press Montreal 2009.

PROCEEDINGS ICM xxxx. Throughout the book, a quote of the form: "Proceedings ICM xxxx" refers to the Proceedings of the International Congress of Mathematicians that took place in the year xxxx. All these Proceedings can be freely downloaded at `https://www.mathunion.org/icm/proceedings` (consulted 20 October 2021).

PROCEEDINGS *of the International Symposium on Algebraic Number Theory Tokyo & Nikko, September, 1955.* Published by The Organizing Committee of the International Symposium on Algebraic Number Theory, Science Council of Japan, Ueno Park, Tokyo October 1956. The entire brochure can be downloaded at `https://www.jmilne.org/math/Documents/TokyoNikko1955.pdf` (consulted 28 June 2021).

Lewis PYENSON. *Neohumanism and the Persistence of Pure Mathematics in Wilhelmian Germany.* American Philosophical Society Philadelphia 1983.

Adolphe QUETELET. Notes extraites d'un voyage scientifique, fait en Allemagne pendant l'été de 1829. *Correspondance mathématique et physique* 6 (1830), 126–148; 161–178; 225–239.

—. Recherches sur le poids de l'homme aux différens âges. Présentées en juin 1832. *Nouveaux Mémoires de l'Académie Royale des sciences et belles-lettres de Bruxelles* VII (1833); 38 pp.

Tibor Radó. *On the problem of Plateau.* Springer 1933.

Gerhard RAMMER. *Die Nazifizierung und Entnazifizierung der Physik an der Universität Göttingen.* PhD Thesis Göttingen 2004.

Michael RAPOPORT, Norbert SCHAPPACHER, Peter SCHNEIDER. *Beilinson's Conjectures on Special Values of L-Functions.* Academic Press 1988.

Anne RASMUSSEN. Internationalismes au début du XXe siècle. In [Audoin-Rouzeau & Becker 2004]; pp. 71–82.

Dietrich RAUSCHNIGG, Donata v. NERÉE (eds.). *Die Albertus-Universität zu Königsberg und ihre Professoren.* Jahrbuch der Albertus-Universität zu Königsberg Band XXIX, 1994. Duncker & Humblot Berlin 1995.

Constance REID. *Hilbert.* Springer 1970.

Kurt REIDEMEISTER. *Knotentheorie.* Springer 1932.

Volker REMMERT. Oberwolfach in the French Occupation Zone: 1945 to early 1950s. *Revue d'histoire des mathématiques* 26 (2020), 121–172.
— & Ute SCHNEIDER (eds.). *Publikationsstrategien einer Disziplin: Mathematik in Kaiserreich und Weimarer Republik.* Mainzer Studien zur Buchwissenschaft 19, Harrassowitz Wiesbaden 2008; pp. 9–51.
— & Ute SCHNEIDER. *Eine Disziplin und ihre Verleger. Disziplinenkultur und Publikationswesen der Mathematik in Deutschland, 1871–1949.* transcript Verlag Bielefeld 2010.

Vera RICH. Boycott of Soviet contacts is for individuals, says NAS. *Nature* 275 (7 September 1978), 3.

Elaine McKinnon RIEHM & Frances HOFFMAN. *Turbulent Times in Mathematics. The Life of J.C. Fields and the History of the Fields Medal.* American Mathematical Society & The Fields Institute 2011.

Adrian RICE. Partnership and Partition: A Case Study of Mathematical Exchange. *Philosophia Scientiæ* 19 (2015), 115–134.

Clara Silvia ROERO (ed.). *Dall'università di Torino all'Italia unita. Contributi dei docenti al risorgimento e all'unità.* Deputazio e Subalpina di Storia Patria Torino 2013.

Laurent ROLLET. Une science sans frontières ? Le Répertoire Bibliographique des Sciences Mathématiques. In *Sciences et frontières. Délimitations du savoir, objets et passages* (Philippe Hert & Marcel Paul-Cavallier, eds.). EME Fernelmont 2007; pp. 255–282.

Rudolf ROTHE. D. F. Seliwanoff †. *Jahresbericht der Deutschen Mathematiker-Vereinigung* 44 (1934), 210–214.

David E. ROWE. Three Letters from Sophus Lie to Felix Klein on Parisian Mathematics during the Early 1880's. Translated from the German. *The Mathematical Intelligencer* 7 (1985), 74–77.
—. Disciplinary Cultures of Mathematical Productivity in Germany. In [Remmert & Schneider 2008]; pp. 9–51.

—. Who Linked Hegel's Philosophy with the History of Mathematics? *Mathematical Intelligencer* 35 (2013), 38–41; 51–55.
—. On Franco-German Relations in Mathematics, 1870–1920. In *Proceedings of the International Congress of Mathematicians, Rio de Janeiro* 2018, vol. 3; pp. 4081–4096.
—. *Emmy Noether — Mathematician Extraordinaire.* Springer 2021.
— & Volkmar FELSCH. *Otto Blumenthal. Ausgewählte Briefe und Schriften II.* Springer Verlag 2019.
— & Mechthild KOREUBER. *Proving It Her Way. Emmy Noether, a Life in Mathematics.* Springer 2020.
— & Robert SCHUMANN (eds.). *Einstein On Politics. His Private Thoughts and Public Stands on Nationalism, Zionism, War, Peace, and the Bomb.* Princeton U Press 2007.

Paul RUBINSON. "For Our Soviet Colleagues": Scientific Internationalism, Human Rights, and the Cold War. In [Iriye et al. 2012]; pp. 245–264.

Philippe SANDS. *East West Street. On the Origins of Genocide and Crimes Against Humanity.* Weidenfeld & Nicolson 2016.

Umberto SANSONI. *Il nodo di Salomone. Simbolo e archetipo d'alleanza.* Electa Milan 1998.

Chikara SASAKI. Science and the Chrysanthemum: The Paradox of Enlightenment in Imperial Japan. *Historia Scientiarum (International Journal of the History of Science Society of Japan)* 11 (2001),24–47.
—. The Emergence of the Japanese Mathematical Community in the Modern Western Style, 1855–1945. In [Parshall & Rice 2002]; pp. 229–252.

Norbert SCHAPPACHER. Max-Planck-Institut für Mathematik. Historical Notes on the New Research Institute at Bonn. *Mathematical Intelligencer* 7 (1985), 41–52.
—. On the History of Hilbert's Twelfth Problem. A Comedy of Errors. In *Matériaux pour l'histoire des mathématiques au XXᵉ siècle. Actes du colloque à la mémoire de Jean Dieudonné Nice 1996.* Séminaires et Congrès 3, Société Mathématique de France Paris 1998; pp. 243–273.
—. Politisches in der Mathematik: Versuch einer Spurensicherung. *Mathematische Semesterberichte* 50 (2003), 1–27.
—. David Hilbert, Report on Algebraic Number Fields (*'Zahlbericht'*) (1897). In *Landmark Writings in Western Mathematics 1640–1940* (Ivor Grattan Guinness, ed.). Elsevier 2005; pp. 700–709.
—. Seventy years ago : The Bourbaki Congress at El Escorial and other mathematical (non-)events of 1936. *The Mathematical Intelligencer*, Special issue ICM Madrid August 2006, 8–15.
—. A Historical Sketch of B. L. van der Waerden's Work in Algebraic Geometry: 1926–1946. In *Episodes in the History of Modern Algebra (1800–1950)* (J.J. Gray & K.H. Parshall, eds.). History of mathematics series, vol. 32, AMS / LMS 2007; pp. 245-283.
—. Rewriting Points. *Proceedings of the International Congress of Mathematicians, Hyderabad, India* 2010; pp. 3258-3291.
—. Panorama eines Umbruchs. Die Modernisierung der Mathematik im Rückblick. Book review of [Gray 2008]. *N.T.M.* 20 (2012), 233–238.
—. Remarks about Intuition in Italian Algebraic Geometry. *Oberwolfach Reports* No. 47/2015[a], 2805–2807.
—. Ideologie, Wissenschaftspolitik, und die Ehre, Mitglied der Akademie zu sein. Ein Fall aus dem zwanzigsten Jahrhundert. *Res doctae*, online server of Göttingen Akademie der Wissenschaften 2015[b] `http://hdl.handle.net/11858/00-001S-0000-0023-9A17-2` (consulted 3 November 2020).

— & Martin KNESER. Fachverband – Institut – Staat. Streiflichter auf das Verhältnis von Mathematik zu Gesellschaft und Politik in Deutschland seit 1890 unter besonderer Berücksichtigung der Zeit des Nationalsozialismus. In *Ein Jahrhundert Mathematik 1890–1990. Festschrift zum Jubiläum der DMV* (G. Fischer, F. Hirzebruch, W. Scharlau, W. Törnig, eds.). Vieweg Braunschweig 1990.

— & René SCHOOF. Beppo Levi and the arithmetic of elliptic curves. *Mathematical Intelligencer* 18 (1996), 57–69.

— & Cordula TOLLMIEN. Emmy Noether, Hermann Weyl, and the Göttingen Academy. A Marginal Note. *Historia Mathematica* 43 (2016), 194–197.

— & Eckhard WIRBELAUER. Zwei Siegeruniversitäten; Die Straßburger Universitätsgründungen von 1872 und 1919. *Jahrbuch für Universitätsgeschichte* 13 (2010), 45–72.

Winfried SCHARLAU. Who Is Alexander Grothendieck? *Notices of the American Mathematical Society* 55–8 (2008), 930–941.

Désirée SCHAUZ. Vergangenheitspolitische Kommunikation im Privaten. Aufzeichnungen und Korrespondenzen des Biochemikers Adolf Windaus (1945–1949). In Petra Terhoeven & Dirk Schumann (ed.s), *Strategien der Selbstbehauptung. Vergangenheitspolitische Kommunikation an der Universität Göttingen (1945–1965).* Wallstein Göttingen 2021; pp. 310–354.

Peter SCHINDLER. *A Short history of the Committee on Freedom and Responsibility in the Conduct of Science (CFRS) and its Predecessor Committees of the International Council for Science (ICSU).* 2017. https://council.science/wp-content/uploads/2017/04/CFRS_history.pdf (consulted 2 August 2021).

Ivo SCHNEIDER. *Archimedes. Ingenieur, Naturwissenschaftler und Mathematiker*. Wissenschaftliche Buchgesellschaft Darmstadt 1979.

Erhard SCHOLZ (ed.). *Hermann Weyl's* Raum – Zeit – Materie *and a General Introduction to his Scientific Work*. DMV Seminar 30. Birkhäuser Basel 2001.

Brigitte SCHROEDER-GUDEHUS. *Deutsche Wissenschaft und internationale Zusammenarbeit 1914–1928. Ein Beitrag zum Studium kultureller Beziehungen in politischen Krisenzeiten.* PhD Thesis Geneva 1966.

—. *Les Scientifiques et la paix, la communauté scientifique internationale au cours des années 20.* Les Presses de l'Université de Montréal 1978.

Dirk SCHUMANN, Désirée SCHAUZ (eds.). *Forschen im "Zeitalter der Extreme". Akademien und andere Forschungseinrichtungen im Nationalsozialismus und nach 1945.* Wallstein Göttingen 2020.

Gert SCHUBRING. The origins and early history of ICMI. *International Journal for the History of Mathematics Education* 3–2 (2008), 3–33.

Arthur SCHUSTER. International Science. Discourse delivered at the Royal Institution on Friday, May, 18. *Nature* 74, issue 5 July 1906, 233–236; issue 12 July 1906, 256–259.

— (ed.). *International Research Council. Constitutive Assembly held at Brussels, July 18th to July 28th, 1919. Report of Proceeedings* ed. by Sir Arthur Schuster, F.R.S., General Secretary. Harrison & Sons London 1920.

— (ed.). *International Research Council. Second Assembly held at Brussels, July 25th to July 29th, 1922. Report of Proceeedings* ed. by Sir Arthur Schuster, F.R.S., General Secretary. Harrison & Sons London 1923.

— (ed.). *International Research Council. Fourth Assembly held at Brussels, July 13th, 1928. Report of Proceeedings* ed. by Sir Arthur Schuster, F.R.S., General Secretary. Harrison & Sons London 1930.

Johannes SCHWEITZER & Thorne LAY. IASPEI: its origins and the promotion of global seismology. *History of Geo- and Space Sciences* 10 (2019), 173–180.

Naoko SHIMAZU. *Japan, Race and Equality. The Racial Equality Proposal of 1919.* Routledge London, New York 1998.

Goro SHIMURA. Yutaka Taniyama and his time. Very Personal Recollections. *Bulletin of the London Mathematical Society* 21 (1989), 186–196.

—. *The Map of My Life.* Springer 2008.

Ulrich SIEG. *Deutschlands Prophet. Paul de Lagarde und die Ursprünge des modernen Antisemitismus.* Hanser Verlag München 2007.

Reinhard SIEGMUND-SCHULTZE. Die Anfänge der Funktionalanalysis und ihr Platz im Umwälzungsprozeß der Mathematik um 1900. *Archive for History of Exact Sciences* Vol. 26, No. 1 (1982), 13–71.

—. Einige Probleme der Geschichtsschreibung der Mathematik im Faschistischen Deutschland – unter besonderer Berücksichtigung des Lebenslaufes des Greifswalder Mathematikers Theodor Vahlen. *Wissenschaftliche Zeitschrift der Ernst-Moritz-Arndt-Universität Greifswald, Mathematisch-Naturwissenschaftliche Reihe* 23 Heft 1–2 (1984), 51–56.

—. *Mathematische Berichterstattung in Hitlerdeutschland.* Vandenhoeck & Ruprecht Göttingen 1993.

—. *Rockefeller and the Internationalization of Mathematics Between the Two World Wars.* Birkhäuser Basel etc. 2001.

—. Maurice Fréchet à Strasbourg. Les mathématiques entre nationalisme et internationalisme, entre application et abstraction. In [Crawford & Olff-Nathan 2005], pp. 185–196.

—. *Mathematicians Fleeing from Nazi Germany. Individual Fates and Global Impact.* Princeton U Press 2009.

—. Opposition to the Boycott of German Mathematics in the Early 1920s: Letters by Edmund Landau (1877–1938) and Edwin Bidwell Wilson (1879–1964). *Revue d'histoire des mathématiques* 17 (2011), 99–125.

—. "Mathematics Knows No Races". A Political Speech that David Hilbert Planned for the ICM in Bologna in 1928. *The Mathematical Intelligencer* 38–1 (2016), 56–66.

—, Henrik KRAGH SØRENSEN (eds.). *Perspectives on Scandinavian Science in the Early Twentieth Century.* The Norwegian Academy of Science and Letters 2006.

Norman SIEROKA. *Umgebungen. Symbolischer Konstruktivismus im Anschluss an Hermann Weyl und Fritz Medicus.* Chronos Verlag Zürich 2010.

Skúli SIGURDSSON. *Hermann Weyl, Mathematics and Physics, 1900–1927.* Thesis Harvard U 1991.

Ana SIMÕES, Maria Paula DIOGO, Kostas GAVROGLU (eds.). *Sciences in the Universities of Europe, Nineteenth and Twentieth Centuries. Academic Landscapes.* Springer Verlag 2015.

Steve (Stephen) SMALE. On the Steps of Moscow University. *The Mathematical Intelligencer* vol. 6, no. 2 (1984), 21–27.

Denis Mack SMITH. *Italy. A Modern History.* U Michigan Press 1969.

Société Mathématique de France (SMF). *Conférences de la Réunion Internationale des Mathématiciens tenue à Paris en juillet 1937.* Gauthier Villars Paris 1939.

Liza SOUTSCHEK, Kärin NICKELSEN. "Zusammenwirken" oder "Wettstreit der Nationen"? Kooperation und Konkurrenz in der deutschen Antarktisexploration um 1900. *NTM Zeitschrift für Geschichte der Wissenschaften, Technik und Medizin* 27/3 (2019), 1—35.

Brigitte STEGER. *Piazza Armerina : la villa romaine du Casale en Sicile.* Antiqua 17. Picard Paris 2017.

Ernst STEINITZ. Algebraische Theorie der Körper. *Journal für die Reine und Angewandte Mathematik* 137 (1910), 167–309.

Brigitte STENHOUSE. Mister Mary Somerville: Husband and Secretary. *The Mathematical Intelligencer* 2020, published online: https://link.springer.com/article/10.1007/s00283-020-09998-6 (consulted 2 April 2021).

Marshall H. STONE. *Linear transformations in Hilbert space and their applications to analysis.* American Mathematical Society 1932.

Arild STUBHAUG. *Gösta Mittag-Leffler. A Man of Conviction.* Springer 2010.

James Joseph SYLVESTER. *The Collected Mathematical Papers of James Joseph Sylvester. Volume II (1854–1873).* Cambridge U Press 1908.

Rossana TAZZIOLI. Interplay between local and international journals: The case of Sicily, 1880–1920. *Historia Mathematica* 45–4 (2018), 334–353.
—. Quelques remarques sur les mathématiques européennes au XIXe siècle : Le cas de l'Italie. *Repères IREM* 110 (Jan 2018), 21–36. Online available at: http://numerisation.univ-irem.fr/WR/IWR18002/IWR18002.pdf (consulted 3 March 2020).
—. Review of [Bongiorno & Curbera 2018]. *The Mathematical Intelligencer* 42 (2020), 75–81

Oswald TEICHMÜLLER. *Gesammelte Abhandlungen. Collected Papers.* Edited by L.V. Ahlfors & F.W. Gehring. Springer 1982.

Vladimir M. TIKHOMIROV. Moscow Mathematics 1950–1975. In Jean-Paul Pier (ed.), *Development of Mathematics 1950–2000.* Birkhäuser 2000; 1107–1120.

Renate TOBIES. Mathematik als Bestandteil der Kultur. Zur Geschichte des Unternehmens "Encyklopädie der Mathematischen Wissenschaften mit Einschluß ihrer Anwendungen". *Mitteilungen Österreichische Gesellschaft für Wissenschaftsgeschichte* 14 (1994), 1–90.
—. *Felix Klein. Visionen für Mathematik, Anwendungen und Unterricht.* Springer 2019.

Cordula TOLLMIEN. "Sind wir doch der Meinung, dass ein weiblicher Kopf nur ganz ausnahmsweise in der Mathematik schöpferisch tätig sein kann ...." Emmy Noether 1882–1935. *Göttinger Jahrbuch* 38 (1990), 153–219.
—. Der Krieg der Geister in der Provinz. Das Beispiel der Universität Göttingen 1914–1919. *Göttinger Jahrbuch* 1993, 137–210. http://www.tollmien.com/pdf/kriegdergeister1993.pdf (consulted 17 May 2020)

Adam TOOZE. *The Deluge. The Great War and the Remaking of Global Order.* Penguin Random House 2014.

Alan M. TURING. The extensions of a group. *Compositio Mathematica* 5 (1938), 357–367.

Laura E. TURNER & Henrik KRAGH SØRENSEN. Cultivating the Herb Garden of Scandinavian Mathematics: The Congresses of Scandinavian Mathematicians, 1909–1925. *Centaurus* 55 (2013), 385–411.

Jürgen & Wolfgang von UNGERN-STERNBERG. *Der 'Aufruf an die Kulturwelt!'. Das Manifest der 93 und die Anfänge der Kriegspropaganda im Ersten Weltkrieg.* 2. erweiterte Auflage mit einem Beitrag von Trude Maurer. Peter Lang Frankfurt a.M. 2013.

Kamil URBANOWICZ & Arris S. TIJSSELING. Work and life of Piotr Szymański. *Proceedings 12th International Conference on Pressure Surges, 18–20 November 2016, Dublin, Ireland.* 2016.

Rudolf VIERHAUS. Wilhelm von Humboldt. In *Wissenschaftspolitik in Berlin. Berlinische Lebensbilder*, Vol. 3 (W. Treue & K. Gründer, eds.). Colloquium Verlag Berlin 1987.

Jonathan VOGES. Eine Internationale der "Geistesarbeiter"? Institutionalisierte intellektuelle Zusammenarbeit im Rahmen des Völkerbundes. In Christian Henrich-Franke, Claudia Hiepel, Guido Thiemeyer, Henning Türk (eds.), *Grenzüberschreitende institutionalisierte Zusammenarbeit von der Antike bis zur Gegenwart*, Nomos Baden-Baden 2019; pp. 355–384.

Henri VOGT. Jules Molk. *Enseignement mathématique* 16 (1914), 380–383.

Aurel VOSS. Die Beziehung der Mathematik zur Kultur der Gegenwart. In *Die Kultur der Gegenwart. Ihre Entwicklung und ihre Ziele* (Paul Hinneberg, ed.), Teil III, Abteilung 1: *Die mathematischen Wissenschaften* (Felix Klein, ed.). Teubner Leipzig 1914; pp. A 1–49.

Bartel Leendert Van der WAERDEN. *Moderne Algebra. Unter Benutzung von Vorlesungen von E. Artin und E. Noether*. Two volumes; first edition. Springer Berlin 1930–31.
—. *Einführung in die algebraische Geometrie.* Springer 1939.
—. *Mathematische Statistik.* Springer 1957.

Oswald VEBLEN. *Projektive Relativitätstheorie.* Springer 1933.

Yuan WANG. *Hua Loo Keng. A Biography.* Translated by Peter Shiu. Springer 1999.

Rolin WAVRE. Les congrès internationaux de mathématiques. In [Lionnais 1948]; pp. 298–303.

Robert WEATHERWALL. The International Education Board. *Nature* 148 (1941), 398–401.

Cornelia WEGELER. *"... wir sagen ab der internationalen Gelehrtenrepublik". Altertumswissenschaft und Nationalsozialismus. Das Göttinger Institut für Altertumskunde 1921–1962.* Böhlau Verlag Wien, Köln, Weimar 1996.

Frode WEIERUD & Sandy ZABELL. German mathematicians and cryptology in World War II. *Cryptologia* 2019. DOI: 10.1080/01611194.2019.1600076

Hans-Georg WEIGAND, William MCCALLUM, Marta MENGHINI, Michael NEUBRAND, Gert SCHUBRING (eds.). *The Legacy of Felix Klein.* Springer Open 2019.

André WEIL. *Œuvres Scientifiques. Collected Papers.* Vol. II, Vol. III. Springer 1979.
—. *Œuvres Scientifiques. Collected Papers.* Vol. I. Corrected second printing. Springer 1980.

—. *Number Theory. An Approach through History. From Hammurapi to Legendre.* Second printing. Birkhäuser Boston, Basel, Stuttgart 1987.
—. *The Apprenticeship of a Mathematician.* (Transl. J. Gage) Birkhäuser Basel, Boston, Berlin 1992.

Paul Julian WEINDLING. *Nazi Medicine and the Nuremberg Trials. From Medical War Crimes to Informed Consent.* Palgrave Macmillan 2004.

Gary WERSKEY. *The Visible College. The Collective Biography of British Scientific Socialists of the 1930s.* Holt, Rinehart and Winston New York 1978.

Hermann WEYL. Über die Definitionen der mathematischen Grundbegriffe. *Mathematisch-naturwissenschaftliche Blätter* 7 (1910), 93–95 and 109–113. Reprinted in *Gesammelte Abhandlungen*, Vol. I, pp. 298–304.
—. *Die Idee der Riemannschen Fläche.* Teubner Leipzig 1913.
—. *Das Kontinuum. Kritische Untersuchungen über die Grundlagen der Analysis.* Verlag Veit & Comp. Leipzig 1918.
—. Über die neue Grundlagenkrise der Mathematik. *Mathematische Zeitschrift* 10 (1921), 39–79. Repr. incl. a postface from 1955 in *Gesammelte Abhandlungen*, Vol. II, pp. 143–180.
—. *The Continuum. A Critical Examination of the Foundation of Analysis.* Translated by Stephen Pollard & Thomas Bole. Thomas Jefferson U Press Kirksville, Missouri, 1987.

Norbert WIENER. *I am a Mathematician. The Later Life of a Prodigy.* MIT Press Cambridge & London 1956.

Edwin Bidwell WILSON. *Aeronautics. A Class Text.* Wiley & Sons New York 1920.

Roger J.A. WILSON. *Piazza Armerina.* Grafton London 1983.
—. *Caddeddi on the Tellaro. A Late Roman Villa in Sicily and its Mosaics.* Annual Papers on Mediterranean Archaeology, Supplement 28. Babesch Foundation 2016.
—. Review of [Steger 2017]. Bryn Mawr Classical Review 2020. See [URL 65].

Shing-Tung YAU (ed.). *Chern. A Great Geometer of the Twentieth Century.* International Press Hong Kong 1992.
—. *Chern. A Great Geometer of the Twentieth Century.* Expanded edition. International Press of Boston, Inc. 2012.

Laurence Chisholm YOUNG. *Mathematicians and their times.* North Holland Amsterdam 1981.

Robert M. YERKES (ed.). *The New World of Science. Its Development During the War.* Libraries Press Freeport, N.Y., 1920.

Oscar ZARISKI. *Algebraic surfaces.* Springer 1935.

Olivier ZUNZ. *Philanthropy in America. A History.* Princeton U Press 2012.

4. WEBSITES

URL 01. `https://edition-humboldt.de` (consulted 2 February 2020).
URL 02. `https://ath.hypotheses.org/a-propos` (consulted 9 November, 2019).
URL 03. `http://ernie.uva.nl` (consulted 3 November, 2020).
URL 04. `https://www.mathunion.org/icm/proceedings` (consulted 10 April 2020).
URL 05. `http://www.indianmathsociety.org.in` (consulted 11 January 2021).

URL 06. https://syrte.obspm.fr/cofusi/index.php?body=Colloque\b\COFUSI2019.html (consulted 19 January 2021).
URL 07. https://rockfound.rockarch.org/natural-sciences (consulted 30 January 2021).
URL 08. https://www.ias.edu/about/faqs#faq4 (consulted 7 February 2021).
URL 09. https://www.math.sciences.univ-nantes.fr/~guillope/LG/JLeray-2006.pdf (consulted 10 February 2021).
URL 10. www.idm314.org (consulted 22 November 2021)
URL 11. https://wolffund.org.il/home-page/ (consulted 22 November 2021).
URL 12. http://www.claymath.org/research (consulted 24 November 2021).
URL 13. https://en.wikipedia.org/wiki/List_of_mathematics_awards (consulted 9 November 2021).
URL 14. https://math.ethz.ch/news-and-events/events/lecture-series/heinz-hopf-prize-and-lectures/committee.html (consulted 26 November 2021).
URL 15. https://www.shawprize.org/the-shaw-prize/about (consulted 30 October 2021).
URL 16. https://breakthroughprize.org (consulted 25 November 2021)
URL 17. https://en.wikipedia.org/wiki/Breakthrough_Prize_in_Mathematics (consulted 26 November 2021)
URL 18. https://breakthroughprize.org/Rules/3 (consulted 26 November 2021).
URL 19. https://www.mathunion.org/organization/imu-secretariat (consulted 26 November 2021).
URL 20. https://www.mathunion.org/outreach/imu-logo (consulted 2 November 2021).
URL 21. https://en.wikipedia.org/wiki/Borromean_rings (consulted 5 November 2021).
URL 22. https://www.birs.ca/cmo (consulted 3 November 2021).
URL 23. https://www.icmihistory.unito.it/timeline.php (consulted 5 October 2021).
URL 24. https://www.icmihistory.unito.it/19111912.php (consulted 5 October 2021).
URL 25. https://www.icmihistory.unito.it/19221936.php (consulted 5 October 2021).
URL 26. https://www.mathunion.org/fileadmin/IMU/EC/Procedures_ICMI_2006-31-12_2012-01.pdf (consulted 5 December 2019).
URL 27. https://www.mathunion.org/icmi/activities/developing-countries-support/capacity-networking-project-canp (consulted 30 September 2021).
URL 28. https://www.canalc2.tv/video/15935 (consulted 3 December 2021).
URL 29. https://council.science/members/online-directory/ (consulted 24 May 2021).
URL 30. https://minorplanetcenter.net (consulted 22 May 2021).
URL 31. https://www.iau.org/science/scientific\b{}bodies/divisions/ (consulted 20 May 2021).
URL 32. https://stratigraphy.org (consulted 21 May 2021).
URL 33. https://www.mathunion.org/activities/imu-commissions-and-committees (consulted 25 May 2021).
URL 34. https://www.mathunion.org/ceic (consulted 13 February 2021)
URL 35. https://blog.wias-berlin.de/imu-journals/ (consulted 11 June 2021)
URL 36. http://www.birs.ca/events/2016/5-day-workshops/16w5003/videos (consulted 23 May 2021).
URL 37. https://awm-math.org/awards/noether-lectures/ (consulted 3 June 2021).
URL 38. https://www.mathunion.org/imu-awards/icm-emmy-noether-lecture (consulted 3 June 2021).
URL 39. https://www.mathunion.org/imu-awards/icm-emmy-noether-lecture (consulted 3 June 2021)
URL 40. https://www.mathunion.org/cwm/events-and-initiatives/individual-events (consulted 31 May 2021),

URL 41. https://www.mathunion.org/fileadmin/CWM/About/CWMAmbassadorslist.pdf (consulted 1 June 2021)
URL 42. https://2018.worldwomeninmaths.org/ (consulted 1 June 2021)
URL 43. https://2022.worldwomeninmaths.org/ (consulted 1 June 2021).
URL 44. https://www.youtube.com/watch?v=WiGf2TapTvQ (consulted 1 June 2021).
URL 45. https://gender-gap-in-science.org/2020/07/01/coordination-meeting-of-the-project-takes-place-remotely-on-july-1st-2020/ (consulted 2 June 2021).
URL 46. https://www.cimpa.info/en/node/10 (consulted 7 June 2021).
URL 47. https://www.mathunion.org/cdc/scholarships/library-assistance-scheme (consulted 5 June 2021).
URL 48. https://www.mathunion.org/cdc/scholarships/past-projects-and-workshops (consulted 5 June 2021).
URL 49. https://www.mathunion.org/cdc/resources/reports-about-mathematics-developing-countries (consulted 2 June 2021).
URL 50. https://www.mathunion.org/cdc/about-cdcabout-cdc/brief-history-commission-developing-countries-cdc-and-its-predecessors (consulted 2 June 2021).
URL 51. https://www.mathunion.org/fileadmin/IMU/Report/Mathematics_in_Africa_Challenges__Opportunities.pdf (consulted 7 June 2021).
URL 52. https://www.mathunion.org/activities/commission-developing-countries-cdc (consulted 2 June 2021).
URL 53. https://www.mathunion.org/membership/imu-bulletins (consulted 3 September 2021).
URL 54. https://www.mathunion.org/ichm (consulted 15 July 2021).
URL 55. https://www.mathunion.org/fileadmin/IMU/SC_Guidelines.pdf (consulted 25 August 2021).
URL 56. https://www.mathunion.org/fileadmin/IMU/Report/SC/2019/structure_committee_final.pdf (consulted 25 August 2021).
URL 57. https://www.mathunion.org/icm-plenary-and-invited-speakers (consulted 20 May 2021).
URL 58. https://genealogy.math.ndsu.nodak.edu (consulted 20 May 2021).
URL 59. https://msc2020.org/ (consulted 20 May 2021).
URL 60. https://zbmath.org/classification/ (consulted 20 May 2021).
URL 61. https://mathscinet.ams.org/msc/pdfs/classifications2020.pdf (consulted 20 May 2021).
URL 62. https://oai.zbmath.org/ (consulted 10 October 2021).
URL 63. http://openscience.ens.fr/MARIE_FARGE/ARTICLES/2017_05_15_BOOK_CHAPTER_FOR_THE_EUROPEAN_COMMISSION/2017_05_15_Chapter_on_publishing_and_peer_reviewing_in_open_access.pdf (consulted 21 March 2021).
URL 64. http://web.math.ku.dk/arkivet/jessen/bjpapers.htm (consulted 16 February 2022).
URL 65. https://bmcr.brynmawr.edu/2020/2020.03.17 (consulted 2 December 2021).
URL 66. http://www.cmat.edu.uy/massera/ (consulted 11 March 2022).
URL 67. https://www.nytimes.com/1983/01/11/opinion/l-one-victim-of-uruguay-s-military-regime-154936.html (consulted 11 March 2022).
URL 68. https://oda.mfo.de/themes/MFO/vendor/pdfjs-dist-viewer-min/build/minified/web/viewer.html?file=https://oda.mfo.de/bitstream/handle/mfo/65/full-text.pdf#page=29 (consulted 14 March 2022).
URL 69. https://www.polytechnique.edu/bibliotheque/fr/lengagement-mathematicien (consulted 24 March 2022).

Index

Abel, N.H., 13
Abel Prize, 266, 312
Ackermann-Teubner, A., 52, 294
Addams, J.L., 93
Agol, I., 339
Ahlfors, L.V., 140, 180, 247
Akizuki, Y., 162, 235
Alexander, J.W., 148, 151, 172
Alexandrov, P., 145, 168, 172, 174, 177, 240, 322
Althoff, F., 39, 40, 49
Anantharaman, N., 298
Antoine, L., 114
Appell, P., 33, 116
Archimedes, 3–5, 27, 270, 271
Arnold, H., 212
Arnold, V.I., 250, 336
Artigue, M., 286, 287
Artin, E., 163, 183, 184, 191, 192, 235, 294
Atiyah, M., 244, 337
Avila, A., 334, 336

Babbage, C., 9–12, 16, 18, 23
Baillaud, B., 105
Baire, R., 53, 54, 75
Baker, A., 330, 337
Ball, H., 82
Ball, J., xii, 249, 267, 287
Bamberger, L., 148
Banach, St., 140, 145, 170, 181, 182, 234, 255
Bannow-Witt, E., 178
Bari, N.K., 146
Bass, H., 39, 275, 276, 286, 287
Baudelaire, Ch., 155, 159
Baxter, A., 206
Behnke, H., 179, 180, 276, 280, 282, 338
Beilinson, A., 258, 259
Beke, M., 94
Bellow, A., 300
Beltrami, E., 22
Benkart, G., 301
Bentham, J., 6, 7, 11, 218
Bergson, H., 82, 110
Bérillon, E., 89
Berthod, A., 175
Bertini, E., 27
Bertrand, J., 23, 24
Besicovitch, A.S., 146
Bessel, W., 8, 9, 12
Betti, E., 22–26, 147
Bhabha, H.J., 221
Bhargava, M., 337
Bianchi, L., 116
Bieberbach, L., 133, 134, 176, 186
Bigelow, J., 211, 213
Birch, B., 263
Birkhoff, Garrett, 143, 209
Birkhoff, George D., 143, 148
Birman, J.S., 298–300
Blaschke, W., 183, 186, 191, 192, 205
Bloch, M., 114
Blondel, Ch., 114
Blumenthal, O., 133, 199
Bohr, H., 177, 179, 180, 187, 229
Bohr, N., 142
Boisbaudran, P.-E. Lecoq de, 31, 32
Bolthausen, E., 338
Bolyai Prize, 266
Bombieri, E., 262
Bompiani, E., 187, 231, 236, 238
Boole, G., 51, 52
Borchardt, C.W., 25, 28
Borcherds, R., 333, 337

N. Schappacher, *Framing Global Mathematics*,
https://doi.org/10.1007/978-3-030-95683-7

Borel, É., 33, 50, 52, 62, 75, 77, 91, 113, 117, 135, 156, 175, 207
Bortolotti, E., 131
Böttinger, H.T., 40
Bourbaki, N., 18, 36, 54, 154, 156, 157, 159, 166–168, 178, 180, 192, 196–198, 213, 273, 282, 284
Bourgain, J., 335
Boutroux, É., 89
Brandi, K., 195
Brandt, H., 164
Brauer, R., 235
Breuil, C., 236
Briand, A., 131
Brieskorn, E., 289
Brill, A., 27, 161
Brioschi, F., 22, 24, 26, 147
Brouwer, L.E.J., 83, 133, 134, 145, 167, 176, 186
Brunn, H., 270
Bush, V., 140, 210

Cannizzaro, St., 30
Cantor, G., 24, 38, 73, 77, 78
Cantor, M., 63, 309
Carleson, L., 213
Carnal, H., 300
Cartan, É., 52, 120, 135, 156, 168, 176, 177, 192
Cartan, H., 120, 154, 197, 198, 229, 246, 255
Cartan née Weiss, N., 120
Cartier, P., 244
Cartwright, M.L., 298
Casorati, F., 24, 26, 147
Castelnuovo, G., 26, 27
Cauchy, A.-L., 12, 18, 54, 88
Cayley, A., 17, 18, 27
Chabauty, C., 197
Chamberlain, J.A., 131
Chandrasekharan, K., 235, 238, 260
Chang, S.-Y.A., 298, 301
Charléty, S., 119
Chasles, M., 24
Châtelet, A., 231, 280
Chazy, J., 156
Chebyshev, P., 29
Chern Medal, 266, 312, 315
Chern, Sh.-Sh., 188, 190–193, 260, 267
Chevalley, C., 154, 156, 157, 166–168, 196, 197, 235, 244
Chisholm Young, G., 55, 56, 86, 87, 106, 135, 175
Chopin, F., 34
Choquet-Bruhat, Y., 300, 301
Chow née Victor, M., 191
Chow, W-L., 169, 170, 188–193
Christodoulou, D., 334
Christoffel, E., 64
Clebsch, A., 24, 25, 27
Coates, J., 337
Cohen, P.J., 244
Coleman, A.J., 304–306
Conant, J.B., 210
Concini, C. de, 337
Connes, A., 263
Conrad, B., 236
Conway, J., 337
Corput, J.G. van der, 175
Costa Lobo, F.M. da, 121
Coulange, F. de, 89
Courant, R., 182, 186, 187, 199, 209, 210
Cramér, H., 119
Crelle, L., 12–14, 25, 28
Cremona, L., 22, 25, 27
Curie, M., 32–34, 110, 143, 189
Curie, P., 32, 143

Dandieu, A., 166
Daniel, Y., 244
Danjon, A., 114
Dantzig, G.B., 211, 212
Darboux, G., 21, 41, 88, 108, 186
Daubechies, I., 267, 293, 298
Davenport, H., 231, 337
Debussy, D., 85
Dedekind, R., 24, 35, 161, 162
Dehn, M., 53, 54, 172
Deift, P., 334
Delage, Y., 105
Delbos, Y., 175
Deligne, P., 258, 259, 330, 337
Delone, B.N., 242
Delsarte, J., 154, 197, 198
Demidov, S.S., 183, 241
Demoulin, A., 107, 116
Denjoy, A., 145, 156, 175, 197
Deruyts, J., 107
Deuring, M., 178–180, 235–237
Diamond, F., 236
Dick, A., 295, 297, 301
Dickson, L.E., 116, 119, 122, 123, 159, 164
Diels, H., 41, 46, 99
Dieudonné, J., 154, 197, 198, 276, 282, 306
Dini, U., 23
Dinur, I., 298
Dirichlet, P.G. Lejeune, 8, 12, 37, 258
Döblin, W., 207
Dobrushin, R., 330, 336

Donaldson, S., 337
Donder, Th. de, 106, 116
Douglas, D., 212
Douglas, J., 180
Drinfeld, V.G., 259
Dubreil, P., 165, 178
Dubreil-Jacotin, M.-L., 178
Duhem, P., 35, 89, 90
Duistermaat, J., 338
Dyck, W.v., 51, 62
Dynkin, E.B., 242, 336

Eckhart, Meister, 80
Eckmann, B., 329, 338
Egorov, D.F., 75, 183
Ehrenfest, P., 51
Ehrenfest-Afanaseva, T., 51
Ehresmann, Ch., 197
Eichler, M., 236
Eilenberg, S., 173
Einstein, A., 33, 86, 88, 110, 119, 133, 148, 163, 201, 269
Eisenstein, G., 12
Emch, A., 93
Enriques, F., 26, 156
Erdős, P., 147, 343
Esclangon, E., 114
Esenin-Vol'pin, A.S., 241
Esnault, H., 298
Esteban, M., 298
Eucken, R., 86
Euler, L., 6, 93

Faltings, G., 257
Faraday, M., 35
Farr, W., 16
Febvre, L., 114, 155, 156, 159
Fehr, H., 62, 136, 177, 276–279
Fejér, L., 177
Fekete, M.-M., 202
Fenchel, W., 179, 180
Fermat, P. de, 236, 257, 259
Ferrié, G., 105
Fichte, J.G., 80
Fields Medal, 3, 27, 140, 162, 169, 180, 228, 229, 235, 243–247, 256, 257, 259, 262–267, 294, 295, 298, 302, 312–316, 323, 326, 328, 330, 332–336, 341
Fields, Ch., 122–126, 135
Figalli, A., 335
Fillmore, C., x
Fleming, W.H., 332
Flexner, A., 148, 151, 193
Florenskii, P.A., 75
Fomin, S.V., 241
Forsyth, A.R., 56
Fourier, J.-B.J., 12, 18, 37
Fowler, R.H., 93
Fraenkel, A., 160, 201, 202
Franchis, M. De, 116, 117
Fréchet, M., 52, 113–115, 117, 118, 124, 156, 181
Freedman, M., 257, 332, 339
Frege, G., 74
Frenkel, E.V., 259
Freudenthal, H., 39, 275, 276, 280, 283–286, 337
Frey, G., 259
Friedlander, F.G., 206
Friedman, M., 209
Friis, Aa., 109
Frobenius, G., 52
Frostman, O., 232, 233
Fueter, R., 177
Fujisawa, R., 63–65
Fuld, C., 148, 149

Gaitsgory, D., 259
Galois, É., 26, 35, 156
Galton, F., 16
Gauss, C.F., 8, 10, 12, 14, 21, 37, 88, 165
Gauss Prize, 266, 298, 312, 314–316, 326
Gelfand, I.M., 242, 246, 336
Genocchi, A., 24
Gentile, G., 130, 155
Geppert, H., 203–205
Geppert, M.-P., 178, 203
Gergonne, J.D., 13
Gerland, G., 57
Germain, S., 10, 264
Gini, C., 130
Glaisher, J.W.L., 107
Godeaux, L., 156, 175
Godement, R., 36
Goethe, J.W.v., 2, 14, 261
Goffman, E., x
Goldstein, C., xii, 298, 337
Goldstine, H.H., 210, 211
Goldwasser, Sh., 298
Gonseth, F., 175
Gontcharov, V.L., 146
Gosse, R., 156
Goursat, É., 197
Gowers, T., 293, 333
Grötschel, M., 269
Grauert, H., 219
Gray, M.W., 299
Green, G., 18

Greenhill, G., 62, 277
Grell, H., 168, 178
Gromov, M., 266, 323
Grothendieck, A., 36, 163, 213, 219, 244, 245, 259, 330, 337
Grötschel, M., xi, 267, 300
Guccia, G.B., 27, 28, 58, 116, 123, 124, 147, 271
Guionnet, A., 298

Hadamard, J., 35, 88, 93, 117, 118, 145, 156, 189, 229, 280
Halbwachs, M., 114, 156
Hale, G.E., 91, 99–102, 105, 108
Halperin, I., 255
Hamilton, W.R., 23, 36
Hardy, G.H., 72, 122, 142, 187, 189, 190
Hartel, W.A.v., 41
Hasse, H., 154, 165, 171, 185, 187, 204, 207, 236–239
Hatzidakis, N., 116
Hecke, E., 49, 79, 183, 192
Heegaard, P., 53, 54
Hegel, G.W.F., 37, 79
Helmholtz, H.v., 88
Henkin, L., 245
Henrici, P., 329, 338
Hensel, K., 160, 171
Herbrand, J., 171
Hermite, Ch., 20, 21, 23, 24, 28, 38, 50, 51, 54, 88, 89
Herschel, F.W., 9
Hesheng, H., 300
Hesse, O., 25
Heyting, A., 179
Heywood, H.B., 117, 119–121
Hilbert, D., 9, 39, 45, 51, 54, 55, 60, 63, 65, 66, 73, 74, 77, 79, 87, 88, 93, 112, 117, 133, 134, 156, 157, 159, 163, 166, 169, 171, 181, 263, 266
Hinneberg, P., 49
Hironaka, H., 162
Hirzebruch, F., 250, 337
Hobsbawm, E., 153
Hobson, E.W., 19
Hodge, W.V.D., 232, 263, 337
Hogbe-Nlend, H., 304, 306
Hohenemser, K., 180
Holden, H., xi, 268
Hopf Prize, 266
Hopf, H., 168, 172, 232, 236–238, 266, 329, 337
Hostinský, B., 177
Hotelling, H., 209
Hoyle, F., 206
Hua, L., 188–193
Humboldt, A.v., 11–14, 148
Humboldt, W.v., 9, 11
Hurewicz, W., 145, 174, 317
Hurwitz, A., 9, 60, 239
Husserl, E., 77
Huxley, J., 221

Ihara, Y., 332, 337
Iorga, N., 195
Ito, K., 337
Iwasawa, K., 235, 337
Iyanaga, Sh., 179, 235–237, 337

Jackson, R.H., 218
Jacobi, C.G.J., 9, 12, 13
Jacobs, A.H., 93
Jaffe, A., 263
Jeltsch, R., 338
Jessen, B., 231
Jimbo, M., 337
Jones, V.F.R., 299
Jordan, C., 35, 115, 119, 197
Julia, G., 154, 177, 192, 204

Kagan, V., 155
Kähler, E., 192
Kahrstedt, U., 195, 196
Kamke, E., 236, 237
Kant, I., 80, 89
Kashiwara, M., 332, 337
Kato, K., 337
Kawada, Y., 304
Kazhdan, D., 323
Keldysh, M.V., 244
Kenig, C.E., 267
Kepler, J., 88
Kerékjártó, B., 156
Kerkhof, K., 133
Keyfitz, B., 298
Keynes, J.M., 98, 127
Khinchin, A., 180
Khot, S., 334
Kim, J.H., 265
Kirwan, F., 298, 337
Klein, F., 25, 27, 38–41, 48–53, 55, 56, 61, 62, 86–88, 117, 123, 147, 155, 167, 180, 181, 275–280
Kline, J.R., 225, 227–229
Klöti, E., 153
Knaster, B., 181
Kœnigs, G.X.P., 106, 107, 115, 116, 118, 119, 135, 279

Kogbetliantz, E., 197, 198
Kohn, R.V., 334
Koksma, J.F., 179, 238
Kolmogorov, A.N., 140, 179–181, 185, 207, 212, 242, 246, 322, 336, 337
Kontsevich, M.L., 323, 336
Kotani, M., 298
Köthe, G., 178
Kovalevskaya, S., 29, 30, 56, 147, 157, 264
Kronecker, L., 24, 28, 35, 39, 50–53, 64, 160, 161
Kronheimer, P., 337
Krull, W., 168, 179
Kummer, E.E., 12, 24, 28, 37–39, 49
Kunugi, K., 231
Kuratowski, K., 36, 181

Ladyzhenskaya, O., 264, 300
Lafforgue, L., 259, 263, 336
La Fontaine, H., 57
Lagarde, P., 41
Lagrange, J.-L., 12, 18
Laisant, Ch., 276
Lallemand, Ch., 105
Lamb, H., 106, 115
Landau, E., 107, 115, 117, 119, 133, 176, 201, 202
Lang, S., 164, 245
Langevin, P., 33
Langlands, R., 255, 258, 259
Laplace, P.-S., 12, 18
Larmor, J., 116, 119
Laugier, H., 197
Lax, P., 332
Lebesgue, H., 75, 114, 117, 145
Lecointe, G., 101, 102
Leelavati Prize, 266, 312, 315
Lefschetz, S., 145, 148, 193, 213
Legendre, A.-M., 12, 13
Leibniz, G.W., 9, 37, 38
Leray, J., 155, 170, 174, 182, 197, 198, 246
Léveillé, A., 175
Levi, B., 199, 200
Levi-Civita, T., 91, 175, 186, 187
Lévy, P., 181
Lie, S., 156, 157, 191
Liebermann, M., 87
Lindemann, C.L.F., 9
Lions, J.-L., 250, 255, 330, 337
Lions, P.-L., 336
Liouville, J., 13, 24, 28, 151
Lobachevsky, N., 89
Lorenat, J., 310
Lovásc, L., 262, 267, 312
Lovelace, A., 9, 10
Lukasiewicz, J., 156
Lusztig, G., 337
Luzin, N.N., 75, 145, 146

Mac Lane, S., 173, 235–238
MacPherson, R., 332
MacWilliams, J., 299
Malgrange, B., 337
Malmsten, C.J., 28
Maltsev, A., 336
Mandelbrojt, Sz., 145, 154
Mandelbrot, B., 301
Manin, Y., 266
Marbo, C., 33, 116
Marc, A., 166
Marchaud, A., 175
Margulis, G.A., 243, 244, 246, 247, 330, 336
Markov, A.A., 52, 207
Marques, F., 334
Massera, J.L., 228, 254, 255
Matveyev, A., 223
Maxwell, J.C., 18, 35
May Prize, 310
May, K.O., 308, 310
Mayer, A., 25
Mazur, B., 256–258
Mazurkiewicz, St., 181
McDuff, D., 298
McMullen, C., 339
Mehrtens, H., 74, 159
Menabrea, L.F., 10
Mendeleev, D.I., 30–32
Menger, K., 168
Menshov, D., 145
Messedaglia, A., 22, 295
Metropolis, N., 206
Meyer, F., 9
Mickiewicz, A., 34
Milnor, J., 289
Minkowski, H., 9, 39, 45, 112, 159
Minsky, M., x
Mirowski, Ph., 212
Mirzakhani, M., 267, 295, 298, 303, 339
Mises, R.v., 93, 134, 156, 175
Mittag-Leffler, G., 28, 29, 38, 94, 95, 107, 115, 116, 119, 122–124, 143, 147, 272
Miwa, T., 337
Molk, J., 51–54
Mommsen, 41, 59
Montel, P., 145, 155, 156
Montgomery, D., 304
Montgomery, H.L., 337
Montucla Prize, 310

Monzie, A. de, 155
Moore, E.H., 159
Morawetz, C.S., 299, 300
Mordell, L.J., 177, 257
Morgan, J., 301
Morgenstern, O., 212
Mori, Sh., 162, 267, 332
Morse, M., 148, 225, 240
Mouchez, E., 44
Moufang, R., 178
Moureu, Ch., 105
Mumford, D., 262
Musil, R., 76, 77, 82

Nachbin, L., 273
Nagata, M., 36
Needham, J., 220, 221
Néron, A., 235
Neugebauer, O., 168, 179, 186, 187
Neumann, B.H., 304
Neumann, C., 25
Neumann, F.E., 9
Neumann, J.v., 140, 148, 182, 190, 206, 209–212
Neumayer, G.B.v., 57
Neurath, O., 156
Nevanlinna Prize, 256, 263, 298, 312, 314–316, 326, 332, 334
Nevanlinna, R., 314
Neyman, J., 145, 209
Nicolai, G.F., 86
Nikolsky, S., 322, 336
Nilson, L.F., 32
Nöbeling, G., 174, 237
Noether, E., 112, 144, 145, 154, 157, 158, 160–163, 165–171, 173, 174, 176, 178, 185, 191, 256, 258, 266, 294, 295, 297, 299–301, 312, 315
Noether, M., 27, 117
Nørlund, N.E., 94, 95, 119
Novalis, 34

Oh, H., 298, 336
Oh, Y.-G., 265
Olech, Cz., 183, 252
Oppenheimer, J.R., 211
Osgood, W.F., 54
Otlet, P., 57

Painlevé, P., 34, 91, 133
Pais, E., 59
Palis, J., 260, 287
Pandharibande, R., 334
Parenty, H., 107
Parimala, R., 298
Parshin, A.N., 242, 323
Pascal, B., 89
Pasteur, L., 89
Peacock, G., 9
Peano, G., 61
Pearson, K., 127
Perelman, G., 257, 315, 323
Perkins, F., 154
Perrin, J., 175
Perry, J., 276
Petersson, H., 192
Petrović Alas, M., 106, 108, 116
Pfister, C., 113
Picard, É., 35, 38, 48, 54, 88–91, 101, 102, 106–108, 115–117, 119, 120, 126, 127, 129, 131, 135, 137
Picone, M., 91, 92
Pincherle, S., 130–132, 134–137, 233, 279
Pirenne, H., 109
Pisier, G., 337
Pisot, Ch., 197
Planck, M., 87
Plofker, K., 298
Poincaré, H., 21, 27, 28, 33–35, 39, 50, 54, 61, 73, 75, 77, 79, 116, 156, 173, 257, 263, 266
Poinsot, L., 13
Poisson, S.D., 12, 13, 18
Pollaczek, F., 207
Pólya, G., 78, 79
Pompeiu, D., 116
Poncelet, J.-V., 13
Pontryagin, L.S., 213, 214, 219, 244, 250, 252, 311
Possel, R. de, 156, 197
Prandtl, L., 40, 127
Pringsheim, A., 53
Prokhorov, Y.V., 244, 336

Quetelet, A., 13–16, 22

Rabin, M.O., 332
Radó, T., 179
Raman, C.V., 71
Ramanathan, K.G., 235
Ramanujan, S., 71, 72, 90
Ramaswami Iyer, V, 71
Ranke, L.v., 49
Rapkine, L., 197
Rapoport, M., xii, 337
Ratner, M., 298, 336
Raynaud, M., 337
Reidemeister, K., 9, 172, 179, 180

Reina, V., 106
Reiten, I., 301
Reuter, G.E.H., 206
Rey Pastor, J.R., 199
Reye, T., 64
Rham, G. de, 232, 243
Ribet, K., 259
Richard, J.A., 77, 79
Riemann, B., 8, 24–26, 54, 133
Riemann Hypothesis, 257, 263
Riesz, M., 142, 175
Robinson, J., 299
Roosevelt, F.D., 154, 188, 202
Rosanov, Y., 336
Rose, W., 141, 142
Rothé, E., 114
Rudin, M.E., 299
Runge, C.D.T., 40
Russell, B., 156

Sadler, D.H., 206
Saint-Raymond, L., 298
Sakellariou, N., 279
Saks, St., 182
Salam, A., 305
Samuel, P., 244
Sato, M., 332, 337
Scanlon, J.C., 299
Schauder, J., 170, 174, 182
Schering, E., 9, 28
Schmidt, E., 134
Schoen, R.M., 332
Scholze, P., 267, 335, 337
Schouten, J.A., 177, 232
Schreier, O., 168
Schuster, A., 41, 44, 45, 48, 99, 101, 105
Schwartz, L., 213, 228–230, 244, 337
Schwarzschild, K., 55
Segre, C., 26, 27
Selberg, A., 228, 229
Selivanov, D.F., 51, 52
Serfaty, S., 298, 334
Serganova, V.V., 298
Sergescu, P., 175
Serre, J.P., 235, 237, 266, 306
Severi, F., 27, 91, 154, 164, 177, 185, 187
Shafarevich, I.R., 241, 242, 244
Shelah, Sh., 266, 332, 343
Shimura, G., 235, 259
Shiryaev, A.N., 207, 336
Shmidt, O.Y., 183
Shoda, K., 169
Siegel, C.L., 193, 236, 239, 246
Sierpiński, W., 36, 181
Simon, B.M., 266
Sinai, Y., 336
Siniavsky, A., 244
Sitter, W. de, 178
Smale, St., 244, 245, 257
Smith, D.E., 276, 277, 280
Smith, H.J.S., 17
Snyder, V., 128
Somerville, M., 9, 10, 14
Sono, M., 161, 162
Soulé, C., 258, 259
Spivak, M., x
Srinivasan, Bh., 300
Steiner, J., 13, 27
Steinhaus, H., 181
Steinitz, E., 160
Stephanos, C., 94
Stiefel, E., 329, 338
Stokes, G.G., 18
Stone, M.H., 147, 179, 182, 187, 205, 225–230, 276, 280, 282
Størmer, C., 232
Stravinsky, I., 85
Stresemann, G., 131
Strubecker, K., 178
Stuyvaert, M., 107
Sudan, M., 263
Suess, E., 41, 45
Suetuna, Z., 169
Sullivan, D., 330, 339
Sullivan, J.M., 269
Sun, D., 191
Swinnerton-Dyer, P., 263
Sylvester, J.J., 17
Szymański, P., 145

Takagi, T., 65, 169, 176, 179, 235
Tamagawa, T., 237
Tamarkin, J.D., 187
Taniyama, Y., 235–237
Tannery, J., 35
Tao, T., 266, 312
Tardos, E., 298
Tarjan, R., 332
Tarski, A., 36, 140, 317
Taussky-Todd, O., 178, 299
Taylor, R., 236, 337
Teichmüller, O., 173, 205
Terradas, E., 177, 199
Thom, R., 3, 285
Thompson, H.W., 223
Thompson, J.G., 330
Thomson, J.J., 18
Thullen, P., 179, 180

Tikhomirov, M., 241
Tillmann, U., 298
Tits, J., 330
Todd, J., 206
Todd, J.A., 140
Tortolini, B., 24
Treccani, G., 155
Trowbridge, A., 110, 139, 143
Turing, A., 173, 206
Turing Award, 256

Uhlenbeck, K., 298–300
Ulam, St., 206
Ullrich, E., 178
Urysohn, P., 145, 168

Vahlen, T., 93
Valiron, G., 114, 177
Vallée Poussin, Ch. de La, 106, 107, 115, 117, 119, 124, 130, 177
Van der Waerden, B.L., 22, 26, 140, 163, 164, 166, 168, 178, 180, 182, 185, 191, 193, 295, 297, 304
Veblen, O., 148, 150, 159, 177, 179, 180, 187, 193, 209, 216, 232
Vergne, M., 298
Veronese, G., 27
Vessiot, E., 156
Villani, C., 336
Villat, H., 114, 119
Vinogradov, I.M., 189, 252, 311
Virchow, R., 44
Vitushkin, A.G., 250, 336
Voevodsky, V.A., 263, 323
Voisin, C., 298
Volterra, V., 26, 91, 100–102, 106–108, 115, 116, 119, 127, 130, 175, 181
Voss, A., 49, 50

Wagner, K.W., 207
Wang, J.-M., 265
Wang, W., 189
Watson, G.N., 177
Wavre, R., 171, 224
Weaver, W., 197, 199, 209, 210, 213
Weber, H., 9, 25, 35, 65, 161
Weber, M., 23, 172
Wedderburn, J., 159
Weierstrass, K., 23, 24, 28, 38, 39, 54, 124, 166, 268
Weil, A., 6, 20, 26, 147, 154, 174, 184, 185, 196–198, 229, 235, 237, 246, 259, 309
Weiss, P.E., 114, 119, 120
Werner, W., 336
Weyl, H., 75, 77–83, 137, 148, 149, 151, 157, 163, 167, 176, 177, 193, 196–198, 204
Weyl née Joseph, H., 77, 79
Wheeler, M.F., 300
Whitney, H., 172, 285
Wiener, N., 117, 177, 189, 206, 207, 212, 213
Wilamowitz-Moellendorff, U.v., 87
Wiles, A., 236, 257, 259, 263, 264, 267, 337
Williamson, G., 335
Wilson, E.B., 126–129, 131
Wilson, W., 109
Windaus, A., 219
Winkler, C., 31, 32
Witt, E., 236
Witten, E., 263
Wolf Prize, 246, 247, 265, 343
Woodward, R.S., 122
Wright, M., 298
Wrinkle, N., 269

Yoccoz, J.-C., 336
Yong, L.L., 310
Young, L.-S., 298, 334
Young, L.C., 175, 181, 182
Young, W.H., 56, 106, 116, 135, 175
Yu, B., 298

Zagier, D., 323
Zaremba, St., 175, 177
Zariski, O., 26, 147, 174, 179, 185
Zassenhaus, H., 192
Zelinsky, D., 235
Zermelo, E., 77, 163
Zuse, K., 207
Zygmund, A., 145

Printed in the United States
by Baker & Taylor Publisher Services